A Natural History of Time

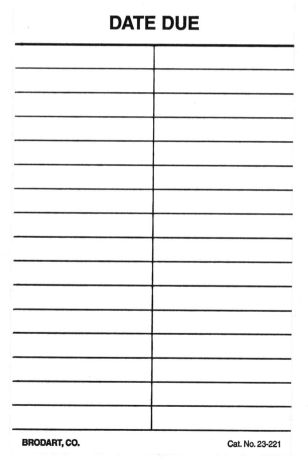

DATE DUE

A Natural
History of Time

PASCAL RICHET

TRANSLATED BY JOHN VENERELLA

The University of Chicago Press   Chicago and London

Contents

Preface

How has the study of nature—from ancient times to our current era—shaped our perceptions of time and of durations of time? According to the Judeo-Christian tradition, the Creation would date back only a few millennia, though the ancient Greeks commonly held that the world had been around for all eternity. Between these two extremes, any among the most various durations of time might be presumed. What I shall attempt to describe here is how natural timescales and the age of the world have been evaluated throughout the ages, how the dark abysses of this "time" ended up being probed—in short, the nature of the long road that had to be traveled before coming upon the billions of years now established by contemporary geochemistry and astrophysics. It was a tortuous road, to be sure, a real game of Snakes and Ladders—with some backtracking included—having as some of its squares the venerable patriarchs of the Old Testament, the precession of the equinoxes, the Passion of Christ, the souls of stars, the shape of the earth, the lives of stones, fossils, the Flood, the paths of comets, the saltiness of the oceans, the primordial heat, the extinction of species, the energy of the stars, sedimentation and erosion, natural selection, radioactivity, the isotopes of uranium, and the atomic bomb. Few problems, indeed, have blended such a large number of different disciplines, raised so many controversies, given rise to so many advances, or offended so many opinions, both secular and religious.

Obviously it is not easy to consider the question of time from beyond the context of all philosophical or metaphysical presuppositions. The temporal dimension of existence masks a familiar drama, the play of life and death; and it is within time that our past is buried. In time, as well, resides our common future, a fact that writers and poets have never ceased to lament, as Pierre de Ronsard (1524–1585) did in *Les amours*: "Time is fleeting, fleeting, my Lady; Alas! it is not time, rather we, who depart, and ere long we shall be felled by the scythe." Beyond how it relates to individuals, this question of time also raises another: that of the place of mankind within the chain of life, or within the cosmos. And finally, birth and death give rhythm to the existence of a species, the earth, and the entire universe. Without chronology there would be no history, and evolution—whether biological, geological, or astrophysical—would remain incomprehensible.

The scale of this irreversibly fleeting time hinges, for us, upon one particular reference point—one special moment—namely, that of the formation of the earth. Because the world was viewed geocentrically for so very long, this choice was not arbitrary, and it has since become quite comfortable in its anthropocentricity. An obvious limiting point in such a consideration is the time of the first appearance of the earliest forms of life. And as a long chain of firmly rooted discoveries dating from antiquity has shown, the age of the earth lends itself much better to precise determination than does the age of the universe.

In order to describe these discoveries, their agents, and their contexts, then, what better place to start than the very beginning of biblical times? However well known this initial moment may be in the West, let us recall here the origins of the world, as the Holy Spirit whispered them to Moses more than three thousand years ago.

When Yahweh had made the heaven and the earth, no wasteland bush existed yet on the earth, and no herb of the field had sprung up yet, because Yahweh had not made it rain on the earth and there was no human being to till the humus.

So Yahweh made a stream emerge from the earth to water the entire face of the humus. Then Yahweh molded the Human from the clay He drew from the humus and blew the breath of life into his nostrils so as to make him into a living being.

Then Yahweh planted a garden in Eden toward the East, and there He placed the Human whom He had molded. . . .

Then Yahweh said [to himself]: it is not good for the Man to remain all alone! I shall make a fitting companion for him! And Yahweh molded from humus all the animals of the field and the birds of the sky; then He led them before the Man to see what he would call them. Whatever name he gave to each one, that would be its name.

All the same, this account of Creation seems curious. It recalls a primitive myth, and not the familiar story evoking the firmament, the trees that produce their fruits, the sea monsters, the winged race, the beasts, and the reptiles of the earth. What, then, had become of all these beings that we are accustomed to seeing born from the divine Word? And the six days of labor followed by the Sabbath rest? Perhaps there is no need to emphasize that this is a noticeably different Genesis that we are recalling here.

According to this second account of Creation, on the fifth day, for example, Elohim [God] created "the gigantic Dragons and all the reptiles with which the waters teem, of every kind, and all the winged birds, of every kind." The next day He commanded: "Let the Earth bring forth animals of every kind: cattle, reptiles, and wild animals of every kind!" And seeing that this was a good thing, Elohim decided finally: "Let us make Mankind in Our image, as a replica of Us, so that they govern the fish of the Sea and the birds of the Sky, the cattle and all the wild animals and all the reptiles that crawl on the Earth!" In this way he created mankind "in the image of Elohim! He created them male and female!" The Creation was complete: "Elohim considered all that He had made and declared that all of it was very good. Thereupon it became evening, then morning: the sixth day."

There are, in effect, two coexisting Genesis accounts of the Creation: the later one precedes the primitive one, contrasting its Elohim, who puts an end to Chaos by means of the Word, with the Yahweh of the original one, who still models the clay with his own hands. Because each has its own characteristic style and vocabulary, these two passages—one according to Elohim, the other, to Yahweh—form just one of the many doublets that exegesis has used to reconstitute the history of the writing of the Pentateuch, the first five books of the Bible. "Once the texts and vocabulary have been properly ascertained and examined in each case, including their language, stylistic characteristics, and individual ideology," summarizes J. Bottéro, whose rendition is quoted above, "then, as geologists can identify a terrain, though it contains great gaps, by its mineral composition and its characteristic fossils, it has been possible to verify that the 'doublets' are coherent within themselves, in autonomous bodies of literature, so that instead of *one* history of ancient Israel . . . , we can single out several histories, written in different times and moods, each with its own language and its own typical way of viewing events."

This comparison between the strata of a text and those of the earth can be taken even further. It was realized during the second half of the seventeenth century that the text of the Bible had a history, at the very same time that it was becoming evident that the earth also retains a memory of its past. This coincidence was not fortuitous. Rather it represented the first signs of a new and powerful sense of history that awakened as a slowly maturing fruit—a fusion of Jewish and Hellenic ideas within the Christian West. It became possible to exploit two types of archives, one human, the other natural, and the two would be found to involve timescales so different as to appear incommensurable. As for nature, the great book to be read was not the Bible; it was the one that the earth itself has never ceased scribbling—in a language that long awaited its own deciphering. Its messages, buried within underground depths, have revealed the immensity of time and the alarming brevity of the human era. The debt to geology was promptly recognized by a common phraseology: whereas the largest of distances are astronomical, unfathomable durations of time have become geological.

Even though the earth's archives and their deciphering make up the essential substance of our narrative, the present volume is far from being a history of geology. First of all, there are already some notable histories, including a recent one by F. Ellenberger. And even more importantly, although the question of natural timescales was a playing field for all "natural philosophers," the noticeable majority of those were physicists and chemists. Because the measure of time is assured by the movement of the stars, and because the age of the earth cannot be completely dissociated from the age of the solar system, some fragrance of the cosmological still hovers about in the shadows of Copernicus, Descartes, Leibniz, Newton, Fourier, and Kelvin. It is one of the ironies of history that the rigorous solution to the problem of time did not come from the world of the infinitely large (the universe), but rather from the world of the infinitely small (the atom), as the radioactivity studies of the Curies and Rutherford revealed. It would not be possible, therefore, to ignore chemistry. However, even though Leibniz and Newton both left their imprint on it, the philosophy of time is not treated here, except in terms of allusions. The same holds for the physical meaning of time and of its irreversibility, to which a vast body of literature has already been dedicated.

A secondary theme of this work is to pay homage to the insight of noted observers, to the ever-alert human curiosity, and to the ingenuity that has tirelessly opened new inroads into labyrinthine depths. A problem as old as humanity found its solution in a totally unforeseen manner, which gives a clear though hardly new illustration of the facts that great discoveries cannot be planned and that the discoverers themselves are often the ones most surprised at the unexpected consequences of their findings. It is quite a challenge to retrace long trains of ideas without distorting them, or to leave aside the dead ends and secondary authors. Even though history is generally the record of the winners, I have not forgotten altogether the losers and those who fought in the rear guard. In this account, which links the stars with the earth and the earth with its fossils, I shall attempt to follow—through the writings of the actors themselves, when possible—the sequences of relevant ideas and of the facts, which often can be even more fortuitous. In short, as the American astronomer Percival Lowell (1855–1916) summarized at the beginning of the twentieth century, "nothing in any branch of science is so little known as its articulation." The purpose of a popularly accessible book, then, is to explain "how the skeleton of it is put together, and what may be the mode of attachment of its muscles." Because "formulae are the anesthetics of thought, not its stimulants," and "symbols tend to fictitious understanding," a few formulas will be mentioned only in the appendix, so as to limit the technical terms as much as possible. For alternative perspectives, one can consult the books of S. G. Brush, F. Haber, P. Rossi, and S. Toulmin and J. Goodfield, which partially cover some of the themes discussed herein.

Finally, this work owes recognition to the many people who have contributed to animating it, from the actors themselves to the authors who have studied them. Although the latter are too numerous to be mentioned in the text, the bibliography will allow the reader to trace any of the sources, whether primary or secondary. All of these sources were found available for consultation thanks to the kind help of librarians of various establishments: the Jussieu Interuniversity Campus, Stanford University, the National Museum of Natural History (Paris), the Geophysical Laboratory of the Carnegie Institution of Washington, DC, the Geological Society of France, and the Catholic Institute of Paris. Thanks are due to the librarians of all of these institutions and in particular to P. Armstrong and T. Noakes at the Special Collections of Rare Books of Stanford University. The author also expresses his sincere gratitude to T. Beardson, G. E. Brown Jr., G. Chevalier, P. Courtial, F. Félin, C. Guilpin, J. Gaudant, D. Goujet, F. Gramain, B. Kolmsee, S. Le Favrais, G. Oudenot, J.-L. and M.-H. Penna, M. Perryman, M.-S. Quinchon, H. S. Yoder Jr., and especially I. Toltsikhin for biographical information on E. C. Gerling and J. F. Stebbins for help in various ways. Much appreciation also to J. Dyon for the illustrations; to C. Darlot, J.-M. Lévy-Leblond, and anonymous reviewers, for their insightful criticism of the manuscript; to the Naturalia and Biologia Association for contributing to this translation; and to the School of Earth Sciences of Stanford University, for an Allan Cox visiting professorship, during the term of which a portion of the book was written.

ONE

Time without a Beginning?

FROM MYTHS TO THE ETERNITY ASSUMED BY THE GREEKS The contrast between day and night and the alternations of the seasons naturally gave rise to
cyclical conceptions of time. From Plato and Aristotle to Hipparchus and Ptolemy,
these were the predominant opinions of Greek natural philosophers. Such conceptions also lent themselves to astronomical measurements of time, which had
important implications for a universe that was assumed to be eternal.

The Origins of the World

Perhaps for its malleability, that humble clay we find sticking to our feet on moist pathways has always had a prodigious destiny within the realm of cosmogony. Has there
ever been a divinity who did not make use of it for modeling
some one or another of his various creatures? For example,
Na'pi, the Old Man of the Blackfoot Indians, found suitable
matter therein, as had Yahweh, for the making of a human.
At the beginning, it was Na'pi who created the animals and
the birds during the course of his numerous wanderings,
and who provided for the placing here and there of the rivers, mountains, and prairies. After the world had been so arranged, he got the idea one fine morning to create a woman
and her child. He outlined their forms in clay, waited several
days, and finally commanded them to rise and walk. Obedi-
ently, the woman and the child followed him to the bank of
a river. It was there that the Old Man presented himself, and
the woman asked him, abruptly: "Will we always live?" The
Old Man was surprised: "I have never thought of that," he
admitted, "We will have to decide it. I will take this buffalo
chip and throw it into the river. If it floats, when people die,

in four days they will become alive again; they will die for only four days. But if it sinks, there will be an end to them." He threw in the chip, and it floated. Gathering up a stone, the woman interjected, "No, I'll throw this stone in the river; if it floats, we will always live; if it sinks people must die, that they may always be sorry for each other." The woman cast in the stone, and the stone sank. "There," announced the Old Man, "you have chosen. There will be an end to them." It was Woman, therefore, who was responsible for Death; but above all else, her desire had been to create a sense of compassion, a trait that the Blackfoot saw lying at the heart of the human condition.

In the fascinating diversity of their expression, the cosmogonical myths testify to this constant need that societies feel to explain life and death, to establish their own origins, to grasp those of the surrounding world, or to organize a pantheon of divinities that animate nature. Few indeed are the societies in which the spectacle created by the sky—the sole source of heat and light, extending as far as the eye can see, over the immense, fertile womb of the earth—has not led people to attribute the origin of all things to majestic, cosmic couplings, or even to the gods themselves. In this spirit, the Mesopotamians left magnificent accounts, collected by J. Bottéro, such as the following Sumerian lyric poem, nearly four thousand years old, which narrates how the first Tree and Reed were born from the coupling embrace of the Sky and the Earth:

The immense platform of *Earth* glittered.
Verdant green was its surface!
Spacious *Earth* was dressed in silver and lazulite,
Bedecked with diorite, chalcedony, carnelian, and antimony,
Adorned with splendid verdure and pastures—
Something of the supreme it had!
What had happened was that august *Earth*, the holy *Earth*,
Had made herself beautiful for *Sky*, the prestigious one!
And *Sky*, this sublime god, drove his penis into
Spacious *Earth*.
He poured at one same time into her vagina,
The seeds of healthy Trees and Reeds.
And, entirely and completely, like an irreproachable cow,
She found herself impregnated with the rich semen of *Sky*!

In a somewhat less crude passage of Genesis, written a good millennium later, Elohim was satisfied simply to order: "Let the Earth bring forth greenery, plants yielding seed, and fruit trees bearing fruit, of every kind, each of them having its own seed in itself, on the Earth." But the relationship that can be recognized between these two texts attests to the permanence of the great questions proposed by myths, which may be different in their forms, and the permanence of the responses they elicit.

As time passed, it became apparent to the authors of antiquity that the cosmogonical accounts were strictly mythical. The Greek historian Diodorus the Sicilian (ca. 90 BC–ca. 30 BC) was conscious of this at the end of the first century BC when he described the origins of mankind in the first pages of his *Historical Library*. Either the earth had existed for all eternity and mankind had always inhabited it, he recognized, in summarizing the opinions of the philosophers and historians of the age, or mankind had made his appearance in a universe that had been created, one that would therefore have a limit to its existence. In the stories that dealt with this creation, every effort was made to relinquish the miraculous. To describe the origin of life, Diodorus related that, according to Democritus (fifth century BC), the sky and the earth had been separate at the time of formation of the universe because the air, which was lighter, had risen to the most elevated of regions, while all the materials that were penetrated with humidity had become concentrated into one lower location. Once they were heated by the fires of the sun, the moist parts of this earth "bubbled up, and appeared as so many pustules wrapt up in thin and slender coats and skins." Finally,

when the births included in those ventricules had received their due proportion, then those slender skins being burst asunder by the heat, the forms of all sorts of living creatures were brought forth into the light, of which those that had the most of heat mounted aloft, and were fowl and birds of the air; but those that were drossy and had more of earth were numbered in the order of creeping things and other creatures altogether used to the earth. Then those beasts that were naturally watery and moist, (called fishes), presently hastened to the place connatural to them; and when the earth afterwards became more dry and solid by the heat of the sun and the drying winds, it had not power at length to produce any more of the living creatures; but each that had an animal life began to increase their kind by mutual copulation.

Cycles of Life and Death

Taking the form of anthropomorphic or supernatural accounts, the myths of "primitive" peoples integrate mankind within the cosmos and give structure to the world by establishing intangible similarities between social order and cosmic order. Myths contribute to the stability of so-cial order by justifying its rules, often attributing their origins to the teachings of gods or demigods; in turn, the observation of ritual pre-scriptions appears to be one of the requirements for the harmonious functioning of society. The myths of late antiquity respond to the same purposes, though they are more like accounts approaching the rational. They represent allegories that symbolize the "immutable physical real-ities, or the permanent verities of On-High," according to the formula of H.-C. Puech. But in either case, a myth must, by its very nature, be timeless. Under the guise of a sequence of events that unfolds in time, "it simulates a genesis, a becoming, in a place where, in fact, there is nothing other than eternity."

Myths are outside of time because nature, above all, is governed by cycles: day alternates inexorably with night, the moon rises and sets, and the seasons succeed one another. There is no visible trace of the ir-reversible: neither beginning nor end can be discerned within the circles that reproduce themselves indefinitely. Human activity linked with the hunt or the fields not only conforms to these rhythms, but it participates in another recurrence, as well: that which leads relentlessly from life to death.

Human generations are like leaves in their seasons.
The wind blows them to the ground, but the tree
Sprouts new ones when spring comes again,

noted Homer (late eighth century BC) in *The Iliad*, with a touch of melan-choly. Another particularly striking example of the cycle is found in the system of counting years practiced long ago by the Egyptians that in-volved starting again with the beginning of each new reign. These cycles of variable duration—ranging from a single day to several decades—had the obvious point in common that they eliminated the sense of succes-siveness in events. "Everything begins over again at its commencement every instant," stressed M. Eliade. "The past is but a prefiguration of the future. No event is irreversible and no transformation is final. In a certain sense, it is even possible to say that nothing new happens in the world, for everything is but the repetition of the same primordial archetypes."

This point of view is perhaps most compelling for those who live among the effects of the extreme manifestations of nature. The nat-uralist Alexander von Humboldt (1769–1859), who took long journeys in South America, reported that in the region of Cumaná, to the south of the Gulf of Cariaco, near Caracas, after a long dry spell followed by repeated earthquakes in 1766–67, "prodigious downpours caused the rivers to swell; the year was extremely fertile, and the Indians, whose frail huts had easily resisted the greatest tremors, celebrated with feasts and dances, in accordance with the ideas of an ancient superstition, the destruction of the world and the coming era of its regeneration." As Eliade also noted, there exists "a conception of the end and the beginning of a temporal period, based on the observation of biocosmic rhythms and forming part of a larger system—the system of periodic purifications (cf. purges, fasting, confession of sins, etc.) and of the periodic regeneration of life."

Within a cyclical framework it is by definition impossible to conceive of history as reflecting a society's evolution. Despite changes of dynasties, or wars and their consequences—which can sometimes be dramatic for individuals—hardly any examples of evolution were recognized in antiquity because of the stability of the institutions and because the ways of life altered very slowly, if at all. In societies where the deep motives of human history are seen in glimpses, at best, anything new disturbs and troubles, because it does not conform to a past that has already been understood. The convenient remedy, then, comes in modifying the cycle so as to integrate the novelty and thus attribute to it some significance. As G. J. Whitrow reminds us, the word *novus* had a sinister ring in Rome, where one "greatly objected to change unless it was thought in accord with ancestral customs, which meant in practice the sentiments of the oldest living senators." This order had scarcely changed at all, in east or west, by the time philosophy and science first took flight. Fear of the unknown is hardly the mark of "primitive" humans alone; by all evidence, there remains much more of it than mere traces in our modern societies.

Chronology according to Diodorus

Independently of their ideas about the nature of time, whether finite or infinite, the Greeks showed very little evidence of any sense of history. The Trojan War of the early twelfth century BC had rather quickly revealed itself to be a myth, in which the memory of heroes was blended with the interventions of the gods. And when, in the fifth century, the germs of a nonreligious history did appear, its purpose was primarily to avoid forgetting the past, not necessarily to understand it. The insatiable curiosity of the Greeks, obvious when one considers their astonishing scientific legacy, also extended to the neighboring countries and the peoples with whom they associated, whether through pacific or bellicose ties. It was that same curiosity that gave rise to the first ethnographers among them. In the grand descriptions inaugurated by Herodotus (ca. 484 BC to ca. 425 BC), history and geography were still mingled together, providing yet another testimony of the perspective still held during the Roman age by the most eminent representatives of the Greek civilization.

As compared with Herodotus, Diodorus the Sicilian was heir to four centuries of hindsight. Upon this basis he became one of the first authors to broaden the scope of history to produce not just the history of one land or regime, but a history of the entire known world. Toward this end he established a composite chronology of antiquity, including all the noteworthy events that had occurred between the Trojan and the Gallic Wars. To justify the utility of writing such "universal histories," Diodorus affirmed that they would serve society's interests. In effect, their authors would "procure to their readers art and skill in politics above the ordinary rate, with great ease and security." But "knowledge gained by experience, though it brings a man to an aptness to be quick in discerning what is most advisable in every particular case," is attended "with many toils and hazards." In contrast, "knowledge of what was well or ill done by others, gained by history, carries along with it instructions, freed from those misfortunes that others have before experienced." And within society, history provides benefits for every age, for every state, for the common harmony:

For young men, it teaches the wisdom and prudence of the old, and increases and improves the wisdom of the aged; it fits private men for high places; and stirs up princes (for the sake of honor and glory) to those exploits that immortalize their names. It likewise encourages soldiers to fight the more courageously for their country, upon the hopes of applause and commendation after their deaths; and as a curb to the impious and profane, it restrains them in some measure, upon the account of being noted to posterity, with a perpetual brand of infamy and disgrace.

Because history increases in value as it increases in scope, it must synthesize on a vast scale, rather than describing only particular episodes. For that reason, Diodorus affirmed, "I determined to compose an entire history, from which the reader might reap much advantage, with little labour and pains; for he who endeavours, to the utmost of his power, to comprehend in his writings the memorable affairs and actions of the whole world (as of one single city), bringing down his history from the most ancient times to his own age, though he set upon a work certainly very laborious, yet he will perform that which, when finished, will be undoubtedly most useful and profitable." But Diodorus scarcely left any illusions for his own readers regarding the final lessons one could draw from such a work: by causing all of mankind to fit into one same pattern, historians acted "as if they were servants herein to the Divine Providence." And Providence, "having marshaled the stars (visible to us) in a most beautiful frame and order, and likewise conjoined the natures of men in a common analogy and likeness one to another, incessantly wheels about every age, as in a circle, imparting to each what is beforehand by fate shared out and allotted for him."

The Stars and Time

That the skies were divine in nature had been postulated ever since the beginning of Greek science. From this premise the idea was derived that the celestial bodies' movements had a determining influence upon terrestrial phenomena. Spectacular illustrations of such influence were the obvious links between the sun's height at zenith and the seasons, or the moon's position and the displacements of water producing the tides. Aristotle himself had averred that "human affairs form a circle, and that there is a circle in all other things having a natural movement and coming into being and passing away." And so, in referring to the cyclical nature of time or the influences of the stars upon human affairs, Diodorus was simply expounding ideas that had already been accepted for centuries. Two thousand years of astronomy had provided sufficient foundations for these influences. Even the great Ptolemy (Claudius Ptolemaeus, ca. 90 to ca. 168), whose geocentric system of the world reigned unconditionally until the time of Copernicus, judged them reliable enough to be a basis for his *Tetrabiblos*, or *Quadripartite: Being Four Books on the Influence of the Stars*, which still constitutes the canon of contemporary astrology. But chronology, quite apart from any idea one could have about history, was at the core of historical thought, and astronomy had long asserted that there were solid bases for it.

Even if only for agricultural interest, keeping track of the alternating seasons was obviously important as a practical aim. From earliest history, the changes in the moon's shape, its different phases, had given a measure of time, and counting lunar phases proved much more practical than counting the passing days. But moving from the lunar month to the year posed all sorts of other difficulties: the periods of revolutions in the earth-sun system and those in the earth-moon system do not relate to each other by an integral ratio. From China to Arabia, and from India to Mongolia, age-old customs induced people to maintain the lunar month for the basis of the calendar, even after these difficulties had been re-solved. Because the year has 11 days more than an integral multiple of the lunar cycle (about 29½ days), the habit developed of inserting an ad-ditional lunar month, in a more or less arbitrary manner, to make the dif-ferent seasons return at the same points of the calendar year. By the end of the third millennium BC, the Babylonians recognized that, for their year of 360 days, a thirteenth month of 30 days needed to be added every 8 years. A better understanding of periods of revolutions led the Greek astronomer Meton in the fifth century BC to define the cycle now bearing his name, based on the fact that 235 lunar months correspond closely to 19 solar years, with 7 additional months inserted for every cycle of 19 years. As for the year having 365¼ days, it was Sosigenes, the Alexandrian astronomer, who promoted its establishment by Julius Caesar in 45 BC. This year reckoning remained in use until the Gregorian calendar reform of 1582, which suppressed three leap years out of every four centuries, thus removing from the Julian year the greater part (10'48") of its excess of 11'14".

The first to measure the year were the Egyptians. Their year was the only one from antiquity that contained 365 days. It was referenced to the floods of the Nile, which marked the alternating seasons quite regularly. The first real astronomical observations must also be attributed to the Egyptians: toward the middle of the third millennium BC they noted that Sothys—the brightest star, now called Sirius—rose shortly before the sun, becoming visible on the horizon on the date corresponding to the summer solstice, the time of year when the waters of the Nile were beginning to rise. But the Egyptians maintained their year of exactly 365 days, causing their calendar year to shift in relation to the sidereal year, in accordance with a cycle of 1,456 years.

Like the Egyptians, those other great pioneers of ancient astronomy, the Mesopotamians, had made early use of the clepsydra, the water clock, as well as of the gnomon, a simple stake driven into the ground that indi-cated the sun's movement throughout the year by its changing shadow, with the shortest and longest shadows at zenith marking the winter and summer solstices. Records on Babylonian tablets show that Venus had already been identified by the beginning of the second millennium BC, but we know nothing about the circumstances of its discovery. The same is true of the discoveries of Mercury, Mars, Jupiter, and Saturn, the other four planets visible to the naked eye. Ancient astronomy con-tinued to be governed mainly by empiricism and progressed gradually until the beginning of the sixth century BC. While the practical needs of Egyptian cults continued to be satisfied with routine observations, the Mesopotamians began focusing, with astrological interest, on the mathematical aspects of the movements of the heavenly bodies. It was in Babylon that the *zodiac* was defined, that is, the band of stars forming the backdrop to the movements of the planets, divided into as many different *signs* (each one characterized by a constellation) as there were months in the year. By around 400 BC the Babylonians had already mathematically described the paths of the sun and the moon in relation to the zodiac.

Including the sun and the moon in the group, the known planets (or "wandering stars") numbered seven. The phases of the moon, which changed appearance with a rhythm of about 7 days, led the Babylonians to subdivide their lunar months into 7-day weeks that were cadenced irregularly in relation to the 30-day months. The complex influence of the skies over the terrestrial world was recognized in the fact that each of the stars bequeathed its name to one of the days of the week. The weeks that overlap the months and years gained ground throughout the Roman world during the last centuries before our era. Oddly enough, it was the church that perpetuated the Babylonian custom of honoring the planets in the names of the days of the week, as is seen in French and the other romance languages (*lun*di, *mar*di, *mer*credi, etc., with the exception of Sunday: *domenica, dimanche, domingo,* or God's day). It is the Egyptians to whom we owe the division of the half-day into twelve hours, though for them the hours varied in duration according to the length of the day; the sixty-minute hour and the sixty-second minute are the direct heritage of the sexagesimal notation of the Mesopotamians. And it was the Greeks who adopted the hours of constant duration. Our division of the day can then be seen, according to the description of G. J. Whitrow, as "the result of a Hellenistic modification of an Egyptian practice combined with Babylonian procedures."

After a long prologue now lost in the mists of time, the Greeks truly entered into the field of science at the beginning of the sixth century BC. From that point on, the progress of physics (from *physis*, nature), astronomy, and mathematics was steady, stymied no more by foreign wars, incessant conflicts among villages, the central position occupied by religious cults, or the crises that frequently pitted scholars against their cities. The contributions to geometry, arithmetic, and astronomy from the school of Pythagoras (ca. 580 BC to ca. 500 BC) at the very beginning of this period are examples. This school, enveloped in mysteries and legends that have led some to say that Pythagoras himself never even existed, took the form of a community of initiates whose beliefs included, among other things, the transmigration of the soul and whose lives were governed by a strict asceticism, both in food and in dress. Pythagoreans asserted, for example, that the earth was spherical, possibly on the basis of analogies with the clearly visible forms of the sun and the moon and because they were aware that the sails of ships disappeared below the horizon after the hulls did; their assertion may even have been based on the aesthetic and philosophical postulate that, because it could not be flat, the earth ought to be spherical. This hypothesis, which made possible the understanding of eclipses, was confirmed by observations of the curved edge of the earth's shadow cast on the moon.

The perfection of the sphere, which was praised by the Pythagoreans, attested particularly well to the existence of an order and a profound harmony within the world, the *cosmos*. The implicit but capital conclusion that ensued was that the cosmos could fall within the realm of human understanding. For Greek cosmology, the fundamental hypothesis from then on was that the stars were spherical in form and that they followed circular trajectories (or movements determined by those of the mobile spheres). The school of Pythagoras, furthermore, left a lasting division of the world into two parts. The first was the world of the stars, incorruptible and eternal, of uniform movement, with the moon's orbit as its boundary—this was the abode of the gods and perhaps that of souls. The second, the realm of life and death, was the *sublunary* world, concentric with the earth; this world, by contrast, was corruptible and animated by uncoordinated movements.

In comparison with the planets, whose paths can be irregular to the point of appearing to backtrack at times, the fixed stars are characterized not by actual immobility, but by constant relative positions as they move collectively in a slow daily rotation from east to west in relation to a fixed point (the *pole*), situated today near the North Star in the northern hemisphere. Once the existence of a second, opposite pole was postulated, this movement could be described as rotation about the polar axis of a canopy bearing the unchanging stars, the *sphere of fixed stars*. As for the sun, its rotations about the earth appeared to be irregular, with the height of its zenith varying according to the season and the direction of its rising and falling course reversing with each solstice. The postulate of circular movements seemed even more reasonable after it was understood that the sun not only rotated daily from east to west about the earth, but also moved each year in a contrary sense along the *ecliptic*, the line along which the signs of the zodiac are distributed. The *equator* was defined as the line of points around the earth where the durations of day and night are equal; the *equinoxes*, as the two yearly intersections of the equatorial plane with the ecliptic. The equator and

Figure 1.1 Apparent rotation of the celestial vault about the pole (northern hemisphere). Photograph by G. Oudenot.

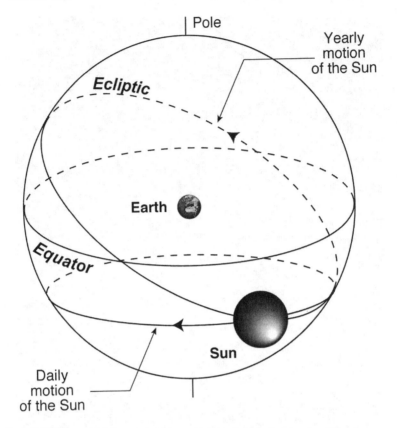

Figure 1.2 Movements of the sun about the earth. The ecliptic, or the plane in which eclipses are produced, is inclined at about 23° in relation to the polar axis.

the *meridian*, the plane determined by the vertical over a place and the line between the poles, combined to provide references allowing one to orient the positions of the stars in terms of their latitudes and longitudes. As the observations of Timocharis of Alexandria illustrated at the beginning of the third century BC, such notions rendered precision astronomy possible.

Of the World and Time

Within this framework, which prevailed until the Renaissance, what were the main conceptions regarding time? We will pass over the divergences that characterized the different Greek schools regarding the origins of the world and mention only the views of Plato and Aristotle, which have exerted immense influence throughout the ages. Plato (428 BC to 347 BC) was a former pupil of Socrates; he founded a school in Athens in 387 BC called the Academy, supposedly because it was built on a plot of land that had been owned by a hero named Akademos. Plato's science and cosmology were principally deductive, if not dogmatic. They nonetheless enjoyed long-lasting authority, possibly because the chiaroscuros of Platonic discourse lent themselves so well to various interpretations.

In resuming the Pythagorean theses that the cosmos is perfect and that it is animated by circular movements, Plato emphasized that the universe resulted from the fertilization of matter by ideas, as directed by the thought of some divine being. Within the universe, the celestial bodies were living entities whose movements were entirely similar to those of other intelligent beings. They were endowed with an immaterial and immortal attribute, a *soul*, which was the origin of their movements. After having "put intelligence in soul, and soul in body," the divine being, necessarily perfect and good, produced "a piece of work that would be as excellent and supreme as its nature would allow." Once the world had been created, "time, then, came to be together with the universe so that just as they were begotten together, they might also be undone together, should there ever be an undoing of them," added Plato. "Such was the reason, then, such the god's design for the coming to be of time, that he brought into being the sun, the moon and five other stars, for the begetting of time."

From an initial moment, that of the Creation, it was thus the movements of the stars within a finite universe that defined a time that was unlimited. According to a renowned formula of Plato, "time limits eternity and circles according to number." The cyclical aspect of time was also emphasized by Plato, with the concept of the cosmic *Great Year*, which had previously been invoked by the Babylonians. The Great Year represented the period after which the stars would find themselves in the same configuration again after having completed an integral number of revolutions: "It is nonetheless possible, however, to discern that the perfect number of time brings to completion the perfect year at that moment when the relative speeds of all eight periods have been completed together and, measured by the circle of the Same that moves uniformly, have achieved their consummation." But the Great Year remained poorly defined, as a result of Plato's disdain for practical matters and because the periods of revolutions of the stars were not yet well known.

Aristotle (ca. 384 BC to ca. 322 BC) had less equivocal ideas. The son of a doctor in the royal court of Macedonia, he entered the Academy at the age of eighteen and maintained his ties with it until Plato's death, completing his education, which until then had been predominantly medical. After spending several years as the private tutor to the young Alexander the Great, he returned to Athens in 335 BC and founded the Lyceum, the seat of the Peripatetic school. Aristotle's encyclopedic work, ranging from cosmology to zoology, was presented in a form more closely resembling academic course notes than didactic treatises. A deductive approach founded upon observation predominated, although it was not free from preconceived ideas. Contrasting with the misty vagueness of the Platonists, it was rational, and it aimed at revealing general, abstract principles, by means of which explanations could be found for concrete, observable phenomena. At the heart of this process was the search for the *four causes*: formal causes (indicating form), material causes (indicating matter), efficient causes (indicating means), and above all, final causes (indicating purpose). Attributing to each thing a profound design, this teleology dominated naturalist thinking until the nineteenth century and constituted one of Aristotle's most lasting legacies.

Aristotle expounded his cosmology in three principal works, *Physics*, *On the Heavens*, and *On Generation and Corruption*. Carrying forward the ideas of the perfection of the sphere and uniform circular movement, he drew close links, like Plato before him, between time and movement: "Not only do we measure the movement by the time, but also the time by the movement, because they define each other." More original was his conception of the finiteness of the universe, attributable to the impossibility of a vacuum, in which movements could not propagate by degrees, and the relation that he established between that finiteness and time: "It is therefore evident that there is also no place or void or time outside the heaven. For in every place body can be present; and void is said to be that in which the presence of body, though not actual, is possible; and time is the number of movement. But in the absence of natural body there is no movement, and outside the heaven, as we have shown, body neither exists nor can come to exist."

If the world is unique and finite, limited by the celestial vault, time, by contrast, is infinite, and Aristotle demonstrates this in numerous ways. Here are three, of different orders. Philosophically, a beginning of time would imply an absence of time before time began; but one cannot say *before* unless one has already supposed the existence of time. Physically, the same conclusion holds, because a movement cannot arise spontaneously: either it has existed for all eternity, or it is the result of the action of another movement, which, itself, has existed for all eternity or is the product of a preceding movement, and so forth. The existence of a celestial world is a third testimony to the eternity of time, because incorruptibility by definition is absolute: we may convince ourselves "that the heaven as a whole neither came into being nor admits of destruction, as some assert, but is one and eternal, with no end or beginning of its total duration, containing and embracing in itself the infinity of time."

The sublunary world, however, placed at the lowest position, at the center of a hierarchically ordered cosmos, is subject to incessant phenomena of generation and corruption. It is composed of the four elements—fire, air, water, and earth—previously favored by Empedocles (490 BC to 435 BC), whereas the stars, the planets, and the spheres that bear them, forming the external strata of the universe, are constituted of a fifth, incorruptible element, ether, or the *quinta essentia* (fifth essence). The movements of the sublunary world, which are determined by the cycles of the celestial world, submit to the same repetitions: generation and corruption already have been produced and will continue to be produced an infinite number of times. The infinity of time appeared to be cadenced, on the one hand, by the majestic, circular courses of the stars, and on the other by the complex cycles governing the corruptible world. The opposition between the two worlds was therefore total. While ever differing from the sublunary world in terms of form and matter, the celestial world still remained the efficient cause of the sublunary. As for the final cause of the universe, that resided in some pure form, or God, an immobile, completely immaterial being that was nonetheless the source of all movement.

Aristotle's four elements represented but a single form of matter that was influenced in differing degrees by two couples of opposing qualities, hot-cold and dry-moist:

$$hot - \text{FIRE} - dry$$
$$\text{AIR} \longrightarrow\!\!\!\!\!\!\!| \longrightarrow \text{EARTH}$$
$$moist - \text{WATER} - cold$$

When blended in pairs, these qualities, with the exception of the opposites, thus form fire (hot and dry), air (hot and moist), water (cold and moist), and earth (cold and dry) and justify the possibility of mutual transformations among the elements. Following its own natural tendency (fire upward, earth downward), each element moves in rectilinear fashion and goes to occupy its "natural" place, within concentric spheres, in the sublunary world: earth at the center, then water, air, and fire. In transforming potentiality into reality and movement into repose, these processes are at the roots of all changes affecting the sublunary world. In actuality, then, the influence of the celestial world implies that this stratification is not fixed, because the elements are in constant, mutual change. Though dry, the earth can rise in an "unnatural" manner above the level of water; fire can burn upon the earth; and water can transmute into air (vapor).

Hipparchus and the Equinoxes

After Plato and Aristotle, serious difficulties arose when variations in distance of the "stars" from the earth became obvious: no longer could simple circular movements concentric with the earth be invoked to explain the planetary orbits. One solution could have been provided by the heliocentric system proposed by Aristarchus of Samos (ca. 310 BC to ca. 230 BC). But because it never proceeded beyond the stage of a rough draft, his system never attracted the sort of calculations that could have made it a matter of practical interest. And above all, it clashed with Aristotle's solid objections to any movement of the earth. Physics would not allow the refuting of this position. For example, if the earth moved, then why would an object thrown up into the air fall back to its original place? And why did the stars not exhibit any parallax? Combinations of various circular movements, not necessarily concentric with the earth, provided a resolution of the difficulty, in a compromise between the physics of the Lyceum and the theory of natural movements. Ptolemy conceived his famous geocentric system in such a manner, where *epicycles*, *equants*, and *deferents* represented geometrical artifices for decomposing any orbit into a combination of circles. *Almagest* (*Al megestos* means "the greatest") was the hybrid title given by the Arabs to *The Great Mathematical Composition of Astronomy*, Ptolemy's work expounding this system. His achievement represented such an important advance that posterity neglected to preserve related works written earlier; consequently, the history of Greek astronomy is based on a few fragmentary quotations from lost texts that have managed to survive the ages.

An early precursor of Ptolemy and his system was Hipparchus (ca. 190 BC to ca. 120 BC), one of the greatest masters of astronomy, about whom we know practically nothing except that he was born at Nicaea (modern Iznik, in Anatolia). No one had "done more to prove that man is related to the stars and that our souls are a part of heaven," commended Pliny the Elder (23–79) in his *Natural History*.

He did a bold thing, that would be reprehensible even for God—he dared to schedule the stars for posterity, and tick off the heavenly bodies by name in a list, devising machinery [the astrolabe, to be specific] by means of which to indicate their several positions and magnitudes, in order that from that time onward it might be possible to discern not only whether stars perish and are born, but whether some are in transit and in motion, and also whether they increase and decrease in magnitude—thus bequeathing the heavens as a legacy to all mankind, supposing anybody had been found to claim that inheritance!

It was on the basis of such measurements that Hipparchus made his great discovery regarding the precession of the equinoxes. This phenomenon, described by Hipparchus in books that today are lost, *On the Movements of the Solstitial and Equinoctial Points* and *On the Length of the Year*, is referred to in several chapters of the present work. According to Ptolemy's account three centuries later, Hipparchus arrived at his result by comparing the position of the star Spica (or Azimech) at the time of the fall equinox with the position measured by the Alexandrian astronomer Timocharis at the beginning of the third century BC: in 160 years this star had increased in longitude by about 2°, although its latitude had remained constant. Hipparchus deduced from this change that the earth's polar axis turns slowly about an axis perpendicular to the ecliptic, after the manner of a spinning top. Because the equinoxes are determined by the intersection of the earth's equatorial plane, which is mobile, and the orbit of the sun, which is situated in the fixed plane of the ecliptic, they move at the same speed and in the same sense as the sun's orbit. In a geocentric system, therefore, the precession of the equinoxes is represented by a slow rotation from west to east of the sphere of fixed stars, and this, because it is superposed upon the diurnal movement, takes place about an axis normal to the plane of the ecliptic.

As an important consequence of this precession, the sun passes through the equinoctial points after a period of time slightly shorter than the time required for an entire revolution. That the *tropical* year is actually shorter than the *sidereal* year was verified by Hipparchus on the basis of his own measurements and of comparisons with observations of the solstices made by Meton in 432 BC and by Aristarchus in 280 BC. The difference between the two types of years, equivalent to about one-one-hundredth of a day, led Hipparchus to affirm that the speed of equinoctial precession was at least one degree per century. But Ptolemy,

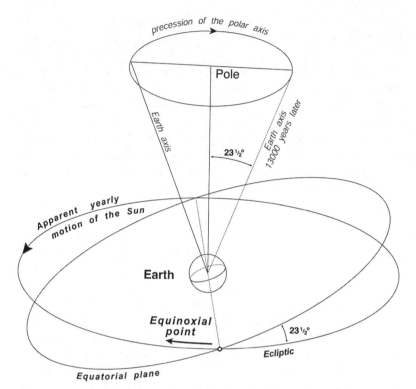

Figure 1.3 Precession of the equinoxes, represented here in terms of a geocentric reference, by the toplike movement of the earth about the pole of the ecliptic. The movement of the equatorial plane causes the sun to return to the equinox slightly earlier than the time when it passes again into its starting constellation.

possibly influenced by the Babylonians' magical numbers, later concluded that the period of precession was 36,000 years (rather than the actual value of 26,000 years, which his astronomical data would have been able to indicate). And so, a new cycle had made its appearance, at the termination of which the cosmos was to find its way back again to the same configuration. The Great Year had received a solid astronomical foundation, although this time it took on an entirely new sense.

The Earth in Slow Labor

The concept of the Great Year had an unexpected implication: the Stoics took it to the point of presuming that each new Great Year presided over a fresh regeneration of the world, in accordance with their belief in inexorable destiny. This assertion only emphasized the vanity of history. The Epicureans, in contrast, postulated the eternity of the world. Following Democritus (ca. 460 BC to ca. 370 BC), their first teacher, they believed the universe was governed by chance, not by gods, and reasoned that the existence of atoms was a sign that matter could not be subdivided ad infinitum. For them, the eternity of space and movement were simply the result of the impossibility of assigning a beginning to time.

Paradoxically, one of the most famous of the Epicureans, the Latin poet Lucretius (ca. 98 BC to ca. 55 BC), asserted the opposite in his epic work *On the Nature of Things*: "If heaven and earth had neither beginning nor origin, and if they have existed since time immemorial, how is it that there was no poet to sing of the earlier events, before the Theban war and the fall of Troy?" Lucretius, accordingly, sang his own conviction: "I believe that the world is new, that nature is altogether recent, and that we are not far from its date of origin." Besides being poorly founded, such propositions were very vague. Although he dedicated numerous verses to denying the existence of centaurs, androgynes, and other mythical beings, and to commenting upon spontaneous generation, Lucretius wrote no more than three verses about the origins of the earth. The eternity of the Peripatetics was not threatened by them in the slightest.

The activity of a teeming nature that so enthralled Lucretius had been seen by Aristotle as the mark of the sublunary world's corruptibility. From this point of view, Aristotle's propositions were scarcely arbitrary, because the ancient world sat in first-row seats before the great works of nature. Seismic destruction, volcanic violence, gullied hillsides, alluvia deposited by torrential rains, the filling in of the seas at the mouths of great rivers such as the Nile—the entire range of the great "remodeling agents" working on the earth's surface deploy themselves in few geographical areas so evidently and so consistently as they do in the eastern Mediterranean basin. A hundred years before Aristotle, the clear-sighted Herodotus had already described the considerable effects of the Nile's flooding and the related accumulation of sediments. Of the territory extending from Thebes to the seaboard, he wrote, "The greater portion of the country above described seemed to me to be, as the priest declared, a tract gained by the inhabitants," adding that he was himself "strongly of the same opinion, since I remarked that the country projects into the sea further than the neighboring shores, and I observed that there were shells upon the hills, and that salt exuded from the soil to such an extent as even to injure the pyramids." And Herodotus drew an important lesson for the future out of this past. "In Arabia, not far from Egypt, there is a long and narrow gulf running inland from the sea called the Erythraean," he remarked, concluding: "Now if the Nile should choose to divert his waters from their present bed into this Arabian gulf, what is there to hinder it from being filled up by the stream, within, at the utmost, twenty thousand years? For my part, I think it would be filled in half the time."

Two centuries later, Polybius (ca. 208 BC to ca. 126 BC), another Greek historian, expressed further interest in the filling in of the seas. Of an aristocratic family of Arcadia, he was a diplomat and military man who finally joined with the Romans. His purpose in writing his *Histories* was to establish "by what means, and under what kind of polity, almost the whole inhabited world was conquered and brought under the dominion of the single city of Rome, and that too within a period of not quite fifty-three years." Because geography was an important factor in the management of politics and wars, it became the object of numerous digressions in the *Histories*, such as the one that dealt with the navigability of the waters in the region of the Bosporus and of the Pontus (the Black Sea). "Those commodities which are the first necessaries of existence, cattle and slaves," Polybius stressed, "are confessedly supplied by the districts round the Pontus in greater profusion, and of better quality than by any others: and for luxuries, they supply us with honey, wax, and salt-fish in great abundance." Within his framework, which was more utilitarian than anything else, Polybius nevertheless demonstrated a certain sense of perspective. He asserted that "the Pontus has long been in the process of being filled up with mud, and that this process is actually going on now; and further, that in process of time both it and the Propontis, assuming the same local conditions to be maintained, and the causes of the alluvial deposit to continue actively, will be entirely filled up. For time being infinite, and the depressions most undoubtedly finite, it is plain that, even though the amount of deposit be small, they must in course of time be filled." After summarizing the question so well, however, Polybius stopped suddenly: because he was a historian, his conclusions applied only to moral or political philosophy, and not to terrestrial dynamics or the implications thereof.

At the origin of Polybius's reflections, in effect, we find Strato of Lampsacus (? to ca. 270 BC), one of Aristotle's successors as head of the Lyceum, and Eratosthenes (ca. 284 BC to ca. 192 BC), the astronomer and geographer who was the first to determine the earth's diameter correctly. Both men affirmed, in treatises that are now lost to us, the predominant role of the slow filling in of the seas and of sea currents in the changes visible on the earth's surface. To be sure, other agents existed, such as those highlighted by the geographer Strabo (ca. 64 BC to AD 25). Strabo was born into a very illustrious family of Cappadocia and was a member of the Stoic school. He wanted to give solid physical and mathematical foundations to a discipline he had encountered after continuing Polybius's *Histories*. Although he reached his goal but very imperfectly in his monumental *Geography*, his criticism of the ideas of Strato and Eratosthenes led him to state the principle of what, much later, was called *actualism*: rather than searching for extraordinary causes for natural formations, actualism brings the discussion "into closer connection with things that are more apparent to the senses and that, so to speak, are seen every day." And what we see is that "earthquakes, volcanic eruptions, and upheavals of the submarine ground raise the sea, whereas the settling of the bed of the sea lowers the sea." As a consequence, "it cannot be that burning masses may be raised aloft, and small islands, but not large islands; nor yet that islands may thus appear, but not continents." Whereas the Bible's psalmist had asked long before, "Sea, what makes you flee? Jordan, why turn back? Why skip like rams, you mountains? Why like sheep, you hills?" Strabo presented a method that excluded any sort of metaphor.

The direct observation of rocks, as was permitted by the digging of canals or the exploitation of quarries (for the Pharaonic construction sites, for example), had early revealed that fossils were ubiquitous. In a late account, Hippolytus Antipope (ca. 170 to ca. 235) reported that the sixth-century BC philosopher Xenophanes of Colophon had observed shark teeth, fish remains, and shells in the soil of Sicily at the end of the sixth century. Xenophanes, who had professed that "nothing comes from nothing" and that "nothing can change," nevertheless boldly concluded that the sea had previously covered not only Sicily, but also all of the continents. A hundred years later, Herodotus was not surprised to discover shells in the mountains, long before the capital importance of this observation had been recognized. According to Strabo, Eratosthenes "says further that this question in particular has presented a problem," Strabo reported in his turn: "how does it come about that large quantities of mussel-shells, oyster-shells, scallop-shells, and also salt-marshes are found in many places in the interior, at a distance of two thousand or three thousand stadia from the sea—for instance—in the neighbourhood of the temple of Ammon and along the road, three thousand stadia in length, that leads to it?" It was primarily in relation to this question that debates about the fluctuating sea level and its causes interested Strabo so much.

The biological origin of fossils seemed to be accepted by most of the ancients, even though Strabo himself had to struggle against a common prejudice that viewed the "little petrifications having the form and dimension of a lentil [Nummulites], which were sometimes found lying upon a bed of debris [also petrified] strongly resembling the peelings of partially-shelled vegetables" as "leftovers from the meals of the workmen who built the pyramids." The seashells in the mountains attested to the continual changes affecting the earth's surface. Doubtlessly recognized by the Greeks because of their profound curiosity about the sea and its actions, as well as about navigation, the link between marine sedimentation and time was later to play a crucial role in the endless discussions throughout the nineteenth century regarding the age of the earth. Although that link constituted a recurring theme in ancient geography, it remained unexploited by the Greeks. In spite of the benefits provided by a half-millennium of clear-sighted observations and well-advised interpretations, Strabo did not, any more than his predecessors, draw specific historical implications from the great sketch that he had drafted so well. The recognition that rocks, valleys, mountains, and shores all retained the traces of an immense history—that of the earth—did not lead the Greeks to attempt to reconstruct that history. Perhaps it was because the endeavor would have been in vain from the outset, since no trace of the beginning of the world was distinguishable.

The lack of any real evolution had been asserted long before by Aristotle, in *Meteorology*, his treatise on natural history:

It is therefore clear that, as time is infinite and the universe eternal, that neither the Tanaïs nor the Nile have always flowed, but the place whence they flow was once dry: for their action has an end, whereas time has none. But if rivers come into being and perish, and if the same parts of the earth are not always moist, the sea also must change correspondingly. And if in places the sea recedes, while in others it encroaches, then evidently the same parts of the earth as a whole are not always sea, nor always mainland, but in the process of time all changes.

The earth, in being wrought by the waters, had obviously been in a constant state of remodeling, but this represented nothing more than ripples passing over a surface that, on the whole, remained unchanged. The unceasing repetition of the same events was what gave time its cyclic nature. "Those whose vision is limited think that the cause of these effects is a universal process of change, the whole universe being in the process of growth," Aristotle had concluded, adding "It is absurd to argue that the whole is in a process of change because of small changes of brief duration like these; for the mass and size of the earth are, of course, nothing compared to that of the universe." For schools that viewed the world as an immutable entity, something that transcended time, the concept that the earth's past could possibly be reconstructed was quite simply out of the question.

TWO

On the Great Book of Moses

SHORT CHRONOLOGIES   Interpretations of the Pentateuch, especially those from the Christian point of view, brought major changes. Time was no longer considered to be cyclical, but linear instead; and the world, no longer thought to be eternal, was understood to have been created by God at a particular moment in time. History ceased being a random sequence of facts and became a succession of events. They were related to each other in a logical manner and within a time frame, the Mosaic chronology established from the Holy Scriptures, into which the entire history of the earth was required to fit. But the renewal of exegesis by Spinoza, Simon, and others in the seventeenth century eventually revealed that the Bible itself had a complex history that was open to critical interpretation.

The Roots of a Long Misunderstanding

Like so many other cosmogonical myths, the Yawhist account of Creation revealed how the breath of the divine had transmuted humus and clay into humanity. By itself, this story could not long have lent itself to literal interpretation; many misunderstandings would have been avoided if this account had not been supplemented by the less metaphorical Elohist description. At the time when the first of these texts was written, in the ninth century BC, the Hebrew nation already had a long history. Dating back to Abraham, Isaac, and Jacob, whose memories were perpetuated by oral tradition, this history had begun a thousand years earlier, when the tribes of Semitic nomads left Mesopotamia, where they had originated, to arrive in the land of Canaan (Phoenicia-Palestine). Around the fifteenth century BC, possibly while searching for greener pastures, one of these tribes, that of the sons of Israel, left for Egypt, where it ended up being held captive. The principal turning point occurred in the thirteenth century BC, when Yahweh incited Moses and his people to flee. The alliance between the Hebrew nation and their new God sealed at the summit of Mount Sinai was a propitious time—with its great kings David and Solomon, Israel dominated Palestine during the tenth century BC—but shortly thereafter disunity took over. Stemming from a schism that occurred upon the death of Solomon, the kingdoms of Israel and Judah both suffered severe defeats: the first in 721 BC by the Assyrians, and the second in 587 BC by the Babylonians under Nebuchadnezzar II. Fifty years later, the Babylonians themselves were conquered by the Persians under Cyrus the Great. The elite Jews who had been exiled into Mesopotamia were then able to return to Palestine, and it was upon their return from exile that a second exposition of Genesis, largely inspired by the Babylonians, was written, and later it was given precedence over the first by the Jewish compilers of the Bible.

So this exile played a considerable role. In particular, the seven-day week of the Babylonians was adopted. The priests of the religion that had in the meantime been codified into Judaism then divided the Creation into six days of labor followed by one day of rest. This allowed them to give a theological foundation to the Jewish practice of the Sabbath, for everybody, not just for the clergy. To these priests, then, we owe the account of the creation of the various families of living beings. The judiciousness of the order in which they caused them to appear troubled the pioneers of paleontology to such a point that Jean-André De Luc (1727–1817), an influential Protestant naturalist, could still maintain in 1809 that "if geology contradicted Genesis, then the latter can be nothing other than fable" and that geology "was necessarily a science to be acquired by the theologians, just as essential as that of the ancient languages." At that time the Babylonian sources of the Bible were unknown; they were discovered in 1834, when excavations in Mesopotamia unearthed thousands of cuneiform tablets that had been buried under the ruins of lost cities. After being brought to light by exegesis, the heterogeneity of the biblical texts was bitterly debated from the seventeenth century on. The naturalists who were then attempting to decipher earth's history using the fossils as a key perhaps did not realize that the text of the Bible had also kept a memory of its past.

Time Takes on a Direction

Without a sense of history, any examination of the past would have incomplete meaning. The sacred texts allow us to observe how this new
sense developed among the Hebrews. One late tradition even attributes
the awakening of our current sense of history to the action of a single
man, Moses. In the ancient world, where the gods tended to be numerous and fickle, and where scarcely any aspects of life or nature escaped
their attention, this prophet broke completely with the age-old customs.
He contracted an alliance with one single divinity, Yahweh, and he
formed it on terms as between equals—even if we might point out here
that one of the parties to the contract was a man, and the other God, the
transcendent Creator of all things. This one God, as J. Bottéro has emphasized, "has nothing to do anymore with the clearing of *one* particular
piece of land, or with *one* man whom He would single out from the rest
to be somewhat like his Peasant, or *one* tree or *one* animal of each kind,
or *one* woman—but with universal realities: Chaos, Water, Heaven, and
Earth; Light and Darkness, the Stars, botanical and zoological Species, the
Human Race." With this alliance, Moses turned his back on those gods
who were nothing more than improved-upon men, and he freed himself
from the ties that bound man with nature and with the cosmos. Within
a nature that was no longer divine, but that sang the glories of its own
Creator, from then on, man became free of his own destiny, responsible
for his own future. It was a relationship of love and no longer one of interest that united man with his God: "You must love Yahweh your God
with all your heart, with all your soul, with all your strength," ordered
Moses in Deuteronomy, the fifth and last book of the Pentateuch.

Israel was destined to remain "the most blessed of all peoples," on
the condition that they scrupulously observe the written law. But the
requirements of this alliance were too rigid for a nation that, like its
neighbors, tended toward idolatry and superstition. In the succession
of misfortunes that crushed Israel, generations of prophets witnessed
the consequences of the recurring breaches of the promises that had
been made to God. "The faithful city, what a harlot she has become!"
exclaimed Isaiah (fl. ca. 746 BC to ca. 701 BC) in the first pages of his book.
In response to this witness of the ruin of the kingdom of the North,
Yahweh avowed, "I am sick of burnt offerings of rams and the fat of
calves. I take no pleasure in the blood of bulls and lambs and goats." For
Yahweh, this sacrificial blood actually did nothing more than sully the
hands of the Jews: "Wash, make yourself clean. Take your wrong-doing

out of my sight. Cease doing evil. Learn to do good, search for justice, discipline the violent, be just to the orphan, plead for the widow." The divine message, which had also been repeated forcefully by Isaiah and other prophets, was not heard. In turn, the kingdom of Judah fell. The chastisement was inevitable: "Who is there to pity you, Jerusalem, who to grieve for you, who to go out of his way and ask how you are? You yourself have rejected me," confided Yahweh to Jeremiah (fl. ca. 627 BC to ca. 587 BC). "You have turned your back on me; so I have stretched my hand over you and destroyed you."

These vicissitudes did not represent repeating episodes of a cosmic drama opposing the forces of good and evil, as M. Eliade has noted. For the prophets, "no military disaster seemed absurd, no suffering was vain, for, beyond the 'event,' it was always possible to perceive the will of Yahweh." As they were confirmed by catastrophes, the prophecies gave religious meaning to political events, and they "revealed their hidden coherence by proving to be the concrete expression of the same, single divine will." Because they gave value to history, the prophecies left cycles to the side and gave a definite direction to time; and it was within such a framework that a religious philosophy of history could slowly be put into place. The most telling of the trials inflicted, that of the fall of the kingdom of Judah, put the Jews into close contact with the Babylonian civilization at the time when Judaism was just coming into being. The consequences were far-reaching for the accounts of the Creation and the Flood, which the new clergy incorporated into the sacred texts, as well as for the genealogical and chronological concerns stemming from the Genesis account that was inspired by Babylonian historiography.

Messianism, that is, awaiting a messenger to be sent by God to abolish sins and establish the kingdom of God, was the second factor that led to the orientation of time. This heralding of a radically new era had truly taken shape with Isaiah in the dark periods before the Babylonian captivity: "A son has been born for us, a son has been given to us, and dominion has been laid on his shoulders; and this is the name he has been given, 'Wonder-Counselor, Mighty-God, Eternal-Father, Prince-of-Peace.'" Isaiah prophesied that he would "extend his dominion in boundless peace, over the throne of David and over his kingdom to make it secure and sustain it in fair judgement and integrity." God could not deny his promise, nor his alliance: a way therefore existed to permit the chosen people to find the path to salvation again, despite the errors of their kings.

Although messianism was insistent before the fall of the kingdom of Judah, it was reduced to secondary importance in the period after the return from exile in Babylon; a new prosperity was burgeoning under Persian control, and during the course of it, Jerusalem and its temple were rebuilt. Under the Syrian domination of the Seleucids, who followed the conquests of Alexander the Great and the rapid disintegration of his empire, a new threat appeared much later, that of Hellenization, which incited the victorious revolt of the Jews under the guidance of Judas Maccabees (ca. 200 BC to ca. 160 BC). It was during this conflict that the latest of the great texts by the Old Testament prophets, the book of Daniel, was written. In this book faith was reaffirmed in an all-powerful God, the master of time, who "uncovers depths and mysteries," and "will set up a kingdom which will never be destroyed." And to complete the description of this ending assigned to history, the book of Daniel even suggested the possibility of a resurrection: "Of those who are sleeping in the Land of Dust, many will awaken, some to everlasting life, some to shame and everlasting disgrace."

A new stage was reached with the book of Daniel, providing a comprehensive tie between the past and the future: the past prepared the way for the future, and the future represented the coming into being of the past. Whether or not the stage was set for the arrival of the Messiah, Pompey the Great took Jerusalem a century later, putting an end to the Hebrew independence. It was in the Roman province of Palestine, under the administration of Pontius Pilate, procurator of Judea, that Jesus of Nazareth was tortured and executed in the year 28 or 29, after three years of public life. "But you (Bethlehem) Ephrathah, the least of the clans of Judah, from you will come for me a future ruler of Israel whose origins go back to the distant past, to the days of old," Micah, a prophet contemporary with Isaiah, had proclaimed. Born opportunely in Bethlehem during the last year of the reign of Herod, the last king of the Jews, Jesus descended through Mary from the tribe of Judah. Was he, as his disciples proclaimed, the Messiah whose coming had been announced by the prophets? Had he come, as Mark proclaimed, "to fulfill the Scriptures"?

The World Acquires an Age

Christianity in its early stages could scarcely be distinguished from the other Jewish sects, at least from the point of view of the Roman powers; in fact, its center was first located in Alexandria, and from there it radiated out to the rest of the Mediterranean basin. By contrast with Judaism, centered exclusively on Israel, this burgeoning Christianity, which was permeated with Greek culture, addressed itself to all nations, to the point where it finally could realize the commandment given to Abraham by Yahweh: "All clans on earth will bless themselves by you." It was Paul of Tarsus (ca. 10 to ca. 63), one of these hellenized Jews, who, in applying himself to converting the pagans, broke abruptly with Judaism, circumcision, and the other ritual prescriptions and, in actuality, founded the *Catholic* (universal) Church. This rupture with the vestiges of cycles was highlighted by Paul himself in his epistle to the Hebrews. Within the framework of the ancient alliance, "the same sacrifices repeatedly offered year after year" could not make those perfect who approached God, because "bulls' blood and goats' blood are incapable of taking away sins." By contrast, the new alliance between God and all humanity was sealed by the blood of Christ and, in fact, was based upon the extraordinary character of the crucifixion within time: "By virtue of that one single offering, he has achieved the eternal perfection of all who are sanctified."

It has sometimes been denied that this first appearance of a sense of history that has just been summarized was specific to the Jews. Some have objected that other ancient religions also contained elements of the messianic. Similar beliefs in a personal relationship between God and man and in immortality may have appeared with Zoroastrianism in Iran before expanding to the neighboring peoples. In addition, ideas of historical recurrence had not entirely disappeared from Judaism, and as the works of the prophets illustrate, the opinions of the masses did not necessarily correspond to those of the elite or of the clergy—whose texts, alone, have been preserved. But it was well in accordance with the direct line of evolution mentioned in the Old Testament that one particular moment, necessarily singular, that of the Crucifixion, marked the rupture with the cyclicity of time. From Genesis to the Acts of the Apostles, Christianity was built, in conformance with a classic formula, upon a succession of historical events mentioned in books considered to be annals. Within a universe that was bounded in time—by the Creation and the Apocalypse—a history that recounted unique events, one marked by majestic, divine interventions, could not be repeated. Generation after generation, the prevailing thought found no better argument in favor of this irreversibility than that of the eschatological perspective imposed by the Last Judgment and the Final Redemption: seated upon his throne of glory, as Matthew first proclaimed in his gospel, the Son of Man would place "the sheep on his right hand and the goats on his left"; the just would receive "eternal life," and the sinners would be relegated "to eternal punishment."

Such a future eternity of well-merited happiness or torment contrasted with the relative brevity of human history. Ancientness, however, was certainly a primary virtue as a hallmark of verity. It was one with which the cults of antiquity were interested in adorning themselves. The long lineage of Moses recounted in Genesis, dating back to Adam, who appeared on the sixth day of the Creation, can be viewed as evidence of such an intention. In the atmosphere of the religious rivalries in contention at the beginning of our era, it was natural for the early Christians to attempt to demonstrate, in their own turn, the depth of their spiritual roots: "Their high antiquity, first of all, claims authority for these writings," the vigorous Tertullian (ca. 155 to ca. 222) declared on the subject of the Scriptures, before rendering glory to Moses for having preceded the most ancient of Greek kings by four hundred years and the Trojan War by one thousand years. And because "all that is taking place around you was fore-announced," the Jewish heritage was not only accepted; it was actually exploited by Tertullian and the other apologists, who represented Christianity as the realization of the prophecies uttered since the account of Genesis, and therefore the fulfillment of Judaism, the most ancient religion.

Flavius Josephus (37 to ca. 100), the Jewish historian and friend of the emperor Vespasian, had compiled the ages of the personages of the Old Testament in his *Jewish Antiquities*. During the second century, Christians began to follow suit. But counting ages and generations from the Mosaic account was not enough; it was also necessary to estimate the intervals between two successive generations. Numerous apologists attempted to perfect these Mosaic chronologies, in manners representing very diverse sensitivities. One of the first to advance along this path was the Assyrian Tatian (ca. 120 to after 173); after the fashion of the anonymous compilers of the Bible, he arranged the four Gospels into one single account, the *Diatessaron*, and then founded the heretical, ascetic sect of the Encratites. Like the works of Tatian, those of Justin Martyr (ca. 100 to ca. 165) were lost, as were those of Clement of Alexandria (ca. 150 to ca. 215), a moralist of pagan origin, and those of Saint Hippolytus (170–235), the bishop of Ostia, who probably died during the course of the first great wave of persecutions. The chronology of Theophilus (? to ca. 190), bishop of Antioch, has come down to us, however; it is centered within a demonstration of the superiority of the new cult, addressed to an eloquent pagan friend. This father of the church, himself of pagan origin, established that at the time of the death of the emperor Aurelius Verus in 169, the creation of the world, such as "Moses the servant of

God recorded through the Holy Spirit," went back "5698 years and the odd months and days" (meaning 5,529 years before the death of Christ). In so doing, Theophilus eliminated all foundations or bases for the beliefs of the impious: "for even if a chronological error has been committed by us, of e.g., 50 or 100, or even 200 years, yet [it is] not of thousands and tens of thousands, as Plato and Apollonius, and other mendacious authors have hitherto written."

Thus, during this period of a few millennia, the divine nature of the revelation as delivered in the Scriptures provided for absolute faith. Julius Africanus (?–232), one of the first Christian scholars, recognized that if the origin of the world could be determined, then the correctness of the Old Testament prophecies, which was constantly being confirmed, ought equally well to permit the prediction of the end of time. In his *Chronicles*, of which only fragments survive, Julius examined the biblical and pagan chronologies within the perspective of millenarianism, basing his reasoning not on the period of one millennium, as given by the Apocalypse for the time separating the Crucifixion from the Last Judgment, but upon a total duration of 6,000 years, deduced from a reading of the book of Daniel. When he stopped his calculations at the third year of the reign of the ephemeral Emperor Elagabalus (or Heliogabalus), 5,723 years had already passed: the end of time was predicted, therefore, for the year 498. But his contemporaries were not intrigued by Julius's perspective.

This art of chronology rapidly reached its first heights with the *Chronographies and Chronological Canons* of Eusebius (265–340), bishop of Caesarea, whose *Church History*, among other works, is very valuable. The *Chronographies* provided an important reference up till the Renaissance, because Eusebius had been inspired by Diodorus of Sicily or by Julius Africanus to juxtapose the history of the Judeo-Christian people with the histories of the peoples of Chaldea, Assyria, Egypt, and Rome, using the form of synoptic tables. He began his universal history with Abraham and ended it at the sixteenth year of the reign of Diocletian, making use of well-known events, such as the birth of Cyrus or the fall of Jerusalem, to collate accurately all the available chronicles. This history was then translated from Greek into Latin by Saint Jerome of Dalmatia (ca. 347–420), who was a hermit in Syria before he retired to a monastery in Bethlehem. Jerome completed the chronology by adding 2,242 years between the Creation and the Flood, then 942 more between the Flood and Abraham, durations similar to those that had already been adopted by Theophilus for these periods. In the face of such consistency as was provided by this comparative method, which associated both sacred and profane chronologies, how could anyone possibly have doubted the age that was subsequently attributed to the world and to mankind?

Why Now?

Because of the importance accorded the Creation, expanding Christianity rekindled an old debate, one that had already involved the finest minds of antiquity in contention. The postulate of eternity, in accordance with Aristotle, gave rise to an impenetrable question: Why should the world even exist? But if one supposed, with Plato, that the world had had a beginning, another enigma presented itself promptly: Why had the world been created at one particular moment rather than at another? The biographer and moralist Plutarch (ca. 48 to ca. 125), who was one of the last great adherents of the Greek cults, presented this old question in such a way that even centuries later Diderot's *Encyclopédie* quipped: "Plato would have it that bodies were already in movement when God decided to arrange them; but Plutarch, wise as he was, made a mockery of this God of Plato's, asking, ironically, whether he had already existed when the bodies began to make their movements. If he had existed," he added,

either he was awake or he slept, or he did neither the one nor the other. One can certainly not say that he did not exist, because he must have existed for all eternity. Nor could one say that he slept, because to sleep for all eternity would be to be dead. If we say that he was awake, we are asking whether something could have been lacking in his beatitude, or whether there was nothing that was lacking. If he had had need of something, then he could not have been God. If there was nothing that he had had need of, then what was the use of creating the world?

The Jewish philosopher Philo of Alexandria (ca. 13–54) offered a response to this question in his *Allegorical Interpretation of Genesis* that Plato himself would not have condemned:

It is quite foolish to think that the world has been created in six days or in a space of time at all. Why? Because every period of time is a series of days and nights, and these can only be made such by the movement of the sun as it goes over and under the earth: but the sun is part of heaven, so that time is confessedly more recent than the world. It would therefore be correct to say that the world was not made in time, but that time was formed by means of the world, for it was heaven's movement that was the index of the nature of time.

Philo thus affirmed that Genesis was but a means for the expression of philosophical verities. For the six days of the Creation, for example, he echoed the school of Pythagoras—which held that "everything was in the number"—by affirming that "we must understand him to be adducing not a quantity of days, but a perfect number, namely six, since this is the first number that is equal to the sum of its own fractions: one-half of itself [3]; one-third of itself [2]; and one-sixth of itself [1]; and it is produced by the multiplication of two unequal factors, 2 and 3."

Philo has been considered a precursor of Neoplatonism, a doctrine that blended Greek philosophy with oriental mysticism and developed in Alexandria during the second and third centuries, under the influence of Plotinus (ca. 205–270) in particular. Neoplatonism proclaimed that everything proceeds from a principle of absolute unity within which no distinctions exist, not even between the knower and the known. In exceeding the confines of his solitary unity, this ineffable principle, God, created by emanation a series of beings who owed their existence and their perfection to him. Intelligence appeared first, containing within itself all possible things; and intelligence, in turn, engendered the soul, the principle of movement, which then produced ideas and the matter that made them material. The engendered being has the tendency to make its way back up toward the perfection from which it emanates, in a *conversion* toward God, which the human soul attempts to effect through ecstasy. In this highly obscure manner, Creation could appear to be an immense "chain of being," maintained through the means of subtle correspondences, commencing from inanimate matter and joining progressively with animate nature, mankind, intelligence, and, finally, God.

Under the influence of Justin Martyr, for example, the Platonic ideas of creation and the immortality of the soul had already been shown to have affinities with the new Christian theology, even to the point of giving one the impression that Plato himself might have studied the Old Testament. Although Neoplatonism represented a reaction against their own nascent philosophy, the first Christian doctors saw these ideas as the only ones originating from the pagan world that would not be irreconcilable with their own dogmas.

Born in Alexandria to a father who was martyred, the exegete Origen (ca. 185 to ca. 254) was so pious at a precocious age that he emasculated himself in order to avoid temptation and calumny. He was especially sensitive to these Neoplatonic ideas in his *On First Principles*, a treatise he wrote to develop an in-depth Christian reflection for the cultured public of his time. Called "the man of bronze" on account of his prodigious capacity for work, Origen drew from the teachings of Clement of

Alexandria and the Neoplatonic masters and finally founded a highly reputed school in Caesarea, to which Eusebius belonged. A well-known passage from Ecclesiastes that recalled the concept of the Great Year struck him particularly: "What was, will be again, what has been done, will be done again, and there is nothing new under the sun!" But even if the world had a beginning in time, Origen recalled, "it is at once impious and absurd to say that the nature of God is inactive and immoveable, or to suppose that goodness at one time did not do good, and omnipotence at one time did not exercise its power." Piety therefore suggested that the divine work had not yet commenced: "He made this visible world; but as, after its destruction, there will be another world, so also we believe that others existed before the present came into being." Although he confessed his ignorance regarding "the number or measure" of these worlds, Origen affirmed that they would not recur in identical manners— unless one were to deny Free Will, something that was unacceptable for a Christian. In effect, it would be inconceivable for the sin of Adam and Eve or the betrayal of Judas to be repeatable.

The theological uncertainties of the age left ample space for the most diverse heresies. Another idea inspired by Plato and defended by Origen was that souls were created by the Word (the Son, subordinated to the Father) and were united to bodies as punishment for their faults. This idea was condemned, just as the idea of cyclical time linked to successive creations was. Few were as penetrating in attempting to put an end to such disorder as Saint Augustine (354–430), the most eminent of the church fathers. Born to a pagan father, who was a Roman civil servant in Numidia (currently Algeria), and a fervent Christian mother, Saint Monica, Augustine led a life of erring ways for a long time. He passed his youth in Carthage, where "the effervescence of shameful amorous liaisons spluttered like boiling oil," after which time he entered upon the assiduous practice of Manichaeanism. Augustine finally converted at the age of thirty-two, after an enthusiastic reading of Plato, while he was teaching rhetoric in Milan near the imperial court. He was baptized in 387 and returned to Africa; he lead a monastic life for a time before becoming, unwillingly, bishop of Hippo Regius (near Annaba, Algeria) in 395. From that time on he lived in the midst of heresies that he combated with his indefatigable pen. Augustine produced a striking analysis of his religious evolution in his *Confessions*, concluding with a meditation on God, Genesis, and time, which appeared to him to be—whether already past or yet still to come—essentially ungraspable. This reflection was then taken up again in *The Literal Meaning of Genesis* and especially in *The City of God*, written while he was bishop in response to those pagans who attributed the sack of Rome by the Visigoths in 410 to gods who were irritated by the development of Christianity. After Augustine died, the Vandals invaded North Africa and laid siege to his town; although he left a doctrine that still today permeates Western thought, nothing was left of his own church.

In more ancient times, some of the apologists had subjugated everything to faith. In writing "Do you not know that faith is the leading principle in all matters?" the bishop Theophilus, for example, had fostered the criticism of pagans such as Celsus, according to whom Christianity wanted only "to gain over the silly, and the mean, and the stupid, with women and children." Augustine professed, on the contrary, that reason had to be cultivated, because it was a gift from God that distinguished mankind from the beasts, and one without which belief would make no sense. One of his most fertile precepts for natural philosophy was his declaration, following Plato, that all things "are good, and altogether very good." Because the world was the work of an intelligent Creator who was intelligible and good, it would submit its secrets to human understanding. But understanding nature was certainly not to be seen as an end in itself; rather, it was to be viewed as a means by which to arrive at knowledge of God. And because any science depends, after all, upon a revelation, it follows that "that which is supported by Divine authority ought to be preferred over that which is conjectured by human infirmity." In this way, the benefit of the doubt goes to the Scriptures: a metaphorical interpretation of the sacred texts is required only when reason irrefutably demonstrates the absurdity of a literal interpretation. Augustine could thus prescribe: "Understand what you may believe; believe what you may understand." Without keeping in mind this precept, which delimited the domains of faith and reason, it would not be possible to understand the behavior of the principal protagonists of the next chapters of the present book. On the subject of the account of Genesis, which is known only by revelation, Augustine continued to follow Philo in tending toward an allegorical interpretation of certain points: the "number six is the first which is made up of its own part." But the sense that was to be given to such days escaped him: "What kind of days these were it is extremely difficult, or perhaps impossible for us to conceive, and how much more to say!"

As for the question "Why now for the Creation?" Augustine finally dealt with it on the basis of analogy. To philosophers who said that "the thoughts of men are idle when they conceive infinite places, since there is no place beside the world," he replied that "by the same showing, it is vain to conceive of the past times of God's rest, since there is no time before the world." Time was therefore not bound to the movements of the stars, as Aristotle had maintained, but rather to the creation of the world: "simultaneously with time the world was made." But it was on the subject of cycles that Augustine broke off clearly with Plato, and it was the Passion, rather than free will, that made him reject the Great Year, or the recurrent creations postulated by Origen. "Far be it, I say, from us to believe this," he declared in *The City of God*, "for once Christ died for our sins; and, rising from the dead, He dieth no more." In conclusion, the only cycles that existed were those in which the sinners found themselves imprisoned: "The wicked walk in a circle," he concluded, citing the Psalms, "not because their life is to recur by means of these circles, which these philosophers imagine, but because the path in which their false doctrine now runs is circuitous."

Aristotle Forgotten, Then Found Again

The Creation, the Passion, and the Last Judgment clearly imposed a sense of history and of time that was neither eternal nor infinite. In putting the finishing touches to the work of Paul of Tarsus, Augustine founded a theology of history. However, little was built upon the bases of the philosophy of science that he had established just as well. With the exceptions of medicine and astronomy, the only two disciplines of practical interest, the study of nature had been an occupation that was primarily subject to the chances of individual initiative: even during their golden age, there never were very many Greeks who applied themselves to this study. When the Roman world dissolved, the flame of science was extinguished, and the Christians did not revive it, being more preoccupied by the Savior than by observation of a sky that could not lead to any discovery that in their eyes would be useful. And even had astronomy's identification with astrology not been sufficient to give it a repugnant aspect, the disagreements among the Greek schools regarding the nature of the heavens evidenced an obvious confusion among the pagans. When these latter ridiculed the Christian indifference to astronomy, Saint Basil the Great (330–379), who was Eusebius's successor at the episcopal see of Caesarea, replied simply: "Do not let us undertake to follow them for fear of falling into like frivolities; let them refute each other, and, without disquieting ourselves about essence, let us say with Moses 'God created the heavens and the earth.' Let us glorify the supreme Artificer for all that was wisely and skillfully made."

The Hellenic cosmologies did in fact have a point in common: they were theologies, as Pierre Duhem emphasized. In expelling God from the stars and fighting astrology, the Christians, in a certain way, secularized the cosmos. And all throughout the Middle Ages, the form the cosmos was considered to have was the one inspired by a literal interpretation of Genesis. On the second day God commanded, "Let there be a vault through the middle of the waters to divide the waters in two," and at the same time, he distinguished "the waters under the vault from the waters above the vault." Following Basil's suggestion, the solution was to suppose the sky to be formed of two distinct envelopes, the internal sphere, that of the fixed stars, and an external sphere, the *Empyrean* of the Greeks, in which, at such a distance, only angels lived; these spheres were separated from one another by a sphere of ice (not by liquid water, for obvious reasons relating to stability). In his commentaries on Aristotle, the Christian John Philoponus (late fifth to middle sixth century?), one of the later philosophers of Alexandria, attacked the old distinction between the corruptible and incorruptible worlds, basing his argument on the differences in the colors of the stars. If the stars were not the same color, Philoponus noted quite elegantly, then they must not have the same composition; and since they do have a single composition, they, just like terrestrial matter, must be susceptible to corruption.

Augustine, himself of the Latin culture, was the church father who was most familiar with secular science. With the exceptions of Plato and the Neoplatonic school, the Latinized Christianity of the high Middle Ages lost the memory of Greek authors little by little. Independently of the attacks that his *Physics* sustained, Aristotle, who had rarely been translated into Latin, escaped total oblivion only on account of his principles of logic. In the eastern Roman Empire, the Academy existed in one way or another until the year 529. When the emperor Justinian finally ordered it to close on the claim of pernicious teachings and paganism, its seven professors made the journey to the court of the king of Persia, from which, however, they returned straightaway. A century later, the Arab conquest began; Alexandria was taken in 641, and the remains of its library disappeared. In less than one hundred years, the Muslim world expanded and, nonetheless, revived the fragments it had gathered from Greek science. As for the natural history of time, no remarkable progress was accomplished during that period.

When the Arab outburst began to weaken, the relay was taken up in the West. In the monastery and cathedral chapters, the Scholastic saw masters and students debating and commenting on the Scriptures and the church fathers. The *quadrivium* was put in place, ensuring that astronomy, arithmetic, geometry, and music would be taught. In an old debate, Abelard undertook the task of emphasizing the importance of reason in relation to faith. His theses had to cede to those of Saint Bernard of Clairvaux, for whom "a pious ignorance was the essential badge of the true Christian," according to the formula of D. C. Allen, but the attitudes continued to evolve. The first universities, which were not subject to direct influence of the clergy, were founded and placed under the administration of the pope. Universities at Bologna, dating from around the end of the eleventh century, and Paris, from around 1150, preceded those at Oxford and Cambridge by several decades; then followed Montpellier, Padua, and Salamanca in the first half of the thirteenth century. In spite of all sorts of troubles during those times, there was an astonishing mobility among scholars and clerics that promoted the exchange and circulation of ideas.

The sciences in these universities were hardly at the level of bare subsistence. Through exchanges with Sicily or Spain, which was partly under Arab domination, new discoveries were made of the treatises from antiquity, such as Euclid's geometry, Ptolemy's astronomy, and the medicine of Hippocrates and Galen. Scholars began translating the works into Latin from the Arabic or Greek, and they learned about later works of Arab scholars and their commentaries on the great authors of antiquity. The encyclopedic work of Aristotle, which depicted all facets of nature, could certainly not have been neglected. But Aristotle rejected the immortality of the soul and free will, and he submitted to an astrological fatalism, expounding a system that was found to be incompatible with Christian doctrine on numerous points. As a result, Aristotle's works were officially blacklisted by the University of Paris at the beginning of the thirteenth century, before being banned by Pope Gregory IX in 1231.

A wiser course than censuring Aristotle appeared to be commenting on his works and interpreting them. One of the greatest minds of the Middle Ages was Albertus Magnus (ca. 1200–1280), a Dominican from Swabia who was entrusted with the task of reconciling Aristotelian teachings with Christian doctrine. Albertus was sent by his order to teach in Paris, where the Place Maubert (Maître Albert) in the Latin Quarter commemorates his name. He had accepted the heavy charge of this compilation only with some reservation. He did it solely "for the honour of Almighty God, the Fount of Wisdom and the Creator, Founder, and Ruler of nature; and also for the benefit of the Brothers, and of any others who read it with the desire of acquiring natural science"; accordingly, the church later made him patron of scientists. In volume after volume, and with the aid of his pupil Thomas Aquinas, Albertus succeeded so well in his task that the University of Paris shortly thereafter made the works of "the Philosopher" required reading for its students.

From New Points of View

The second age of Scholasticism was thus dominated by Aristotle. It was recognized, however, that numerous questions found no satisfactory answers in Peripatetic physics. In order to avoid arbitrarily limiting the omnipotence of God, for example, it was necessary to reject the impossibility of a void and of the plurality of worlds that were demonstrated by this physics. Much more bothersome was that reason seemed to be in agreement with Aristotle regarding the immutability of movement and the consequent eternity of the world. In order not to find the Scriptures contradicted, faith had to impose itself firmly on the discussions that endlessly ensued: 27 of the 219 theses (not all of which came from Aristotle) that were solemnly condemned in 1277 by Etienne Tempier (ca. 1210–1279), the bishop of Paris, touched more or less directly on the question of limitless time. Finally, because the Ascension of Christ remained beyond any sort of interpretation, Thomas Aquinas had to conclude that it indeed constituted a mystery.

Hardly triumphant, Scholastics were truly struggling under the attacks of the first partisans of an "experimental" science, such as the English Franciscans Robert Grosseteste, Roger Bacon, and William of Ockham. Bacon (ca. 1219 to ca. 1292), in his *Opus majus*, exposed to the pope his conceptions of science and his propositions for reforming education; there he clearly recorded how one could avoid the failings that threatened Scholasticism and that would leave it, in the final account, with a bad reputation. He claimed that "there are two modes of acquiring knowledge, namely, by reasoning and experience. Reasoning draws a conclusion and makes us grant the conclusion, but does not make the conclusion certain, nor does it remove doubt so that the mind may rest on the intuition of truth, unless the mind discovers it by the path of experience." And to complete this recourse to experience, Bacon pointed out that there was another aid that was indispensable: "the things of this world cannot be made known without a knowledge of mathematics."

The glorification of the Lord nevertheless remained at the heart of this new development. One was expected to contemplate the divine work and discover the essential mysteries in such a way as to complement the teachings of theology. According to Bacon, "one science is the mistress of the others, namely, theology, to which the remaining sciences are virtually necessary, and without which it cannot reach its end." This science was therefore at the service of the church, which needed to assure its triumph over the Antichrist and to promote the work of evangelization, because of the glory it spread for the Creator. This, at last, would permit man to control nature. In discussing "another alchemy, operative and practical," in opposition to his speculative form, Bacon thus perceived that science "not only yields wealth and very many other things for the public welfare, but it also teaches how to discover such things as are capable of prolonging human life for much longer periods than can be accomplished by nature."

Without really passing from their precepts into practice, Bacon and his colleagues nonetheless introduced a considerable mutation: the Greek philosophy of nature was going to give way by degrees to a true natural science. Pierre Duhem (1861–1916), the theoretical physicist, played a considerable role in the recent illumination of the acquisitions of this medieval science. Beginning with a study of Leonardo da Vinci's science inquiries as expounded in his notebooks, Duhem undertook to rediscover the forgotten roots of modern science by researching its deepest foundations. As a fervent Catholic who was exposed to the hostility of the radical left dominating the scientific environment of the Belle Époque in France, Duhem found it fulfilling to discover that the religious context of the Middle Ages had far from stifled human curiosity. He even dared to assert that, by rejecting the principles of Peripatetic physics, "the excommunications pronounced in Paris, on 7 March 1277 by Bishop Etienne Tempier and the doctors of theology, were the birth certificate of modern physics"! A monumental work ensued, entitled *Le système du monde* (*The System of the World*), in which Duhem expounded the thoughts of numerous authors from antiquity and the Middle Ages. Two of them merit brief consideration here, Jean Buridan and Nicole Oresme, because these two great figures thought nature should be understood not by the expedients of astral influences or final causes, but by means of objective, mechanical explanations.

A Freedom to Speculate

The name of Jean Buridan (ca. 1300 to after 1366) of Picardy is connected with the old dilemma of the ass that died of hunger and thirst because it could not decide between the bucket of oats and the pitcher of water. We know very little about Buridan. He was rector of the University of Paris in 1328 and 1340, but—in spite of the fable that was spread by the poet François Villon in the following century—he could not have been one of the lovers of Queen Jeanne de Navarre (she died in 1304) who were thrown into the Seine in sacks weighted with stones in the early hours of the morning, after a night of dissipation and pleasure in the tower of Nesle. Because he rejected Aristotle's theory of natural movements, Buridan should instead be known for his revival of the notion of *impetus*, developed from the ideas of John Philoponus, which prefigured the principle of inertia: "Since the creation of the world," he declared, "God has stirred the heavens with the very same movements as those that stir them at present; he has thus impressed *impetuses* upon them, by means of which those continue to be moved in a uniform manner. And because they do not encounter any resistance that opposes them, these impetuses are never destroyed nor weakened. According to this conception," he concluded, "it is not necessary to suppose the existence of intelligences that would move the celestial bodies in a suitable manner."

Buridan also left some penetrating observations on mountain relief, owing to his journeys in the Massif Central and to Mount Ventoux, in Provence. In particular, he described the slowness of erosion and its different actions on hard rock in comparison with soft rock. He also advanced the idea that earthquakes formed mountains; and because of the heights of such mountains, he concluded that "by natural means, it would be impossible for there to have been a Universal Flood, that is to say, for the entire earth to be covered by waters, even though God would have been able to do it by supernatural means." The role of erosion permitted Buridan, at last, to consider the important problem of the earth's place within the universe. In order for it to be a natural place, according to Aristotle and his commentators, the center of gravity of the earth had to coincide with that of the universe. Now, the earth's centers of volume and of gravity cannot be coincident, by reason of the nonuniform distribution of land masses and seas. By expanding the areas of land, solar heat had caused an imbalance of mass. Erosion and the seas, then, as mobile agents, came to reduce the land mass by degrees. Buridan deduced from these facts that a slow displacement of the land was occurring all around the globe: "I admit, therefore, in conformance with the preceding conclusion that, in ten thousand years, the ocean must have advanced ten leagues on the eastern side, covering over such an expanse of land as previously was habitable, and that an equal expanse of land must have been abandoned on the western side." Buridan thus invoked displacements of the sort that extended over hundreds of millions of years.

We find a similar viewpoint in the work of the Norman Nicole Oresme (ca. 1320–1382). Shortly after the invention of the mechanical clock in the thirteenth century, this pupil of Buridan expounded the analogy, which since has become a classic, between the movements of the heavens and the mechanism of this instrument: "No one would say that the absolutely regular movement of a clock happens casually without having been caused by some intellectual power," he asserted; "just so must the movement of the heavens depend to an even greater degree upon some intellectual power higher and greater than human understanding." Close for a long time to Charles V, for whom he was a tutor, Oresme was dean of the Cathedral of Rouen in Normandy when this lettered king asked him to translate several of Aristotle's treatises into French. With the assistance of numerous commentaries and discussions, Oresme completed the translation of *De cœlo* in 1377 in the best spirit of the Scholastic. Titled *Le livre du ciel et du monde* (*The Heavens and the World*), this was one of the first scientific works in the French language, and as such it required Oresme to introduce hundreds of words that were new or technical (for example, *angular, oval, longitude, meridian, solstice, motor, gravity,* etc.); others that were not new came into more frequent usage (*imperceptible, rare, duality, intelligence, symbol, observation, distinguish, design,* etc.). He was thus one of the most productive authors of neologisms that later passed into other languages, including English.

But for our purposes *The Heavens and the World* is of interest for other than philological reasons. In a highly significant long passage about the movement of the earth, Oresme concluded that the earth's apparent fixity in space certainly does not prevent one from assuming that our planet actually rotates about itself. If so, then, would there have been any disagreement about the celebrated passage of the Bible in which, at the request of Joshua, Yahweh stopped the sun in its course, in order to allow Israel to triumph over the Amorites? Oresme did not think so: "One could answer the sixth argument, which concerns the reference in Holy Scripture about the sun's turning, etc., by saying that this passage conforms to the customary usage of popular speech as it does in many other places, for instance, in those where it is written that God repented, and He became angry and became pacified, and other such expressions which are not to be taken literally." But nothing, said Oresme, prevents one from thinking that the earth turns about itself, nor is there anything that demonstrates that such a movement does occur. In so resorting to the precepts of Augustine, Oresme rejected the metaphorical interpretation in order to let the Scriptures have the last word; even so, setting aside the verities of faith, he recalled, in a Latin manuscript, "I indeed know nothing except that I know that I know nothing."

From Reform to Revolution

Far from attracting reproaches from the church, *The Heavens and the World* was cause for Oresme's promotion to bishop of Lisieux, another town in Normandy. For Buridan, envisaging periods ranging from tens of thousands to hundreds of millions of years did not involve any sort of unfortunate consequences. He certainly was not providing an outline for any sort of dating, but the view he expressed did indicate that he was not limiting his thought to the Mosaic framework. For the needs of his reasoning, Buridan could calmly state: "I also conjecture that the world has existed in perpetuity, as Aristotle appears to have understood it." The addition of the stylistic clause "even if it is false according to our faith" allowed him to continue right on: "I suppose that it is the celestial bodies that move and direct the world below here, without the need of any miracle." But such liberty to speculate could not weather the Reformation: Faith would soon have its reasons, to which Reason would have to submit.

A full century after Oresme, the thesis that the earth could rotate was taken up again by a young Polish canon. After studying Greek, Latin, philosophy, and medicine in Krakow, Bologna, and Rome, having landed a doctorate in canonical law at Ferrara in 1503, Nicolaus Copernicus (1473–1543) obtained his office in Frauenburg, in eastern Prussia, through the help of an uncle who was a bishop. Copernicus did not like the Ptolemaic system at all. Because circular movement was natural, and because linear movement was against nature, he thought that the earth, with its perfect spherical form, ought to have a trajectory that was equally perfect, that is to say, circular. In 1506 he postulated that the sun was at the center of an immobile sphere of fixed stars and lighted with its own light the other stars that turned about it, in uniform movement, by virtue of their form alone.

Copernicus was a mathematician, not an astronomer, and accordingly he was guided by aesthetic considerations. In fact, the heliocentric system, which he finished working out around 1530, gave no better account of observable facts than did Ptolemy's. Moreover, the problems that were raised by his system were serious: for example, because the earth itself became a planet, the division between the celestial and the sublunary worlds was disrupted, and gravity could no longer be interpreted as a natural movement that was directed toward the center of the universe. Further absurdities abounded in the eyes of the faithful readers of the Bible. As explained by Thomas S. Kuhn, if the earth was not unique, then other similar bodies ought to exist in the universe, and from the certainty of God's benevolence the implication would ensue that these also were inhabited. But how could such living beings have descended from Adam and Eve? How would they have inherited the original sin that had resulted in man's lot of hard toil? How would they have been made aware of the Savior's existence? And how would they have been able to benefit from redemption? And if the universe was infinite, as some would soon propose, where would God's throne be?

Copernicus was exhorted in 1536 by the archbishop of Capua to publish his work, at the archbishop's expense, but he waited seven years before doing so, and during that time one of his friends arranged to have a summary of it produced. Copernicus dedicated his *De revolutionibus orbium clestium* (*On The Revolutions of Heavenly Stars*) to Pope Paul III, and it is said that he died on the same day the first copy of his work was brought to him. Without Copernicus's knowledge, a preface had been added to his book. It was written by the Protestant theologian Andreas Osander (1498–1562), who presented the new thesis as a purely mathematical device, one that was devoid of any physical reality. Even though Copernicus had been inspired primarily by aesthetic considerations, it is doubtful that he would have agreed.

Copernicus prided himself, in his own preface, for having written a book of mathematics for mathematicians: dry and free from any sort of sensationalism, it was read by nobody except the specialists. In the meantime, the Reformation was launched in 1516 by Martin Luther (1483–1546). In a society where the diffusion of ideas through the press only rendered religious conflicts all the more implacable, the scholars' place was at the front line. From the point of view of the Reformation, Copernicus's yet unpublished ideas represented, straightaway, the symbol of those detestable metaphorical or allegorical interpretations that had deviated Christians from their faith and cut them off from the Bible. In 1539 "there was mention of a certain new astrologer who wanted to prove that the earth moves, and not the sky, the sun, and the moon. This would be as if somebody were riding on a cart or in a ship and he was standing still while the earth and the trees were moving." About the ideas of this fellow who "wishes to turn the whole of astronomy upside down," Luther claimed: "Even in these things that are thrown into disorder I believe the Holy Scripture, for Joshua commanded the sun to stand still and not the earth [Joshua 10:12]."

The initial reactions from the Catholic Church to Copernicus's ideas, as we have seen, were not negative. The Council of Trent, which began meeting in 1545, reaffirmed the primacy of the Church's teachings in relation to the Scriptures, correctly observing that heresies had always had a reading of the Bible as a point of departure. Without raising any serious controversies, the heliocentric theory could thus be used to put into practice the great calendar reform instituted by Pope Gregory XIII. As the new calendar took effect, the fifth of October in 1582 was followed immediately by the fifteenth, in order to eliminate the difference that had accumulated between the calendar year and the solar year since the time of Julius Caesar.

The Industry of Chronology

It was at the timid beginnings of the Copernican revolution that a question much more intimately connected to the biblical account returned to the foreground. This was something that affected the date of the Creation and, more generally, the Mosaic chronologies. For Protestants, it represented a relatively weak link in their doctrine, as it would have called those foundations into question again if the link was invalidated. After all, Luther himself, in defending the primacy he accorded the biblical texts, had affirmed that God was incarnated into the Word just as Christ was into the flesh. The golden age of chronology that ensued is associated most frequently in the modern Anglo-Saxon world with the name of James Ussher (1581–1656), an Anglican archbishop of Armagh, in Ireland—and with the thick irony surrounding his claim that the Creation took place at the beginning of the night preceding October 23 in the year 4004 BC. Ussher draws his dubious celebrity from the fact that in 1701 his conclusion was inserted into the King James Bible as a marginal note that was maintained in subsequent editions even into the early twentieth century. Although, like the majority of his colleagues in chronology, Ussher was a great scholar, he was only one among many masters of an industry that prospered until the end of the eighteenth century.

After the works of Theophilus of Antioch, Julius Africanus, and Eusebius, interest in chronology had not waned even slightly. It increased to the extent that an eschatological sentiment began to develop and that many sought to discover the hand of God in historical events or to foretell the day of the Last Judgment. Astronomical recurrences were researched in attempting to identify noble moments within the cycles, during which the world might have been created (such as when the sun was rising at the summer solstice, or during one of the equinoxes in one or another of the constellations). This intense interest is the reason why the concept of the Great Year was resolutely condemned in proposition

6 of Etienne Tempier's 219 propositions issued in 1277. But these chronological methods not only showed an abuse of mathematics; they also
demonstrated a complete lack of any critical sense. An interminable
compilation written by an Oxonian astrologer, John Ashenden (? to ca.
1379), testifies to this continuous effort on the part of the most famed of
scholars—from the Venerable Bede (673–735) to Bacon and Grosseteste.
Ashenden reported dates that various individuals had arrived at for the
Creation ranging from 3752 BC to 5530 BC.

In the years before the Reformation, the return to antiquity that marked the beginnings of the Renaissance was accompanied by a new reading of the biblical texts. Even before the advent of printing, the critics
became vigilant, wanting to rid the manuscripts of the errors and modifications introduced by copyists and translators. There were three principal versions of the Old Testament, each one corresponding to a different
historical period. A Greek translation called the Septuagint was produced
in Alexandria, it was believed, during the third century BC for the use of
the Hellenized Jewish community dispersed around the Mediterranean,
who did not know Hebrew. The Vulgate, the Latin translation revised at
the end of the fourth century by Jerome, was the text of the church. And
of course, there was the Hebrew text, which had been reviewed at various
times by rabbis since the great rewriting in the second century AD. Naturally, some differences became evident when scholars began to compare
these texts. Augustine, the opponent to Jerome in one epistolary controversy, was the first to express uneasiness: "I wish you would have the
kindness to open up to me what you think to be the reason of the frequent discrepancies between the text supported by the Hebrew codices
and the Greek Septuagint version." But these divergences were lost to
view later on. Beside the serious inconsistencies that were found a good
millennium later, the points of dispute between these versions could be
seen as practically anecdotal, such as those concerning what kind of
wood had been used to construct Noah's ark. For an age in which the
conflicts seemed to intensify, however, none of the disagreements were
trivial, because they automatically brought into question the Holy Spirit,
who had inspired Moses. From this point of view, the Hebrew text
was not of much use for certain scholars, because even though it was
difficult to deny the competence and the seriousness of the doctors of
Jewish law, it was fundamentally unpleasant to consult them for the
purpose of resolving questions of Christian doctrine.

Within this context chronology was completely renovated by Joseph-
Juste Scaliger (1540–1609), the son of a celebrated doctor and Italian
philologist who settled in 1525 at Agen, in southwestern France, and who

embraced Protestantism at an early age. He wrote the main part of his work in Touraine while in the service of a French ambassador before accepting a chair at the University of Leiden in 1593. There, upon his death, he was lamented by admirers as "the divine offspring of a divine father," the "perpetual dictator of letters," and "the greatest work and miracle of nature." Whereas the enterprise begun by Diodorus, Eusebius, and Jerome was essentially pedagogical, Scaliger turned it into a full-scale reconstruction of the past. He considered that a chronology ought to take advantage of all available sources and that it was actually all the more necessary to make use of secular documents, for otherwise some passages of the Old Testament would remain obscure. For the episodes regarding the exile in Babylon, for example, the pertinent chronology was certainly not that of the captives.

In this way, Scaliger accomplished an immense task, which was aimed at comparing, contrasting, and harmonizing the ancient chronologies (Hebrew, Assyrian, Babylonian, Egyptian, Greek, and Roman) with the more recent ones of the Vandals, the Ostrogoths, and the Visigoths. In order to date all the significant events that were registered from the Creation to the Middle Ages, he sifted through the astronomical bases of fifty different calendars. He compared calculations made in terms of calends, ides, and decades (in the sense of periods of ten days); made by archons, ephors, and aediles; in the context of Pythian and Centennial games; within Attic years, Syrian years, and embolismic years. The breadth of the deciphering he accomplished still fascinated the romantics a good two centuries later. Scaliger produced two major works, *On the Correction of Chronology* and the *Repertory of Dates*, which completely updated the efforts of Eusebius and Jerome, established a coherent and enormously vast historical framework, rectified numerous errors, and clearly identified historical lacunae.

After Scaliger, the world of scholars was occupied for a long time with the fever of chronology. Among Catholics, the Jesuits were particularly active, such as the French Denis Peteau (1583–1652) or the Italian astronomer Giovanni Battista Riccioli (1598–1671). Among Protestants, the challenge was taken up by Ussher, by the English writer John Marsham (1602–1683), and by the Dutch humanist Gerardus Joannes Vossius (1577–1649). These efforts were further supported by new documentation, such as the endless chronologies brought back from China by the Jesuits, who reveled in the ancientness of the peoples they had evangelized. In the nineteenth century, François-Auguste-René Chateaubriand (1768–1848) noted that all peoples were troubled by the same shortcoming: "Man feels within himself a principle of immortality and shrinks, as it were, with shame, from the contemplation of his brief existence. He imagines that by piling tombs upon tombs he will hide from view this capital defect of his nature, and by adding nothing to nothing he will at length produce eternity." As would be shown by the *Histoire de l'astronomie ancienne* (*History of Ancient Astronomy*) of Jean-Sylvain Bailly (1736–1793), the unfortunate mayor of Paris who was decapitated during the Revolution, these exorbitant pretensions to ancientness were all reduced to nothing by a rigorous examination of some of the calendars of antiquity, where sometimes a "year" represented only a single solar day.

A Critique of Biblical Reason

In a debate that was contaminated by the defense of sectarian interests, there was unfortunately no sign that the various chronologies would converge. "I myself have gathered more than 200 different calculations, of which the shortest counts only 3483 years from the creation of the world to Jesus Christ, and the longest counts 6984," claimed Huguenot Alphonse des Vignoles (1649–1744), who was one of the first directors of the Academy of Science in Berlin, in the preface to his monumental *Chronologie de l'histoire sainte* (*Chronology of Sacred History*). At any rate, it would appear that the chronologies could not possibly converge when the Hebrew text and the Septuagint differed by fifteen centuries in the amount of time that elapsed between the Creation and the birth of Abraham. Feeling that a certain lassitude was appearing, the Cistercian Paul Pezron (1639–1706) acknowledged it in the preamble to his *Antiquité des temps rétablie et defenduë contre les Juifs & les Nouveaux Chronologistes* (*Antiquity of Time Reestablished and Defended against the Jews and the New Chronologists*), published in 1687: "Time, which consumes all things, and which seems to want to commit everything to an eternal oblivion, has practically robbed man of any knowledge of his own duration and ancientness. This is true to such a degree that, after all the care we have taken, in our days, to discover its extent and to find out how many centuries have elapsed between the origin of the world and the coming of the Messiah, not only have we not come any closer to the truth, but we have actually gone quite a bit farther from it."

As time passed, the date of the Creation took on more importance, since by itself it summed up all of the disagreements. But the parties who were so greatly at odds with one another did agree on one capital point: the brevity of time. With respect to the antiquity of the earth, whether the Creation had occurred six thousand or eight thousand years before made scarcely any difference: the history of the earth was necessarily confused with that of humanity. Without encountering any sort of opposition, and apart from all religious presuppositions, the timescales that were established were disproportionately brief. Returning like a leitmotiv in the study of nature, these short durations had profound consequences for the interpretation of fossils, as well as for the role that the Flood had played. And to the antiquarians of time, the inconsistencies of these ages could not be ignored. They justified the pursuit of a chronological investigation that lasted nearly until the French Revolution and was complemented by the archaeological excavations first undertaken at the beginning of the eighteenth century. Whereas atheists felt a certain satisfaction in seeing some justification for their impious theses in these unyielding difficulties, others had reactions that finally led to an entirely new scrutiny of the biblical texts.

The first of the new examinations, because it was radical, assured a great success for its author, Isaac de La Peyrère (1594–1676), a gentleman from Bordeaux. A sincere but critical Calvinist, La Peyrère found his attention caught by a remark regarding sin in Paul's Epistle to the Romans that made him think that Adam could not have been the first man. In order to prove his thesis, he embarked upon a literal reading of the Bible, which he pushed to the extreme. Noticing a number of lacunae, transposed passages, chronological inconsistencies, and even irresolvable contradictions, he questioned how it could it have been otherwise: "To these deluges, fires, and devouring times, add that gross ignorance, which in several ages has overrun the whole world, more powerfully than fire, than water, than time itself, which hath swallowed, blotted out, and defaced the memory of things past"! Following a long theological discussion on sin, La Peyrère demonstrated, in a book published anonymously at Amsterdam in 1655, how the belief that humanity had begun with Adam and Eve resulted in nothing but absurdities. For example, if God had not previously created other men, how could Cain, the laborer, have had available to him a weapon with which to kill his brother? And how could he have been able, afterward, to take a wife in order to perpetuate his race? Accepting such stories, La Peyrère assured, was equivalent to believing that all Italians had descended from a certain *Italicus*, and all French from a *Francus*, and it made him think about the naïveté of "the two-penny Doctor, who told the sick man he had eaten an Asse, because he saw the dorsers[1] standing under the bed!"

---

1. *Dorser* refers to "a basket carried on the back or slung in pairs over the back of a beast of burden," according to the *Oxford English Dictionary*, s.v. "dorser," "dosser."—Trans.

The Bible's account was therefore only the history of the Jews, God's chosen people, and not of all humanity. Before the Jews, as La Peyrère explained, God had created the *preadamites*. These had been abandoned by the Creator but had escaped his punishment, because the Flood had been limited to Palestine, where the only Gentiles to have been drowned were those who had mingled with the Jews. La Peyrère's book was greeted with jubilation by impious persons and quickly ended up being burned in public. It resulted in repercussions for its author (whose anonymity was fragile) that were not all happy. After being arrested at Mechlin, in Belgium, and thrown into prison, La Peyrère was able to breathe fresh air again through the intervention of his protector, the prince of Condé, but only on the condition that he renounce Calvinism and Preadamitism at once. He went to do so in person before the pope, who received him, it is said, with benevolence. As for the sincerity of his renouncement, the epitaph that honored him leaves some room for doubt:

Here lies La Peyrère, a good Israelite
Huguenot, Catholic, and finally Preadamitist.
Four religions pleased him at the same time;
And his indifference was so extraordinary,
That after the eighty years he had for making a choice;
The good man departed, and never chose a one.

Perhaps he was not really such a fool as he was believed to be.

In large numbers the apologists of that time hastened to push the shadowy preadamites back into limbo. But the questions that La Peyrère had raised did not disappear with the same move. Some of the passages in Deuteronomy, such as the eulogy of Moses and the account of his death, had long demonstrated that the text could not have been entirely from the hand of the first prophet. The Manichaeans, for example, had already made this assertion in their attacks on the fathers of the church, but these were merely details; in no way did they call the main point back into question. By the middle of the seventeenth century, there was a long-standing suggestion that these problematical passages were not isolated examples, and even that the Pentateuch might have been drafted by a later prophet who had equally been inspired by God. The blistering attack by La Peyrère prepared some minds for new scrutiny of one of the dogmas that had been accepted, by the church as well as by the synagogue, since the earliest times of our era. They concluded, along with our preadamitist historian, not only that the Pentateuch was not entirely the work of the hand of Moses, but that it had been written by several different authors of several different periods.

One of the first steps accomplished in this direction was taken by Baruch Spinoza (1632–1677), a Dutch philosopher. For him, only a proper and well-reasoned examination would permit one to understand the Bible without drawing from it a teaching that it did not profess. Putting this principle into practice, Spinoza was able to establish clearly, in a treatise published secretly in 1670, that the Bible had a "history" and that as it was written and revised by men, this history had become confused with that of its authors. Although Willem van Blyenbergh, a Dutch theologian, judged that Spinoza's book was "filled with curious, but abominable, discoveries, the science and research of which could only have originated in hell," the book was in fact responsible for suggesting the foundations of modern exegesis. In France, a scholarly priest of the congregation of the Oratory, Father Richard Simon (1638–1712), began making efforts to demonstrate the importance of tradition in the interpretation of biblical texts. What Simon proved in his *Histoire critique du Vieux Testament* (*Critical History of the Old Testament*), published in 1678, was that the sacred texts had indeed been written by different authors of different times and that all too often they were corrupted or represented mediocre translations of sources that had since disappeared. Simon furthermore confirmed that "everything we know about the chronology of the Bible is still insufficient to give us exact knowledge of the number of centuries that have elapsed since the creation of the world." Shortly after it was printed, Simon's book had a sensational impact: the energetic Bishop Bossuet had all copies of it seized to be turned into pulp, even before Simon was accused of degrading the Bible to the level of the works of Homer or Aristotle.

But the march of the critic was making good progress. Even Voltaire was to have his hand in it at one point, felicitously, although it was Jean Astruc (1684–1766) who finally identified the Yahwistic and Elohistic sources at the origins of the two versions of Genesis. Astruc, a physician specializing in venereal diseases, was also a consultant to Louis XV; a professor at Montpellier, then later at the Collège de France; and an amateur naturalist. His *Conjectures sur les mémoires originaux dont il paroît que Moyse s'est servi pour composer le livre de la Genèse* (*Conjectures on the Original Memoirs Moses May Have Used to Compose the Book of Genesis*) was published in 1753, shortly after Georges Louis Leclerc, Count of Buffon (1707–1788), published his *Theory of the Earth*, the first wide-ranging attempt to recount the past of our planet. This coincidence, by no means

Figure 2.1 After the Flood: abandoned shells, future fossils. From Parkinson, *Organic Remains of a Former World*, vol. 1, frontispiece.

fortuitous, was exceeded by one that followed immediately after: at the very time when La Peyrère, Spinoza, and Simon were laying the foundations of biblical history, Robert Hooke, the English physicist, and Niels Stensen, the Danish anatomist, were striving to do the same for historical geology. A new sense of history was thus beginning to take shape, extending broadly across all the disciplines.

In spite of the development of biblical studies throughout the nineteenth century, particularly in Germany, the new theses of Simon and

Astruc were not easily accepted. As the *Biographie universelle* demonstrates, around the middle of the nineteenth century there were still some who castigated Simon for the dangerous principles he had promulgated. Even so, during that same time, the six days of the Creation had already been relegated to the status of a metaphorical interpretation. The celebrated paleontologist Georges Cuvier (1769–1832) was an inveterate Protestant who saw nothing more in the Mosaic chronology than a history of humankind. On this account, "there is no reason to doubt but that the book of Genesis was composed by Moses himself," he wrote, adding that it sufficed "to read it to perceive that it was composed partly of fragments of former works." But in his opinion, the fact remained that these sacred texts were of divine inspiration. As such, credit still needed to be accorded them by his science. A millennium and a half after Augustine, the debate between Faith and Reason had still not been definitively settled.

THREE

Genesis as Viewed through the Prism of Natural Philosophy

REASON IN THE SERVICE OF FAITH   The scientific revolution stemming from the astronomical and cosmological work of Copernicus, Kepler, and Galileo had important implications for the examination of chronology. While Halley's pioneering geophysical investigations provoked controversies, Newton misused the powerful resources of his new physics in his quantitative interpretations of the Mosaic chronology.

The Renaissance of the Soul

A general feature of great discoveries is that they tend to wipe clean the slate of the past: once Copernicus's system was introduced, the scientific revolution quarantined medieval science into oblivion. Even though the earth had lost its central place, the humblest of all positions at the very center of the corruptible world, the universe remained finite and bounded by the sphere of the fixed stars. No empirical observations had raised any questions about this sphere, for the simple reason that Copernicus had established his system starting from data that were scarcely different from those used by Ptolemy. The scientific revolution resulted, in effect, from the more mature approaches that Bacon and his colleagues had begun using as they sought correspondences, whether empirical or mathematical, among observable facts and experimental results, the validity of which could then be tested by phenomena. From this point of

view, the revolution actually preceded the great progress of instrumentation: the telescope, the thermometer, and the microscope were the fruits
of this revolution, not the causes.

No one ever encouraged breaking off from Scholasticism and its
commentaries with more contagious enthusiasm than Petrus Severinus
(1542–1602), a Danish doctor. "Go forth my sons," he ordered;

sell your fields, your homes, your clothing, your titles, burn your books, purchase
shoes! Conquer the mountains! Explore the valleys, the deserts, the seashores, the deep
folds of the earth! Observe the different species of animals and plants! The classes of
minerals, the properties and the origins of each and every thing! Do not be ashamed
to learn attentively, by heart, the Astronomy of the peasants, the Philosophy of the
earth. Lastly, buy coal, build ovens, be attentive, and cook without repugnance. It is
in this manner, and in no other, that you will achieve a knowledge of bodies and their
properties.

Behind these precepts, however, can be found some less familiar attitudes, which continued throughout the long period of transition that accompanied the Renaissance. In many respects, these represented an obvious regression, and as such they affected progress in the study of nature
and time. In order to recognize them, we need only go back to Severinus.

"These would be weighty tasks if they did not promise profitable re-
turns for your labor," he had continued. "Because they contain obvious
explanations of hidden properties, they reveal the sources of actions
and the manners of those actions, the times for which have been determined beforehand. They show the agreement and harmony of all nature.
They constitute you, you who have been adorned with these riches, and
confirmed by this abundant experience, as the philosophers, legitimate
interpreters, and ministers of nature."

There was one great German astronomer who, beyond the shadow of
a doubt, exceeded all the others as interpreter and attentive minister of
these hidden properties and harmonies of nature. He was not content
merely to describe the movements of the stars, their rotations about
themselves, or their revolutions around the sun; he had the innovative
idea of searching for the *physical* causes of these movements. Although
he postulated the existence of magnetic interactions between celestial
bodies, and of attractions between bodies proportional to their distances,
he was not able to achieve his aim, in particular because the principle of
inertia was not known at the time. The old hypotheses remained: the sun
and the planets were not inanimate bodies; an inert body's movement
was analogous to life for matter; and the earth therefore had a soul,

which was in tune with the harmony of astral configurations. Thus our astronomer took inspiration from Virgil when, in 1619 in *The Harmony of the World*, he compared "the bosom of the earth to *the thighs of a wife*, and indeed a *joyful wife*, that is, one who perceives what is happening to her with pleasure and helps her husband with suitable motions. All these things are signs of life, and suppose a soul in the body which experiences them. For it would not be easy for the sun, destitute of suitable troops, to invade this citadel of the bowels of the earth, without the co-operation of some kind of soul, seated within, to collude with the enemy and open the gates to him."

Somewhat backward on this point compared to Buridan or Oresme, this great scientist who credited the stars with souls was Johannes Kepler (1571–1630), the founder of modern astronomy. In his mind, such reminiscences of mythical, cosmic loves were not incompatible in the least with his unshakable Christian faith. When his own work led him to abandon the circle, the figure that for the Greeks was divine, in his descriptions of planetary orbits, it made way for the Holy Trinity to take possession of the universe: the Father occupied the central place, that is, the sun, the point of reference through which light was projected and whence the movements of the planets were governed; to the Son was associated the sphere of the fixed stars; and the Holy Spirit expanded throughout the rest of space. As for the laws that govern planetary movements, these expressed musical harmonies, which the human mind could comprehend while communing with God.

In his quest for inexpressible harmonies, and in his conviction that a world devoid of souls would have been an abomination, Kepler was certainly not alone. "Heaven is man, and man is heaven, and all men together are the one heaven, and heaven is nothing but one man," proclaimed the indefatigable Swiss doctor and chemist Paracelsus (1493–1541), who was resistant to all sorts of tradition and eager for all sorts of knowledge. For this great master of theosophy and his numerous followers, nature was supposed to be read directly, page by page, in the search for the invisible forces that animated it. In nature, man, the *microcosm,* was the image of the *macrocosm*: the body corresponded to the visible world; the senses, to the astral world; and the soul, to the celestial world. Both alchemy and medicine and both astronomy and astrology were often pursued concurrently by the same person, and they merely represented different ways to understand the microcosm starting from the macrocosm, to discover the subtle relationships that, like the quintessence, provided for these correspondences between the material and the immaterial.

This philosophy, which signaled a spectacular resurgence of Neoplatonism, had its theorists in the Florentine Academy, a group of scholars, such as Pico de la Mirandola, close to Prince Lorenzo de Medici at the end of the fifteenth century. The Christianized Neoplatonism of these lettered men, however, was not familiar to the greater public, whose view of nature continued to rest on a shallow foundation of ancient, magical beliefs. While societies were decimated by epidemics, fascinated by death, threatened by the advances of the Turks, and surprised by the discovery of the New World, the Reformation, and the collapse of beliefs and institutions, and while legends about the great fears of the year 1000 found their places retrospectively, the course of nature did not appear to have been disrupted. As A. Koyré noted on the subject of Paracelsus's philosophy, nature is neither a system of laws nor a system of bodies governed by laws; it is an unflagging, omnipotent force, because it is everything, present in all beings without being reduced to them. To sum up, everything appeared to be possible through this inexhaustible, vital force: "Magic is natural because nature is magical." It was at that time that witches proliferated and ended up being burned; Kepler's mother was herself accused of witchcraft and barely escaped the stake. Nature was being examined through eyes that were clouded over by transcendencies dating back to the Middle Ages. It was this world, imbued with various occult qualities, that presided over the beginnings of the scientific revolution, which, in turn, caused its demise.

From the Stars to the Nativity

The astronomers, practically by definition, were astrologers, especially those under engagement by sovereigns. Even so, the Dane Tycho Brahe (1546–1601), in service to his king for a time and then later to Emperor Rudolphe II in Prague, exemplified better than anyone else the renewed importance now being accorded to measurement. Using precision unequaled until the telescope appeared, Tycho observed and followed the stars in a systematic way, no longer fragmentary—a way that had scarcely been used since Ptolemy. The system of the world he left did not survive him. It was intermediate between those of Ptolemy and Copernicus in that the earth kept its central place and had, as satellites, the sun, around which the other planets turned, and the moon. Tycho left us, above all, a very precise catalog of some one thousand stars, and he made two capital observations. In 1572, within the constellation Cassiopeia, a star arose that was visible in broad daylight, surpassing even the most brilliant stars in brightness. Then it weakened and disappeared, after undergoing considerable variations in color. Tycho proved that this was a new star, a *nova*, and not some atmospheric phenomenon. Secondly, he concluded that the comet that appeared in 1577 had come from a distance that was at least six times greater than the distance between the earth and the moon. Because they were incompatible with the conception of immutable heavens, these two observations soon played a significant role in opening the sphere of the fixed stars outward toward the infinite. And in the end, it was on the basis of the positions of the planets detected by Tycho that Kepler was able to formulate his great laws.

Kepler was an ecumenical Christian, albeit an intransigent one. He shifted back and forth between the Catholics and the Lutherans all through the troubles that preceded the Thirty Years' War. He was obliged to interrupt his studies of philosophy and theology in order to go to Graz, Austria, to teach mathematics. There, in 1596, he published a little book, the *Mysterium cosmographicum* (*The Secret of the Universe*) in which he affirmed that he had penetrated the most profound secret of the universe. In the Copernican system, which he was among the first to adopt, the radii of the circles traversed by the six planets corresponded to those of spheres nested successively, one within the other, in such a way that one could inscribe within them the five fundamental, regular polyhedra. Proceeding from Mercury to Jupiter, these were the octahedron, the icosahedron (twenty faces), the dodecahedron (twelve), the tetrahedron, and the cube. Related to the essays on geometry and on harmonies in Plato's *Timaeus*, this was actually the first attempt at a construction of the universe on the basis of mathematical principles, from which all arbitrariness was to be excluded. In the eyes of its author, it confirmed the profound, divinely inspired harmony that determined the physical world.

After taking refuge in 1600 near Tycho in Prague, Kepler was charged with giving an account of the apparently rebellious movement of Mars. Unfortunately he had to resign himself to disillusionment, because the precise observations of Tycho could not be reproduced on the basis of circular trajectories; the beautiful construction of his *Mysterium cosmographicum* collapsed. After about ten years of laborious calculations, Kepler resolved the difficulties and published his first two laws in his *Astronomia nova* (*New Astronomy*). There he explained that the orbit of Mars is actually an ellipse, with the sun at one of the foci, and that Mars travels more slowly along this ellipse where its distance from the sun is greater (or more precisely, the line joining Mars and the sun sweeps out equal areas in equal intervals of time). Yet another decade of efforts finally allowed Kepler to extend his conclusions to the other planets and to formulate his third law in his *Harmonice mundi* (*The Harmony of the World*), which established that the square of the period of a planet's revolution is proportional to the cube of its average distance from the sun ($T^2/D^3 =$ constant). As a spectacular illustration of the validity of these laws, the planetary tables that Kepler published shortly before his death proved to be a hundred times more precise than those they replaced.

Kepler's entire work was painted over a backdrop of service to the divine; no other scientist could have better applied himself to putting astronomy to the service of sacred history and its chronology. Following a long tradition, Kepler in his turn attempted to date the Creation. Because "God did not start the motions [of the stars] at random," he asserted in his *Mysterium Cosmographicum* that "if the year of Christ is taken as the year 5572 of the universe (though it is generally and by very sound authority reckoned as 5557), the creation will come to an illustrious combination of stars at the start of Aries." More originally, he considered it his duty, above all, to establish the precise date of the birth of Christ. In this way he would "silence the Godless Epicurean scoffers and scorners of Christianity of whom this day there is not an inconsiderable number among Jews and Christians in large towns at court and in commercial cities," who were capable of bringing up incorrect calculations of the years so as to give free rein to their hatred of religion. Had the star of Bethlehem that attracted the magi been a nova? Or did it represent a conjunction of Jupiter, Mars, and Saturn? When did the lunar eclipse that preceded Herod's death actually occur? Kepler compared these considerations with historical and biblical texts and affirmed, contrary to the opinions of Scaliger and other scholars, that the date of the Nativity was actually five years earlier than the beginning previously established for the Christian era. This conclusion has since been accepted. But another, even more illustrious, scientist did not find as much success in exercises of this sort.

The Infinite Universe, Created Anew

While Kepler was formulating his laws, Galileo Galilei (1564–1642) was studying falling bodies in Venice. His work was interrupted in 1609 when the rumor spread that a marvelous invention had been made in the Netherlands: by properly arranging two glass lenses, one could create a *telescope*, which made it possible to see objects at enormous distances. Galileo was a highly curious scientist, as well as a capable artisan, so he set to work immediately; the exceptional ninefold enlargement that he obtained prompted him, at first, to boast of the great military value of his new instrument to the doge. When he later had the inspiration to turn his telescope to the skies, rather than toward the horizon in search for enemy sails, what he discovered was a new universe. With its four satellites, Jupiter proved to be another center of movement within the sphere of fixed stars. Even more amazing—countless, previously unknown stars suddenly burst forth when the somber immensity of the celestial vault was scrutinized. The infinity of the universe was being revealed. It had been foreseen by Nicholas of Cusa (1401–1464), a mathematician, diplomat, and cardinal, and sustained by Giordano Bruno (1548–1600), the former Dominican who was burned at the stake for pantheism and heresy against the Trinity. The sphere of fixed stars, though still maintained by Kepler, was disappearing, along with the medieval sphere of ice. Moreover, far from the perfectly smooth surfaces that had been imagined for the moon and the sun, the tormented relief of the moon and the spots of the sun now betrayed the intrinsically corruptible nature of these two bodies. The ancient view of the cosmos as finite and immutable had already suffered blows from Copernicus. Galileo now finished the job by completely shattering it.

While the Protestants were gradually becoming accustomed to heliocentricism, the Curia's hostility toward Copernicus's ideas was increasing in Rome, being further nourished by the disordered support that Bruno had brought to it and by the repercussions that this system was beginning to produce, even outside of strictly scientific circles. In 1617, a year after Copernicus was put on the Index, Kepler, his staunchest advocate, joined him. A decade later, this condemnation did not prevent the very persistent Jesuits from trying anew to win him over to their faith. But—the story is too long and complicated to be told in detail—an implacable driving force was at work. Galileo had affirmed that "since two truths cannot contradict one another, it follows necessarily that this [heliocentricism] and the Scriptures are in perfect agreement," but one cardinal had answered that it would be better for Galileo, rather than trying to convert the theologians, to restrict himself to the mathematical aspects of astronomy. Isolation and the abjuration of heliocentricism finally broke the spirit of Galileo, authenticating his trial as an eminently durable symbol of religious obscurantism.

For René Descartes (1596–1650), this movement of the earth was of extreme importance: "I confess that if it is false, all the foundations of my Philosophy are also false, because they are demonstrated by those in a manner that is manifest," he wrote to his friend, Father Marin Mersenne, in 1633. Furthermore, Galileo's condemnation had the accessory effect of frightening him: "I had set myself the proposition of sending you my *Monde*" (a draft of his unpublished work, *The World*), he confided to Mersenne, but "I have sent inquiry, these days, to Leiden and Amsterdam, as to whether there were any copies at all to be found of Galileo's System of the World.... They wrote me back that it was true that it had been printed, but that all the copies of it had been burned in Rome, all at the same time, and that he was condemned to some fine. This is something that has astonished me so strongly that I have almost resolved to burn all my papers, or at the least, not to let them be seen by anyone." Though he was audacious, Descartes nonetheless professed five years after his retirement to Holland that "to live unnoticed is to live in happiness." Soon, however, the austere Calvinists there would accuse him of atheism—even though he had never ceased producing new proofs of the existence of God. In order to avoid any sort of polemic, he decided in 1633 to keep his *Monde* to himself, for in it he had undertaken to deduce all natural phenomena from a few fundamental principles.

This was only a temporary deferral. After producing other classic works, such as *Dioptrics*, in which he derived the law of the refraction of light, and *Geometry*, in which he threw the study of geometry into upheaval by stating that any figure can be associated with an algebraic equation, Descartes resolved to publish his *Principles of Philosophy* in 1644. His purpose was to expose "everything that is most general in Physics, namely, the explanation of the first laws or Principles of Nature, and the way in which the Heavens, the fixed Stars, the Planets, the Comets, and generally all the universe is composed." Clearly, this Cartesian cosmology was in line with the traditions of Philoponus or Buridan, and not with that of Kepler. With the principle of inertia being formulated correctly for the first time, his point of departure marked the disappearance of the circle as a natural movement: as long as no action was exerted, a body at rest would remain at rest, and a body in movement traversed a rectilinear trajectory in a uniform manner. Descartes asserted furthermore that every action is exerted by means of collisions (though he formulated the laws incorrectly) and that collisions had, in the distant past, gradually put an end to the primordial chaos by organizing matter, which was fluid and continuous, into the form of adjacent *vortices*, which established themselves mutually and expanded infinitely out into space. In a nature that abhorred a vacuum, it was shocks that corrected the centrifugal tendencies conferred to material particles by their inertia and gave them closed, though not perfectly circular, trajectories.

Each vortex could be the seat of a solar system, a star forming when volatile matter was accumulating at its center. Light represented a material

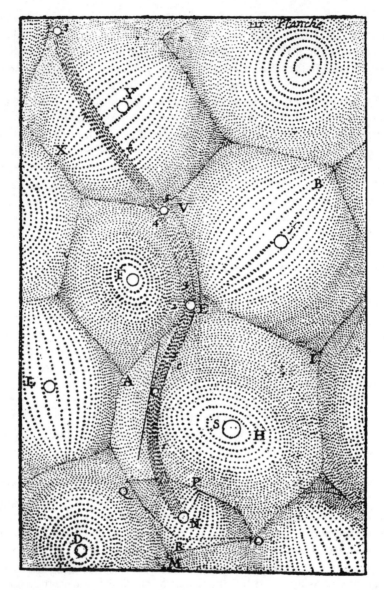

Figure 3.1 The famous vortices of Descartes: sun (*S*), fixed stars (*F, L, D*), and a comet beginning its journey (*NCEV*...) when the star *N*, initially situated at the center of its vortex, is "pressed" too much by the adjacent vortices, which end up removing it. From Descartes, *Principes*, plate 3.

vibration, transmitted afar, whose source was the seething of the star rotating at the center of its vortex. Some other opaque material could come to condense at the surface of the star, forming spots analogous to those on the sun, and end up by covering over the star, which would then become a planet and find itself caught up by another vortex. Jupiter with its four satellites, and the earth with the moon, were illustrations of the fact that "in the great vortex which forms a heaven having the Sun at its center, there are other smaller ones which we can compare to those I have often seen in eddies of rivers, where they all follow the current of the larger vortex which carries them, and move in the direction in which it moves." And from time to time, as in the case of dead stars, coagulations of matter were ejected from their original vortex and continued on, in an erratic course determined by the centrifugal pressures relating to the neighboring vortices: these were the comets.

As Pierre-Simon de Laplace (1749–1827) noted, Descartes had the immense merit of substituting "for the occult qualities of the Peripatetics, the intelligible ideas of motion, of impulse, and centrifugal force." Aristotle and Scholasticism had to be left behind once and for all. For all sorts of phenomena, from light to gravity (due to the thrusting of vortices of subtle material upon terrestrial bodies), a mechanical description had become necessary, and Descartes had the audacity to extend it even to the world of the living, with his vision of the "man machine." Many have blamed him, the author of *Discourse on the Method of Rightly Conducting the Reason, and Seeking Truth in the Sciences*, for neglecting Galileo and Kepler and, even more so, for trampling underfoot the reasonable doubt that he had introduced into the heart of scientific methodology. Nonetheless, he affirmed "that the matter of the heaven does not differ from that of the earth; and that even if there were countless worlds in all, it would be impossible for them not to all be of one and the same [kind of] matter." In so doing, Descartes displayed such boldness that some were still holding it against him in the middle of the nineteenth century. And above all, the old cosmogonical myths had had to give way, before his *Principles*, to a cosmology that basically remained the only theoretical description of the evolution of the universe until the Big Bang concept.

In particular, Descartes's theory of the earth, the first such theory, fit into this grandiose picture that some qualified derisively as "romance physics." This theory formed the point of departure for most of the further speculations on the subject. The earth appeared to Descartes as a crusted-over star, upon which had come to be deposited successive layers of matter: metals, water, clayey earth, and air. It had adopted a layered structure covering a still-hot core of solar matter and a crust of opaque

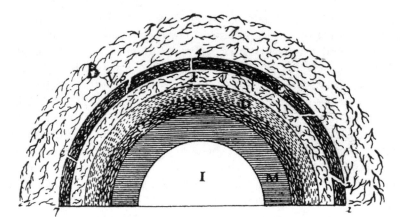

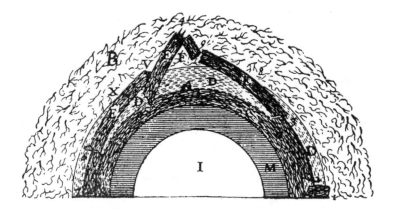

Figure 3.2 The earth in section, according to Descartes, before and after the formation of the mountains. *I*: primal, originally subtle, matter; *M*: opaque body, similar to the crust of sunspots; *D*: water; *B* and *F*: air. From Descartes, *Principes*, plate 15.

matter. Under the influence of the sun's heat, the layer of water had, over time, provoked the fissuring, then the subsiding, of the layer of clayey earth, and had thus given birth to the mountain chains and to the seas surrounding the plains.

Describing a Creation in which the role of God had been at best accessory, Descartes concluded, "Nonetheless, mindful of my insignificance, I am affirming nothing: but submit all these things both to the authority of the Catholic Church and to the judgment of men wiser than I; nor would I wish anyone to believe anything except that which he is convinced of by clear and irrefutable reason." Without explicitly comparing his account to that of Genesis, Descartes had actually offered an explanation for the Flood with his masses of water escaping from vast, subterranean reservoirs, although he ignored erosion, sedimentation, fossils, and time, among many other things. "But only in the space of many years could the particles of body D have been reduced to the two types described a short while ago, and all the shells of body E be formed," he remarked in passing. The brief Mosaic chronology had not been shaken in the slightest by this Cartesian outline that was devoid of any timescale.

Academicians and Virtuosi

As a member of the wealthy nobility, Descartes lived peacefully off his private income before departing for Sweden, where he died prematurely, a victim of the rigors of the climate, shortly after having enlisted the patronage of Queen Christine. His situation was doubly classic for an age in which scholars who were not Jesuits or in a position to enjoy personal revenues had to seek the patronage of rich protectors, upon whom they would reflect prestige. Science developed slowly in the universities, where professors had neither the leisure nor the means required to practice the experimental method. It was most effectively pursued in colleges directed by religious figures where all new knowledge remained a source of glorification of the divine work. With the obvious exception of heliocentricism, not only did science suffer little from their highly vigilant censure, but as J. L. Heilbron has noted, "the single most important contributor to the study of experimental physics in the seventeenth century was the Catholic Church, and within it, the Society of Jesus." It was indeed among the Jesuits that physics was taught the best; and especially in France it was through their schools that the elite of society passed, such as Descartes and Buffon.

In the seventeenth century the number of scientists increased noticeably: informal groups were formed, networks were created, and science became the principal object of the new societies of the erudite. Founded in Rome in 1609, the Accademia dei Lincei (Academy of the Lynxes) was the first scientific academy. In Paris, the cell of Father Marin Mersenne became a center of exchanges, well before Jean-Baptiste Colbert founded the Royal Academy of Sciences in 1666. The scholars, closely supervised by the king, were divided into various grades, and they each received a salary from the state, whose resolute engagement made it possible for them to undertake ambitious expeditions. From 1699 to 1739 Bernard le Bovier de Fontenelle (1657–1757) was secretary of the Academy. He had already met unequaled success in 1686 with his *Conversations on the Plurality of Worlds*, where he treated "philosophy in a very un-philosophical manner," and his reputation would later give a special luster to his *Histoires*, which served as introduction to the *Mémoires* published by the Academy's members.

On the other side of the Channel, another sort of inquiring mind began to distinguish itself, that of the virtuoso. Roaming the world on his own, or keeping up a vast correspondence, this was the sort who "values a *Camelion*, or *Salamander's* Egg, above all the Sugars and Spices of the West and East-Indies and wou'd give more for the Shell of a *Star-fish*, or *Sea Urchin* entire than for a whole *Dutch* Herring Fleet," quipped the essayist and mathematician Mary Astell (1668–1731), in her *Essay in Defense of the Female Sex*. When a group of scientists founded the Royal Society of London, shortly after the Stuarts had been restored, and had its statutes approved by King Charles II in 1662, they became a society that no longer had to depend upon a monarch who could find himself short of money; they funded themselves instead by the subscription revenues from their members alone. The society members were not limited by qualifications: besides the scientists, eminent aristocrats and rich virtuosi were invited.

To accompany the movement, the first periodical publications appeared. In 1665, one year before the Academy, the *Journal des Sçavants* (*Savants' Journal*) was created; it was "a succinct and condensed selection of all the most astonishing occurrences in Nature, and of what has been done or discovered to be most curious, within the Arts and Sciences." More of a scientific gazette than a journal, it gave a large place to theology. By contrast, theology was excluded from the discussions of the Royal Society of London, as well as from the columns of its rich *Philosophical Transactions*, which began publication shortly before the *Journal des Sçavants*. Both periodicals dealt with diverse topics of discussion. For example, after a half century of its publication, one could still find in *Philosophical Transactions*, printed side by side, speculations by the astronomer Halley on "the Infinity of the Sphere of the fixed Stars," a new recipe for "making Sugar from the Juice of the Maple Tree in New England," and "an Account of a boy who lived a considerable Time" (three years) without eating or drinking.

But it was not easy to get into the habit of publishing rapidly. In a climate of suspicious competition, scientists hesitated to expose themselves, often preferring to advance a concept under the mask of a cryptogram, so as to be able to establish a priority tersely. For example, the physicist Robert Hooke (1635–1703) put forth "Ceiiinosssttuu" in 1676 to announce the law of elasticity bearing his name (the anagram unscrambled is "ut tensio, sic vis," that is, "as the extension, so the force"). Through the long years, it was Hooke who animated this heterogeneous Royal Society, the initial enthusiasm of which had quickly slackened to the point where its finances were continually unstable. The son of a respectable minister on the Isle of Wight, Hooke had been troubled since infancy with delicate health and the feeble constitution of a hunchback; it was even doubted that he would make it through early childhood. He showed a talent for painting, so when he was orphaned at the age of thirteen, he entered apprenticeship with a renowned portraitist. The paint fumes aggravated his headaches, which were already too frequent, and so the adolescent had to change his path. He resumed his scholarly studies, demonstrating brilliance in mathematics and ancient languages, though in the end it was music—a chorister's scholarship—that opened the doors to Oxford for him in 1653. The group of scientists who formed the core of the Royal Society met there regularly, and Hooke did not miss the chance to make himself appreciated by them, through the inventive fever that had animated him since youth.

Starting off as assistant to the chemist Robert Boyle, Hooke conceived and constructed apparatus of all sorts. In the end he was engaged by the Royal Society, in 1662, as curator of experiments, charged with the task to "bring in every Day three or four Experiments of his own, and take care of such others, as should be mentioned to him by the *Society*." Hooke worked wonders in this impossible mission of factotum scientist; he was engaged in experiments on the refraction of light, the force of gravity, elasticity, animal respiration, the physical properties of air, meteorology (for which he conceived a thermometer, a barometer, an anemometer, and a hygrometer), and others. And nothing stimulated his imagination so much as the innumerable constraints imposed upon him by his duties, to which those of librarian and head of the "repository of rarities" were soon added. It was through these offices that Hooke later established himself as one of the founders of geological science. It was also at the Royal Society, of which he was made a member in 1663, where he crossed paths for the first time in 1672 with the younger Isaac Newton (1642–1727), who was then also on the verge of becoming a member (and whose memoir on optics Hooke had hastily condemned). Hooke pirouetted from one subject to the next with brilliant perspicacity, even though his approach lacked analytical depth, while Newton, in contrast, would not leave a problem until he had exerted his utmost powers of penetration upon it. A fruitful relationship between the two physicists ensued, though they finally turned bitter toward each other; Hooke referred to Newton as the "veryest knave in all the Ho," whereas the Newtonian camarilla in return labeled him unjustly as an embittered quarreler.

Newton

As an infant, Newton was both posthumous and premature—saved in extremis by Providence, he thought. He was born to a well-to-do rural family of Lincolnshire in the same year that Galileo died. Owing to his fragile health, his childhood was not a particularly happy one. His mother had remarried and had him raised by his grandmother; Isaac did not see his mother again until he was eleven years old, when his stepfather, who was a minister and had been supervising his education, died. Newton was very adroit with his hands, but to all appearances he would have been incapable of taking over the family business, and so he entered the university. He was admitted to Cambridge in 1661 and just eight years later became the second holder of the only chair in science, which had been well endowed by Henry Lucas. Meanwhile, the year 1666, an annus mirabilis, marked the conclusion of a period that began in 1663 when Newton acquired a book on astrology at a fair. Stopping short before a figure of the heavens, he purchased a book on trigonometry; because he did not understand its proofs, he continued on to Euclid, then to Descartes, the Englishman John Wallis, and the other great authors of the time. With hardly a transition after this learning stage, he established infinitesimal calculus under the name of the calculus of *fluxions* (derivatives), along with some other mathematical feats; he conceived his theory on light; and he presented his first reflections on gravitation.

Newton spent the greater part of the year of miracles at home. In June of 1665 it had "pleased the Almighty God," as was recognized locally, "in his just severity to visit this towne of Cambridge with the plague of pestilence." According to a famous anecdote first told by Voltaire, it was on his own estate that a falling apple had put him on the track of his gravitational theory. As Kepler and Galileo had demonstrated, the laws that describe the movements of the stars and the falling of bodies were simple ones. This force that attracted the apple must have extended much farther than had been thought, perhaps also keeping the moon in its orbit about the earth, as well as the planets about the sun. But the calculations for the speed of the moon's fall toward the earth were inconclusive. Newton abandoned the attempt. However, Hooke had proposed the idea that a planet's movement was the result of the equilibrium between a centrifugal force and a central force of attraction (one that acted between center and center). Hooke further believed, independently of Newton, that this force was an inverse function of the square of the distance between two bodies, but he was not able to solve the problem mathematically. Through an exchange of letters in 1679 and 1680, he discussed these concepts with Newton, who then got back to work, putting into action his vast mathematical resources and his awareness of the importance of Kepler's laws. He found, first of all, that for any law of central force that maintained "the Planets in the Orbs, the areas drawn from them to the sun would be proportional to the times in which they were described." Subsequently, he demonstrated that it was through the use of a law of attraction, which was an inverse function of the square of the distance, that the other two laws of Kepler could be deduced.

But Newton had the habit of keeping his discoveries to himself. When the young Edmond Halley (1656–1742) came to consult him about the movement of a planet subjected to a force of attraction inversely proportional to the square root of distances, he was surprised to learn that Newton had already solved the problem. It was Halley who urged him to divulge his theory: if Newton would only draft it, Halley would take care of the rest. Newton allowed himself to be convinced. He developed his views more explicitly and laid the foundations for the theory of dynamics that had eluded him for so long. Among other things, his theory distinguished the notions of mass and weight and demonstrated that masses, when considered in regard to his law of attraction, could be supposed to be concentrated at the centers of their bodies. Humphrey Newton, his longtime, homonymous assistant, remarked at one time that "he seldom left his chamber except at term time, when he read in the schools, as being Lucasianus Professor, where so few went to hear him, and fewer that understood him, that oftentimes he did, in a manner, for want of hearers, read to the walls."

As a response to Descartes and his *Principles of Philosophy*, Newton's three books of *The Mathematical Principles of Natural Philosophy* were published, at Halley's expense, in 1686 and 1687. Their theme was the single governing force for the falling of bodies, for the tides, and for the movements of the planets and comets. It was a veritable system of the world, built upon this central force of attraction, which was proportional to the mass of objects and inversely proportional to their distance from one another. And this force exerted itself throughout the utmost depths of space: void and space were becoming absolute in themselves. One of the most spectacular triumphs of the Newtonian system was provided by Halley. He calculated in 1705 that the comet of 1682 had an elliptical trajectory, rather than a parabolic one, as Newton had thought; accordingly, he predicted that it would return in 1758. Because the comet was faithful to its appointment, it has borne Halley's name ever since.

Meanwhile, Newton eventually wearied of Cambridge. In 1696 he obtained the important office of warden of the mint in London. Although he was still far from having published all his works, he had at that time accomplished the essential part that could be considered properly scientific. Quite well versed in theology, he concealed his highly unorthodox opinions about the Trinity, whose existence he denied. But he did no such thing in regard to his contempt for Catholicism, letting it be known that he considered the pope an Antichrist and the devotion to saints as idolatry. He became a foreign member of the French Academy of Sciences in 1699 and president of the Royal Society of London in 1703, which he directed with a firm hand until his death; he was even knighted by Queen Anne in 1705. In short, the authority exerted by Newton was awe-inspiring. He had already permitted himself in 1695 to blurt the order to John Flamsteed, the astronomer royal, who was on bad terms with Newton's friend Halley: "I want not your calculations but your Observations only!" And when Newton denied his rival Gottfried Wilhelm Leibniz (1646–1716) any sort of credit relating to the invention of infinitesimal calculus, very few at the time knew that Newton himself had written the report for the commission of the Royal Society of London, appointed on his request in 1712 to settle the differences, and that the favorable commentary on said report, published in the *Philosophical Transactions*, had also come from Newton's pen.

Newtonian physics was not exempt from difficulties, though. If one excluded Descartes's concept of action through contact, recalling the mechanism of billiard balls that transmit their movements to one another, how could the force of gravitation come to exert itself instantaneously, all the way to the cores of stars? Newton refused to answer the question, using the celebrated maxim "hypotheses non fingo" (I make no pretense to hypotheses). And he added, "for whatever is not deduced from the phenomena is to be called an hypothesis; and hypotheses, whether metaphysical or physical, whether of occult qualities or mechanical, have no place in experimental philosophy." There was a whole philosophy of experience behind Newtonian physics, and this found its place and established itself slowly, even in England.

Gravitation smacked of heresy, particularly on the Continent, where, despite Newton's denials, people feared a return to the occult qualities that were so dear to the Peripatetics. In the long struggle between the Cartesians and the Newtonians, the foremost propagandist of the new ideas was Voltaire (1694–1778), while he was sharing life with Gabrielle-Émilie Le Tonnelier de Breteuil, marquise du Châtelet (1706–1749), who translated Newton's *Principles* into French and whom Frederick the Great nicknamed "Venus-Newton." At the Château de Cirey in Champagne, in the company of a woman who loved physics, and "verses and the wines of Champagne" as well, Voltaire, the Prince of Enlightenment, confided tranquilly: "My serious occupation is to study Newton, and to attempt to reduce that giant down to the measure of my fellow dwarves." Though he was entirely devoted to his latest passion, he did not completely forget his earlier ones: "Newton, here, is the god to whom I sacrifice," he recognized, "but I also have chapels for other, subordinate divinities." His admiration for this man who measured his laughter and ever disdained the company of women was tinged with regrets when he said, daringly, "I wish Newton had done vaudeville; I would have appreciated him more. A man who has but a single talent can become a great genius; someone who has several is more likeable!" While still in the care of Madame du Châtelet, Voltaire published a complete exposé of the new physics in his *Eléments de la philosophie de Newton* (*Elements of Newton's Philosophy*) in 1738. He proclaimed the downfall of Descartes in this work before a public that was still hostile to its hero: "These vortices, pressed one against the other / Moving about without space and piled one upon another, without any sort of order—/ These learned phantoms are disappearing before my eyes."

The Lessons of Distant Voyages

In the third book of his *Principles*, Newton wrote about an object he had found worthy of his theory of gravitation, the earth itself. The remote origin of this interest came from observations made in French Guiana in 1672–73, on the initiative of the recently established Paris Academy of Sciences. For only a short time had telescopic viewings or pendulum clocks allowed taking the positions of stars or measuring time with considerably increased precision. As a consequence of this progress, new fields of study had opened up. In order to make comparisons, it even became necessary to know the precise latitudes and longitudes of the modern and ancient observatories, such as those of Ptolemy or Tycho

Brahe. A vast ensemble of physical and astronomical measurements figured among the priorities that were outlined by the Academy in Paris in 1667. Because these were to profit geography and navigation, as well as the age-old problem of determining longitudes at sea, this first scientific expedition to be truly dedicated to such a program, and not just to the collecting of exotic specimens of flora and fauna, had the advantage of receiving financial support from the king and from Colbert, his general comptroller of finances.

In the end, the expedition chose French Guiana as its destination. It had been a French possession since 1664, it was close to the equator in the southern hemisphere, and communications with it were not too difficult. And so Jean Richer (1630–1696), a junior astronomer of the Academy, set off on 8 February 1672 from La Rochelle on a ship of the French West Indies Company; after falling ill, he departed Cayenne on 5 May 1673, leaving his assistant, who finally died there. The most spectacular result from this expedition was the first precise measurement of the distance between the earth and the sun. Beginning with observations of Mars made simultaneously in Cayenne by Richer and in Paris by Dominique Cassini (1625–1712), the founder of an illustrious line of astronomers, the distance was found to be 21,600 terrestrial radii. Another important discovery, however, escaped the attention of the general public. Richer had noticed with surprise that the pendulum of his clock needed to be shorter at Cayenne than at Paris in order to beat the same seconds. If the length of the pendulum was kept constant, the difference in the period corresponded to nearly two and a half minutes per day. What was the significance of this curious fact?

Richer's observations, which Halley later confirmed, did not intrigue Newton at first. Although Hooke had already thought that the earth exhibited a bulging, in 1680 Newton was "most inclined to believe it spherical or not much oval." If the pendulum swung slower at the equator than at Paris, he finally supposed it must be because gravity was less strong there as a result of the greater centrifugal force. He concluded, following Hooke, that the earth ought to bulge at the equator, as had been observed for Jupiter through the telescope. Regarding this point, which was consistent with Descartes's cosmogony, Newton concluded that the earth had initially been a fluid mass in rotation and that it had maintained its bulge as it solidified. In order to calculate the combined effects of gravity and centrifugal force (maximum at the equator, minimum at the poles), Newton limited himself to the case of a hydrostatic equilibrium between two channels with polar and equatorial axes. Using calculations that his contemporaries could but poorly follow, he found that the flattening of the earth ought to be 1/230. Pushing his analysis a step further, Newton could then account for the precession of the equinoxes: this movement, which had escaped explanation ever since the times of Hipparchus, was a consequence of the equatorial bulge. The earth turns on an axis that is inclined with respect to the ecliptic, so this bulge, in effect, creates an asymmetry of mass in relation to the sun. Because the force of gravitational attraction from the sun is stronger upon the part of the bulge that is closest, a coupling is established, which tends to drive the plane of the equator into the plane of the ecliptic, producing the top-like movement of the axis of the poles in accordance with a period of twenty-six thousand years. If instead the earth were perfectly spherical and homogeneous, the axis of its rotation would be fixed, and there would be no precession whatsoever.

These demonstrations given in the *Principles* contradicted the fact that the lengths of arcs of meridians, as measured in France between Dunkirk and Collioure, indicated a lengthening of the earth along the direction of the polar axis. Those results were first published in 1718 by Jacques Cassini (1677–1756), then confirmed several times, until in 1736 they triggered a serious battle in France between the "flatteners" and the "lengtheners" of the poles. Among the latter, some refused on principle to accept that the figure of the earth could possibly be dictated by a theory originating from a foreigner, while others, following the example of Jean-Jacques de Mairan (1678–1771), attempted to find some theory within Cartesian physics that would uphold their own affirmations. To cut the controversy short, the Academy of Sciences resolved to have measurements made for one degree of the earth's meridian near the poles and one degree near the equator. Under the direction of Pierre-Louis Moreau de Maupertuis (1698–1759), the surveys done in Lapland (at a latitude of 66°N) in 1736–37 gave a length of 57,438 toises (1 toise = 1.95 meters), that is, 378 toises more than the value measured around Paris. And so when Bouguer and La Condamine came back in 1744–45, after their ten years of adventures and misadventures in Peru, with the 56,753 toises they had measured near the equator, there was nothing left to be said. The earth's flattening, thus measured to be 1/215, accorded remarkably well with Newton's calculations. After praising Maupertuis as "the flattener of worlds—and of the Cassinis," Voltaire marked the return of the geometers with a sally of wit that has since become a classic: "You have confirmed, by going to places filled with troubles / What Newton understood without having to leave home." Buffon—as we shall see—made good use of this result.

The Age of Salt and of the World

The findings of an expedition undertaken shortly afterward by Halley echoed Richer's. Halley was not merely the astronomer royal who had been honored by a comet, or the midwife to the *Principles* of his friend Newton, into whose shadow he had been relegated by posterity. Halley also had an enterprising mind, active throughout the course of his long life. He successfully blended a profound intuition, a sharp sense of observation, and "a boldness, often successful, because it was always lucid," according to the expression of his colleague Mairan. Astronomy, mathematics, and physics were not the only disciplines to profit from his talents. Halley had an equal inclination for antiquity, and he published one of the very first tables of mortality (by which the prices of life annuities could be rigorously evaluated); he also invented a diving bell, which he used himself for studying the absorption of sunlight by water; and he was assigned diplomatic missions for Queen Anne. And the reason why we meet with him here is that he also undertook what was probably the first oceanographic cruise, making him one of the founders of meteorology and geophysics. Of special interest is that one can follow, almost step by step, the train of ideas that led Halley to establish, over the years, a link between the salt of the oceans and the age of the world.

The son of a well-to-do London merchant, the young Halley, a brilliant pupil, was attracted by mathematics and eventually seduced by astronomy. When he was barely twenty years old, he resolved to take advantage of the new astronomical methods. His ambition was to prepare a catalog of the stars of the austral skies, to complement the works then in progress in Europe for the northern hemisphere and those just recently undertaken by Richer for the southern hemisphere. Having his father as a liberal financial backer and King Charles II as his protector, he succeeded in setting up his expedition in less than four months. He left England in November 1676, sailing on a ship of the British East Indies Company, along with an aide and armed with sextant, quadrant, and telescopes. His destination was Saint Helena, the southernmost British possession at the time; there he took quarters (at his own expense) with the governor of the island and installed his observatory on a site overhanging that of the future tomb of Napoleon. All through his sojourn of more than a year, Halley was irked by unfortunately cloudy skies and abundant rains. The catalog published upon his return nonetheless included 265 stars that were not observable from England. His reputation in astronomy was established from that time on, and Halley was elected a member of the Royal Society of London at the precocious age of twenty-two. He was one of its most active and highly respected members: "he was generous," noted Mairan, "and his generosity was exercised even at the expense of vanity, from which scientists are not exempt any more than other men are, and which they display, perhaps, even more comfortably."

During the long months of navigation, Halley was intrigued by the interactions among the winds, the climate, and atmospheric pressure. In 1686 he calculated how pressure varies as a function of altitude and published the first map of the trade winds, attributing the causes of those winds to the convections induced by the sun's heating of the atmosphere (his explanation was only half correct, because it did not take into consideration the effects of Coriolis forces, discovered 150 years later). The importance of the clouds and the rains did not escape his notice, and Halley proceeded to measure the rates of evaporation of salt water, so that he could determine how these rates depended upon the winds and the sun's heat. His results, extrapolated at the Mediterranean, indicated to him in 1687 that "the quantity of the evaporations of water, as far as they arise from heat" was three times greater than the influx of water from rivers, and that this was the cause for the strong Atlantic currents coming through the Straits of Gibraltar, which compensated for the losses from evaporation. From that point, he concluded that the lakes that had no effluents, such as the Caspian Sea, the Dead Sea, and Lake Titicaca, ought to have increased in salinity over time and that the salt slowly introduced by tributaries ought to have become concentrated, because it was the water alone that evaporated.

Halley, who was a shrewd observer, had the remarkable idea of recognizing in this phenomenon the very first natural chronometer, but he waited until 1715 to suggest putting it into practice. "On the Cause of the Saltness of the ocean, and of the several lakes that emit no Rivers; with a Proposal, by means thereof, to discover the Age of the World" was the title of his memoir. Now if evaporation is the "true reason of the Saltness of these Lakes," thought Halley, "'tis not improbable that the ocean itself is become salt from the same Cause, and we are thereby furnished with an Argument for estimating the Duration of all things, from an Observation of the increment of Saltness in their waters." The solution, then, was simple, at least in principle: "I recommend it therefore to the Society, as opportunity shall offer, to procure the experiments to be made of the present degree of saltness of the Ocean, and of as many of these lakes as can be come at, that they may stand upon record for the benefit of future ages." With this idea of deducing the age of the world from the current salinity of the oceans and its rate of increase, Halley had not uttered his last word. Being a rigorous scientist, he considered the possibility that the level of salt in the oceans, and perhaps that of some of the lakes, might not have been negligible at the beginning of time. In that case, the world would certainly be younger than the method would indicate. But that possibility did not trouble him, because his intention was "to refute the ancient notion some have of late entertained of the eternity of all things; though perhaps by it the world may be found much older than many have hitherto imagined." This enigmatic conclusion did not reveal what he was really thinking, nor by how much he might have thought the Mosaic chronology fell short.

The Impact of the Flood

Although he did not, perhaps, "push skepticism to the extreme," as the nineteenth-century physicist Arago claimed, Halley was scarcely a religious zealot. Newton "could not bear to hear anyone talk ludicrously of religion," J. Conduitt, his nephew by marriage, noted, and he was "often angry with Dr. Halley on that score." According to Mairan, Halley "had no fear of going against the common opinion, and he had no scruples against conceiving or proposing hypotheses, or conjecturing in accordance with his observations and his particular ideas." The relative timidity of his conclusions regarding the age of the world might therefore be surprising, as might the fact that twenty-eight years elapsed between his first conclusions on the salinity of lakes and the publication of his idea to use salinity as a chronometer. The fact is that Halley was a voice crying in the wilderness: not only was no salinity measurement made, but the principles of his method were rapidly forgotten.

History does not tell us whether the threat of censure prompted Halley to keep his insights on chronology to himself for so long. It does, however, attest to his fear regarding the theory that he put forth to attempt a physical explanation for the Flood. Halley was struck by the precision of the biblical account: he thought that one could be "fully assured of [the fact] that such a deluge has been; and by the many signs of marine bodies found far above the sea, it is evident that those parts have been once under water: or, either that the sea has risen to them, or they have been raised from the sea." As one who was familiar with the skies and the heavens, Halley was in a position to bring forth an original solution that would also permit him to resolve the problem of the fossils perched upon mountain peaks. He remarked, first, that the Flood could not possibly be explained by rains: the level of the seas would not rise more than 22 fathoms (about 40 meters), even if all the annual rainfall of the rainiest place in England fell, over the entire earth, in one single day. "The Almighty generally making use of natural means to bring about his will," Halley sought the agent that would have been able to provoke the rising waters described in Genesis. He rejected the thought of resorting to caverns or other subterranean reservoirs, which numerous authors made use of to vary, each in his own way, the total quantity of water present upon the surface of the globe. As a scientist, he considered that the Flood could have had as its origin "the casual shock of a comet, or other transient body." Under the effect of this impact, the axis of the poles and the diurnal rotation of the earth would have been changed abruptly; the chaos resulting from the immensely powerful ebbs and flows of the seas would have swept everything away and produced the accumulation of materials, previously uprooted by the waves, upon places where opposing waves would have annihilated themselves against one another, and so the long mountain chains would have been formed. We understand, consequently, why "the earth seems as if it were made out of the ruins of a former world."

It was on 12 December 1692 that Halley presented these "considerations about the Cause of the universal Deluge." After discussing them with "a person whose judgment he had great reason to respect," however, Halley added on the following day, in a prophetic fashion, that, as this event had occurred before the appearance of humankind, it could not have been known to us except by "Revelation." In returning to his initial conclusions, Halley nonetheless contradicted his postulate that man had witnessed the Flood. We know neither who this provider of wise counsel was, nor the private context in which this discussion took place. The fact remains that these ideas had to wait thirty-two years before being published in 1724–25, because of Halley's fear that by "some unguarded Expression he might incur the Censure of the Sacred Order." Nonetheless, by describing the effects of a possible shock to the earth from "a comet, or other transient body," Halley inaugurated another new field of speculation, which has never flourished so well as in our own times. His first follower was William Whiston (1667–1752), Newton's protégé and successor at Cambridge. Whiston, who was somewhat less prudent than Halley, had successfully integrated comets into his *New Theory of the Earth*, published in 1696, as we will see later. But in 1724, a year we have already visited with Halley, there was a great, new dispute causing murmurs in the scientific world on both sides of the Channel. This would have been the last stroke for Newton if he had not died just when the controversy was really about to burst.

Commotion about Dates

Rarely has there been such a famous person who has succeeded so well as Newton did in concealing vast portions of his own work. First of all, in Cambridge, it appeared that he passed very long and mysterious days in a laboratory next to his apartments. "He very rarely went to bed till *two* or *three* of the clock, sometimes not until *five* or *six*, lying about *four* or *five* hours, especially at spring and Fall of the Leaf, at which times he used to employ about 6 weeks in his Elaboratory, the Fire scarcely going out either Night or Day, he sitting up one Night, as I did another," reported Humphrey Newton, "till he had finished his Chymichal Experiments, in the Performance of which he was the most accurate, strict, exact: What his Aim might be, I was not able to penetrate into, but his Pains, his Diligence at those set Times, made me think, he aimed at something beyond the Reach of humane Art and Industry." When Newton had to abandon his laboratory and forsake alchemy as he was leaving Cambridge in 1696, he conserved thousands of pages that had been darkened with his steady writing throughout the years. Today their interpretation is still a matter of controversy among scholars. Were these enigmatic pages simply lecture notes, or did they betray an alchemical mind behind Newton's reflections, just as Kepler's thought had found its roots in mystical harmonies?

Meanwhile, in London, Newton was able to pursue another ineffable quest until his last days, one that he began during his very first period in Cambridge and which he jealously guarded with the same secrecy: his establishment of a *Chronology of Ancient Kingdoms*. Only a few who were close to him, such as the philosophers Henry More and John Locke, knew beforehand of Newton's passions for chronology and for exegesis of the Old Testament prophecies. The "republic of letters" discovered them late, in circumstances that provoked a violent controversy. At the origin of an affair in which some of the protagonists put on such intricate "masks," there should be no surprise in finding a certain noble Venetian abbot of eclectic talents, Antonio Conti (1677–1749), whose province was the salons and the courts of Europe. This abbot had taken it upon himself to mediate the priority controversy between Newton and Leibniz regarding the invention of infinitesimal calculus. In 1716, through a common friend, the princess of Wales, he became close enough to Newton to have him submit an abstract of about twenty pages recording the principal dates that he had established. As time passed, it became more difficult for Conti, who liked to put on a good appearance, to respect his promise of keeping the abstract's existence a secret. He ended up showing it to a few scholars, one of whom was the French Jesuit Etienne Souciet (1671–1744), who was among the top specialists in the field; even so, the complete mystery about these new lights on ancient chronology persisted. In response to the questions that were asked of him, Newton answered only that the key to his method was to be found in the principle of the retrogradation of the equinoctial points. This was enough for Souciet, a man bound up with the established principles, to understand the sense of the challenge: he set to work taking it up, immediately and quietly.

In 1724 the abstract finally slipped from Conti's hands and fell into those of Nicolas Fréret (1688–1749), one of the most qualified members of the Academy of Belles Lettres and Inscriptions. Even though some of his own private remarks about religion could scarcely be considered scholarly, the young academician could not believe his eyes: between the times of the Creation and Alexander the Great, a half-millennium of world history had been canceled by Newton's long strokes of the pen. Fréret translated the abstract, adding a few observations of his own invention—highlighting in particular that what had been called the Age of the Argonauts and the lengths of the generations ought to have been sufficient to arouse the reader's suspicions—and he asked a Parisian publisher to print it anonymously. The publisher wrote Newton in a manner that was, to say the least, cavalier, inquiring about some corrections to be made to his abstract before publishing its translation into French. Of course, a veto arrived from London, but the presses had already done their work: on 11 November 1725, Newton received from Paris, free of charge, a copy of his *Abstract of Chronology*. The height of insults!—it had been published as a supplement to Humphrey Prideaux's *History of the Jews and Neighboring Nations*, also translated from English, whose chronological framework was inconsistent with Newton's.

"Newton was piqued by this proceeding; he could not, without suffering, see that someone would have wanted to stifle this child of his leisure, practically before it was even born; his paternal tenderness had been aroused, and he responded sharply to the observations; in this response one did not feel any sort of the Stoic phlegm that would have sat so well upon a philosopher," his translator commented later. The controversy took on an international dimension when Newton wrote in the columns of *Philosophical Transactions* that Conti was a traitor who had made a farce of his promises of confidentiality. Conti defended himself by appealing to the various services he had rendered to the Newtonian cause and then concluded, with some foresight as well as a touch of haughtiness:

I like these sorts of study quite a bit; but they do not disquiet me in the slightest, and deep down, I do not value the subject of it any more than the quadrille or the hunt; it all comes down to the same thing when it is examined dispassionately; and besides, I am persuaded that, if we except fifteen or twenty problems, which are useful to the arts and to the customs of society, all the rest will perhaps one day be disregarded, like certain questions of the Scholastics, or questions of the void, of atoms, of time, or of the perfection of the universe, etc., that Mr. Newton disdains.

As for Father Souciet, the publication of this *Abstract* delivered him from the silence with which he had surrounded his own work. Beginning in 1726, he attempted to cut the ground from under the feet of his adversary by publishing *Cinq dissertations contre la chronologie de M. Newton* (*Five Dissertations against Newton's Chronology*), dedicated to Abbot Conti, in which he demonstrated that the Englishman had indeed been wrong about the 530 years that were missing from his account. Newton held his ground firmly, put his own manuscript in order, and at that juncture died at the age of eighty-eight on 20 March 1727. "He lived in honor from his compatriots and was buried like a king who had done well to his subjects," noted Voltaire, who was in London for the solemn funeral, having settled there just after his release from the Bastille the previous year. When Newton's executors examined the papers left by the deceased, they were not transported by enthusiasm: Newton's chronology did not really inspire them, any more than his alchemy had. To the underhanded satisfaction of Whiston, who was impatient to see the public ruin of his previous master's system, *The Chronology of Ancient Kingdoms, Amended* was finally published in 1728, first in London, then in Paris.

The Apotheosis of Chronology

Beyond its anecdotal context, the publication of Newton's *Chronology* illustrates the passions that ancient history still aroused around the beginning of the eighteenth century. Perhaps from the heights of his celestial spheres Newton had been able to achieve for the ancient quest of time the same sublime understanding as for the organization of the world, the figure of the earth, and the precession of the equinoxes. For him the issue was not literary at all; it aimed, rather, at the very foundations of science and history. Historical truth, in effect, like time and space, was absolute. Like the church fathers, Newton thought that "the real evidences of Christianity were historical: the narration by witnesses of true events and the demonstrations of veritable prophecies," noted F. E. Manuel. "This religious and secular history of the world, an apology conceived in the grand dimensions of the *City of God*, Newton wrote himself, with his own hands; he did not leave it to others."

And Newton wrote his apology as he did his other works, without concessions or facilitations. An excerpt from his *Chronology* allows today's reader a glimpse of his style, while at the same time reviving somewhat the discussions that had animated the period's guardians of antiquity:

Now since *Erathosthenes* and *Appollodorus* computed the times by the Reigns of the Kings of *Sparta*, and (as appears by their Chronology still followed) have made the seventeen Reigns of their Kings in both Races, between the Return of the *Heraclides* into *Peloponnesus* and the battle of *Thermopylae*, take up 622 years, which is after the rate of 36 1/2 years to a Reign, and yet a Race of seventeen Kings of that length is no where to be met with in all true History, and Kings at a moderate reckoning Reign but 18 or 20 years a-piece with another: I have stated the time of the return of *Heraclides* by the last way of reckoning, placing it about 340 years before the battle of *Thermopylae*, which makes the Taking of *Troy* eighty years older than the Return, according to *Thucydides*, and the *Argonautic* Expedition a Generation older than the *Trojan* war, and the Wars of *Sesostris* in *Thrace* and death of *Ino* the daughter of *Cadmus* a Generation older than that Expedition.

As for the new proofs that Newton had advanced in order to establish his chronology, the first was to reduce the average duration of the reigns in early history. Newton made use there of the indisputable exactness of the Greek, Roman, and modern archives. For example, the average duration of a reign in the French and English dynasties lasted between 19 and 20 years. Thus, Newton noted, "the *Greek* Chronologers, who follow *Timaeus* and *Erathosthenes*, have made the Kings of their several Cities, who lived before the times of the *Persian* Empire, to Reign about 35 or 40 years a-piece, one with another; which is a length so much beyond the course of nature, as not to be credited." But it was especially with the second proof that Newton demonstrated the full measure of his genius. In the precession of the equinoxes, whose secret he had penetrated, what he had seen was the first lamp suitable to illuminate the truth about history. Because the equinoctial points have a uniform retrograde movement of one degree every 72 years, it was actually possible to date an ancient event based solely upon astronomical observations.

Newton thus noted regarding the expedition of the Argonauts, when Castor and Pollux departed to capture the Golden Fleece in company with some fifty other Greek heroes, that the astronomer Chiron had fixed the spring equinox at the fifteenth degree of Aries. This positioned the summer solstice at the fifteenth degree of Cancer. Similarly, one year before the Peloponnesian War, Meton had fixed the summer solstice at the eighth degree of Cancer, which meant that seven times 72 years, or 504 years, had elapsed between the voyage of the Argonauts and the beginning of the Peloponnesian War, rather than 700 years, as Greek history had recorded. In combining his two different types of proofs, Newton concluded that the expedition of the Argonauts took place in 909 BC, and not in 1400 BC, as had previously been believed. And so it was that the world became a half-millennium younger, even if it did receive the same age, in the final account, as had been given earlier, by Father Peteau, for example.

In basing his work on an astronomical phenomenon whose physical explanation had been established, Newton believed he could reduce the inevitable uncertainties of history. Even so, there was a surprising amount of opposition to his method, particularly on the part of the Jesuits, who themselves had little to learn about the subject of astronomy. Despite Newton's critical approach to his sources, the precision of his quotations, and the attention he brought to a multiplicity of proofs, controversy raged. The faithful Halley, among the very first to take part in the debates, took the defense of his late friend, restricting himself to a justification of the astronomical hypotheses. Perhaps this cautious support reflected some doubt on his part. Voltaire, in France, naturally became the zealous and eloquent advocate of Newton's system of history. He strived to make its argumentation comprehensible, putting his talent all the more willingly toward the service of this cause, for as he was defending it, he was also attacking his favorite enemies, the Jesuits! But Voltaire's support did not soften the derision of the critics, who were skeptical on principle about chronology. As was summarized in the preface to a book published anonymously by Jean-Baptiste de Mirabaud (1675–1760), the permanent secretary of the Académie Française, "At least, what has been said about this will be useful in informing readers of how little depth they need bring to the chronologists' calculations, and to convince them of the feebleness of their efforts, as well as of the uselessness of their research." Abbot Conti, for his part, was not completely indifferent to history and chronology. In Italy he became close to Giambattista Vico (1668–1744)—at that time an obscure Neapolitan author—who was on the verge of achieving status as one of the innovators of historical science with his *Principles of the New Science concerning the Common Nature of Nations*, published in 1725, a work that later influenced generations of historians.

For their part, chronology scholars established that the ancient texts gave no sort of authority to the firm conclusions that Newton had made. Dismissed from his chair at Cambridge in 1711 for Arianism, Whiston had long been ready to struggle on this terrain against his former master, who had not defended him and who never admitted his inferiority in regard to exegesis. It was for fear of killing Newton, he claimed, that he had attacked him only after his death. And so, Whiston asserted that Newton had interpreted Hipparchus incorrectly and that the expedition of the Argonauts had actually taken place one hundred years earlier—not later—than had been thought. And, more than happy to enlist the support of one Englishman, Whiston, to attack another, Fréret accused Newton of having confused different types of reigns, for example, hereditary, elective, and collateral, in preferring his reigns of 19 or 20 years over the traditional, but more realistic, periods of 33 years. "I will dwell no longer on Newton's work. I believe that I have already destroyed its fundamental principles," was the cruel conclusion of his *Défense de la chronologie*, published posthumously in 1758. The astronomer Bailly pronounced the final absolution in 1775: "The idea of regulating chronology by the ancient determinations of equinoctial and solstitial points was fine and great, worthy of a man of genius; but Newton erred in the application he made of that, and the system resulting from it has fallen, because it is contrary to the facts."

Newton's handsome edifice was thus undermined at numerous points even before it was fully expounded. The great philosopher nonetheless declared: "I have drawn up the following Chronological Table, so as to make Chronology suit with the Course of Nature, with Astronomy, with Sacred History, with *Herodotus*, the Father of History, and with itself; without the many repugnancies complained of by *Plutarch*." Such pretentiousness was bound to irritate his adversaries. "How weak, how very weak, the greatest of mortal Men may be in some Things, though they be beyond all Men in others," deplored Whiston. "Sir *Isaac*, in Mathematicks, could sometimes see almost by Intuition, even without Demonstration.... And when he did but propose Conjectures in Natural Philosophy, he almost always knew them to be true at the same Time," he persisted, before concluding: "yet did this Sir *Isaac Newton* compose a Chronology, and wrote out 18 Copies of its first and principal Chapter with his own hand, but little different one from another, which proved no better than a sagacious Romance."

How would Newton, who detested controversy, have reacted to these pitiless criticisms? Constantly harassed by his own excesses of pride and introspection, he remained, in truth, an enigma to himself: "I do not know what I may appear to the world; but to myself I seem to have been only like a boy playing on the sea-shore, and diverting myself in now and then finding a smoother pebble or a prettier shell than ordinary, whilst the great ocean of truth lay all undiscovered before me."

FOUR

Nature's Admirable Medals

THE STORY TOLD BY FOSSILS   Ever since the Renaissance, fossil discoveries had raised major questions about whether they were "sports of nature" counterfeiting animal remains or the actual remains of dead animals, as Nicolaus Steno and Robert Hooke asserted. Although the Great Flood could explain the latter hypothesis, the enormous numbers of fossils deposited in rocks found even at the tops of mountains were obviously inconsistent with the brief Mosaic time frame.

Newton's God

In his controversy with Abbot Conti, Newton had advised the public that his treatise "was the fruit of his vacant hours, and the relief he sometimes had recourse to, when tired of his other studies." But his chronological research was no game at all; it was directly involved with principles at the roots of his entire scientific work. Just as one of the purposes of natural philosophy was to discover a limited number of axioms that could be used to resolve the apparent complexity of the world, the purpose of exegesis was to decipher the obscure symbols that punctuated the biblical account, in order to understand more profoundly the sense of the divine word. Physics was one of the instruments available in this quest, and the essential truths being investigated could be shown, according to Newton, to bear the imprint of an elegant simplicity. In disclosing the magnificence of the movements of celestial bodies, which were governed by the mere force of gravitation, natural philosophy did not disrupt the biblical account in any way. On the contrary, it proved that "this most beautiful system of the sun, planets, and comets, could only proceed from the counsel and dominion of an intelligent and powerful Being" whose work was not subject to the laws of gravitation until its completion.

In reality, this indisputable testimony to the existence of a Master was merely a point of departure. Because "the light of the fixed stars is of the same nature with the light of the sun," it could be affirmed that "if the fixed stars are the centres of other like systems, these, being formed by the like wise counsel, must be all subjects to the dominion of One." In developing this argument based on the existence of a Design, Newton added that "lest the systems of the fixed stars should, by their gravity, fall on each other," the Supreme Being has "placed those systems at immense distances from one another." And so God recovered his role as the great organizer, from which Descartes had zealously removed him. The inevitable conclusion was that "this Being governs all things, not as the soul of the world, but as Lord over all." And in a burst of enthusiasm nearing the mystical, Newton proceeded to describe the essence of the divine Master: "He is eternal and infinite, omnipotent and omniscient; that is, his duration reaches from eternity to eternity; his presence from infinity to infinity; he governs all things, and knows all things that are or can be done. He is not eternity and infinity, but eternal and infinite; he is not duration or space, but he endures and is present. He endures forever, and is everywhere present; and, by existing always and everywhere, he constitutes duration and space."

The progress of instruments that revealed endless numbers of new stars, along with the fact that a universe of merely finite measure when submitted to gravity would necessarily collapse into its own center, led Halley and Newton to recognize the immense dimensions of the universe and to arrive at a more concrete sense of the infinite. But these new conceptions dealt with space alone, not with time. The important discovery that light had a speed could possibly have opened the way to a connection between these two dimensions. In 1676 the Danish astronomer Ole Römer (1644–1710), while he was professor of mathematics to the dauphin and a member of the Academy of Sciences in Paris, proved that light is not transmitted instantaneously. His observations of eclipses of Jupiter's first satellite caused him to conclude that light had a finite speed and that it traversed the distance of the diameter of the earth's orbit about the sun in 22 minutes (giving a speed of 215,000 kilometers/ second, which agrees respectably with the currently accepted value of 300,000 kilometers/second). Though one can quickly calculate from this result that the light emitted from the sun takes about 8.5 minutes to reach the human eye, neither Newton nor his contemporaries could possibly have imagined the time required for light to travel from the most distant stars.

With his physics, Newton launched a movement that took a century to dominate science. His chronological, or prophetical, labors, in contrast, which revealed his old-fashioned ways of thinking, were concealed—like the drunken nakedness of Noah—by well-meaning admirers during the nineteenth century. The early-nineteenth-century physicists Jean-Baptiste Biot and Laplace, for example, attributed them to an episode of severe depression Newton experienced in 1693 and then, later, to old age. As a matter of fact, Newton subscribed enthusiastically to the old ways of thinking, viewing the perfection of nature as a sign of the benevolent hand of God and as proof of design by a Creator. In this way he represented a transition, by himself alone, among several types of attitudes that coexisted through long periods; he was an exemplary case, as one who upheld the temporal framework that still dominated at the beginning of the eighteenth century. Fossils would later play a major role in the repudiation of that framework, so we'll backtrack here to take up their history.

"Fossils" versus "Minerals"

When there were still debates about whether or not fossils were of organic origin, when the idea of spontaneous generation was still largely accepted, and when concepts of evolution were still unknown, one could hardly have foreseen the importance of these curious fragments of matter buried deep underground that had so intrigued the ancients. Actually, the distinctions among fossils (in the modern sense of the term), minerals (the elementary constituents of stone), and rocks have only recently been properly recognized. As late as 1742, Antoine-Joseph Dézallier d'Argenville (1680–1765) stated, following the Latin etymology, that fossils were "in general, all things that are enclosed within the bowels of the earth, and that one finds in the excavations made therein." The difficulty arose from the fact that fossils, minerals, and rocks all have the same obvious point in common: they are *stones*. Certainly one could distinguish gems and metals (including sulfides, such as pyrite), but for a long time, there were no utilitarian or secondary criteria for distinguishing the various fossils, other than listing them in alphabetical order. As a consequence, we cannot today determine whether many of the minerals listed in the ancient or medieval lapidaries are fossils, minerals, or rocks. The chemistry of Aristotle, in its blending with Neoplatonism, had the additional effect of submitting the processes of the formation of stones to astral influences, and naturally, the starry shapes of certain "figured stones" (what we would call fossils) supported this idea. From these influences also came the justifications for the various magical, occult, or medicinal virtues that had been attributed to stones since time immemorial. The physicians of the time, who were often the first mineralogists, took pains to exploit these virtues.

The long durability of ideas is perhaps best demonstrated by a book written by Albertus Magnus around the middle of the thirteenth century. The seventh of some twenty treatises dedicated to natural history, *De mineralibus* (*On Minerals*) was perhaps the most original, for the simple reason that, after long seeking a work on the subject by Aristotle that had actually never been produced, Albertus had to resort to his own talents and write one himself. He had traveled widely, making observations in mines, in particular, and he was familiar with alchemy. Nevertheless, Albertus—the philosopher and theologian who believed that a principle that disagrees with well-informed experience is not really a principle—followed the custom, in the descriptive part of his book, of classifying minerals according to alphabetical order and indicating their various uses. Thus, *alecterius* allowed one to excite carnal desire; *chrysolitus* was used to chase off ghosts; *magnes* (magnetite) was used to give nightmares to adulterous women; *melochites* (malachite), to protect cradles; and *orites*, to avoid pregnancy or to provoke a miscarriage; whereas *crystallus* (quartz), taken in powder form with honey, would maintain a woman's lactation.

But *De mineralibus* had another importance, which extended well beyond any of the catalogs drawn up by the lapidaries. As the mineralogist D. Wyckoff pointed out, its plan was a modern one, amounting to an exposition of general principles that preceded the individual descriptions. Following Aristotle's ideas, Albertus made lengthy efforts to find the material, efficient, formal, and final causes of minerals. He did so within a coherent framework that, though not in agreement with modern conceptions, accounted for the forces exerted in all natural environments. Everyday experience displayed numerous examples of such forces: petrifying fountains slowly accomplishing their work, stalactites growing in caves, minerals appearing on the walls of mines, algae surrounding themselves with solid crusts, stones forming in the kidneys or the gall bladder, and even stones condensing within the atmosphere, to fall on the ground in the form of meteorites. There was therefore no doubt about the existence of a power called the *vis lapidifica*, a force that was capable of petrifying things. What remained to be understood was how this force operated. Here, in the following pages, we shall examine a number of portraits that allow us to contemplate the processes of trial and error that gradually led observers to notice the connections between fossils and their beds, giving rise to concepts that were to prove very productive for the consideration of timescales.

The Middle Ages of Fossils

Even though they expressed themselves in Arabic, the first authors of interest to us here were actually Persian. Foremost among them were the Brothers of Purity and Sincerity, a half-secret confraternity of scientists formed during the tenth century, who drew up an encyclopedia at Basra (now in Iraq) that the Ishmaelite sect drew upon, in part, for the syncretism of its doctrine. One of the passages of this encyclopedia describes "how the mountains are formed, and the seas, and how malleable clay is made into stone, and how the stones break up to become pebbles and sand," and so forth. In short, the Brothers of Purity explained with a remarkable clarity of vision the great effects of erosion and sedimentation; and in so doing, they introduced a very important notion, but in such an incidental way that it attracted hardly any attention at the time: "Then the seas, through the causes of the force of their waves and of the intensity of their agitating and seething, deposited these sands, clays, and pebbles upon their bottoms, layers upon layers, over the courses of time and through the ages." But in spite of this direct association between time and the depositing of layers, the Brothers of Purity raised no question of timescales throughout the rest of their descriptions. Time remained unlimited and cyclical, and the presence of the organic remains within the layers was not discussed.

The conclusions of the celebrated Avicenna (980–1037) regarding the formation of rocks ought to be mentioned concurrently with the observations of great phenomena that were made by the anonymous sages of Basra. The prince of doctors, as he was called, also the vizier to a Persian sovereign, had begun practicing at the age of sixteen. One of the numerous treatises he wrote, the *Canon of Medicine*, was still used as a university teaching manual in the seventeenth century. Avicenna was not only a physician but also a philosopher, a cosmologist, and an epistemologist. He ordered the known sciences into eight principal and seven subordinate classes. The science of minerals was fifth on the list of the principal sciences. Avicenna is interesting to us here on account of a short text he wrote around 1022 regarding the formation of stones and mountains, which he inserted into his large *Book of the Remedy*. As the following passage demonstrates, for him there was no doubt whatsoever as to the organic origin of fossils: "Mountains have been formed by one [or other] of the causes of the formation of stone, most probably from agglutinative clay which slowly dried and petrified during ages of which we have no record.... The more probable is that petrifaction occurred after the earth had been exposed, and that the condition of the clay, which would then be agglutinate, assisted the petrifaction. It is for this reason that in many stones, when they are broken, are found parts of aquatic animals, such as shells, etc." Over the course of a few decades, Avicenna observed for himself the solidification of the muds that produced rocks, a process that was designated by the Latin term *conglutinatione*. He also described the precipitation of minerals within water; this process was called *congelatione*. And so, his text was distributed in the West at the very beginning of the thirteenth century under the Latin name *De congelatione et conglutinatione lapidum* (*On the Solidification and the Precipitation of Stones*), around the time when Albertus Magnus was writing his *De mineralibus*.

In reality, fossils occupy only a very small portion of Albertus's work. The principal remark he made about them was that "it seems wonderful to everyone that sometimes stones are found that have figures of animals inside and outside. For outside they have an outline, and when they are broken open, the shapes of the internal organs are found inside." Albertus clearly had observed fossilized shells with careful attention, for example in the limestone from the Paris basin that had been used to build so many monuments. He also retained the lessons of Avicenna, who had said "that the cause of this is that animals, just as they are, are sometimes changed into stones. For he says that just as Earth and Water are material for stones, so, animals, too are material for stone." Albertus could interpret the transformation naturally enough within the framework of his theory of minerals: "in places where a petrifying force is exhaling, they change into their elements and are attacked by the properties of the qualities [hot, cold, moist, dry] which are present in those places, and the elements in the bodies of such animals are changed into the dominant element, namely Earth mixed with Water; and then the mineralizing power converts [the mixture] into stone, and the parts of the body retain their shape, inside and outside, just as they were before."

In following the trail blazed by Avicenna two hundred years earlier, Albertus Magnus thus described fossilization around the middle of the fourteenth century in a way that would hardly be worthy of remark in our own time, if we were to leave aside the chemical aspect of the transformation. One could object that Albertus was much more ambiguous in another passage, where, perhaps reflecting another common opinion, he wrote: "That it is the viscous and unctuous moisture which gives coherence to the material of stone is indicated by the fact that the animals called shellfish are very commonly produced with their shells in stone."

Leonardo the Great

The progress of the science of fossils would appear astonishingly linear if we were to skip over a period of time equal to that separating Albertus from Avicenna, to go to decipher the strange notebooks kept by a certain Tuscan painter at the very beginning of the sixteenth century. Ever since his apprenticeship in Florence, this artist had scrutinized the surrounding landscapes; his *Virgin of the Rocks* and his *Virgin and Child with Saint Anne* have enigmatic, rocky backgrounds. As architect and engineer, he directed excavations of canals and constructions of roads for years. As it happened, the geology of the area was such that many of the upturned soils were rich in fossils. Our curious observer understood that these fragments had a history to relate, one that needed only to be decoded. This man of incomparable talents was, of course, Leonardo da Vinci (1452–1519).

For example, on folio 10 of what is now called the Codex Leicester, after deciphering his celebrated retroverse handwriting, we read:

That in the drifts, among one and another, there are still to be found the traces of the worms which crawled upon them when they were not yet dry. And all marine clays still contain shells, and the shells are petrified together with the clay. From their firmness and unity, some persons will have it that these animals were carried up to places remote from the sea by the deluge. Another sect of ignorant persons declare that Nature or Heaven created them in these places by celestial influences, as if in these places we did not also find the bones of fishes which have taken a long time to grow; and as if we could not count, in the shells of cockles and snails, the years and months of their life, as we do in the horns of bulls and oxen, and in the branches of plants that have never been cut in any part.

Besides, having proved by these signs the length of their lives, it is evident, and it must be admitted, that these animals could not live without moving to fetch their food; and we find in them no instrument for penetrating the earth or the rock where we find them enclosed. But how could we find in a large snail shell the fragments and portions of many other sorts of shells, of various sorts, if they had not been thrown there, when dead, by the waves of the sea like the other light objects which it throws on the earth? Why do we find so many fragments and whole shells between layer and layer of stone, if this had not formerly been covered on the shore by a layer of earth thrown up by the sea, and which was afterwards petrified? And if the before-mentioned deluge had carried them to these parts of the sea, you might find these shells at the boundary of one drift but not at the boundary between many drifts. We must also account for the winters of the years during which the sea multiplied the drift of sand and mud brought down by the neighboring rivers, by washing down the shores; and if you choose to say that there were several deluges to produce these rifts and the shells among them, you would also have to affirm that such a deluge took place every year.

Although Leonardo may not have observed the actual petrifaction of worms and shells, he seems to have found it obvious that the soil and the silt were deposited in layers, and he even remarked that the accumulation of these layers would provide a measure of time. Furthermore, he showed perfectly well that the Flood was completely foreign to the account given by the rocks. But when the discussion of such a passage, or of an aria of Mozart, would only expose the modern commentator to ridicule, it is better to remain silent and admire, and to remember that this "painting" was merely one among the many that Leonardo sketched to depict the activity of our planet.

Fishes Bearing Shells

In large part, the height of Leonardo's vision was undoubtedly connected to the long times he invoked incidentally in some of his other notes. Without an understanding of extensive durations, one would be condemned to observation without comprehension or to theorizing on unsound bases. Another astonishing observer, the engaging clay potter Bernard Palissy (ca. 1510–1590), demonstrated this point unwittingly. Born into a rural family of Périgord, in southwestern France, he developed the talents of painting and glassblowing and also engaged in portraiture and geometry. Then he undertook, as P. A. Cap synthesized, an excursion around France, "practicing glassmaking, portraiture, and surveying at once, but observing, as well, the topography all around, the accidents of the soil, the natural curiosities, traversing the mountains, the forests, the riverbanks, visiting the quarries and mines, the caves and caverns, and, in a word, inquiring everywhere of Nature herself the secrets of the marvels she offered for his admiration and study." Afterward, Palissy started a family and settled at Saintes, near La Rochelle, where he became an ardent Calvinist. When he discovered ceramics there in 1539, it was a revelation.

While isolated in his province, Palissy launched upon an unbridled quest to discover the secrets of glazings on terra-cotta. The history is well known: after sixteen years of fruitless efforts, ruined, ill, exposed to the hostility of his acquaintances but borne along by his faith, Palissy was reduced to burning the wooden floors and furniture in his kiln so as not to surrender prematurely. Providence was watching; there was light at the end of the tunnel and his perseverance was repaid: the "rustic ceramics" soon came to seduce the constable of Montmorency, through whose influence their creator, Palissy, received an inventor's patent and was able to escape the worst of the religious persecutions. In Paris, where he had been called by the Queen Mother for the decoration of a grotto at the Palace of the Tuileries, the self-taught Palissy, famous for his ignorance of Latin and Greek, went to teach among the distinguished minds of the capital everything that he had learned through his long life of curious observation. In 1580, eight years before religious conflicts landed him in the Bastille, where he ended his days, the contents of these courses were published as *Discours admirables* (*Admirable Discourses*). The discourses had the form of animated dialogues in which one interlocutor, a certain Theory, tending naturally toward elevated ideas, had to be called back into line by the other interlocutor, Practice, behind whom Palissy was not completely successful in disguising himself.

Theory wanted to know, in particular, what Practice thought about stones, "for some say that they were formed at the time of Creation, and others say they grow everyday." Obviously, Practice was not familiar with the great authors of antiquity, with Aristotelian thinking, or with Neoplatonism. And Palissy "had not studied astrology in order to contemplate the stars"; he had had "no other book than the sky and the earth, which is known to all, and it is given to all to know and to read in this beautiful book." Concerning the *vis lapidifica* (petrifying force), therefore, Palissy could simply read within his great book that the waters, as they wore down the rocks, became charged with mineral substances, which they then deposited farther along; the stars played no role whatsoever in the actions of this congealing water.

As the discussion continued and moved to fossils, Theory declared straightaway that "shellfish, which are petrified in many quarries, were born on the very spot, while the rocks were but water and mud, which since have been petrified together with these fishes." This opinion, which Palissy was among the first to put forth, was followed by a very important remark, uttered by Practice, which he made on the basis of his painstaking efforts of species identification. What he observed was that some petrified animals belong to species that nowadays live only in the tropics: "Some are found in Champagne and in the Ardennes that are similar to certain purple shells, whelks and other large snails, which kinds are not found in the Ocean Sea, and are not seen, except through sailors who often bring some from the Indies and Guinea." Practice made another observation, also quite remarkable, that certain fossils do not correspond to any known species at all: "And although I have found petrified shells of oysters, cockles, hard-shell clams . . . and all kinds of other snails that live in the Ocean Sea, still I have found . . . some kind whose like is unknown to us, and none are found except they are lapified."

But Palissy was not able to comprehend the importance of these two discoveries, which he had mentioned only incidentally in the course of his theme. In the Bible, he had read that "Moses gives witness that in the days of the Flood, the abysses and floodgates of heaven were opened, and it rained for forty days, which rains and abysses brought the water on the land, and not the overflowing of the sea." Because he therefore rejected the rising of the seawaters upon the lands for an explanation of the presence of fossils in the mountains, Practice had to attribute some sort of terrestrial origin to these fossils. He made an awkward attempt to justify such an origin by proposing that the land, with its various expanses of water, actually harbored greater numbers of fishes than do the seas: "If you had thought about the great number of petrified shells that are found in the earth, you would know that the earth produces hardly fewer fish bearing a shell than the sea: including in it rivers, fountains and brooks." And with the all-too-brief durations of time that were inherent to his framework, Palissy was not capable of imagining any reason for the disappearance of species other than what today we call overexploitation by man: "for it is certain that animals and fishes that are good to eat are so closely pursued by man that in the end he causes them to loose their seed."

The Birth of Mineralogy

In his modern approach to the petrifying actions of natural waters, Palissy was preceded slightly by Georg Bauer, known as Georgius Agricola (1494–1555), who was a friend of Erasmus, the archetype of Renaissance erudition. Agricola began his studies by pursuing the classics: his first work was a Latin grammar, produced in 1520. Philosophy and medicine followed, which he studied between Germany and Italy. After obtaining a doctorate at Ferrara in 1526, he was named physician and apothecary at the Bohemian town of Joachimsthal, then the center of an important mining district. From the miners he treated, Agricola developed an interest in mines. Then he studied minerals and began publishing his observations on this topic in 1528. Neither his relocation in 1534 to Chemnitz, another mining center in Saxony, nor the numerous political and diplomatic responsibilities that he accepted interrupted his various research activities, which also touched on religion, historiography, and medicine. His masterpiece, *De re metallica libri xii*, appeared posthumously, illustrated with about one hundred plates that are remarkable for their attention to detail. That book is considered an essential reference work on the mining and metallurgical activity of this period. With regard to the study of time, however, two other treatises are of greater relevance, *De ortu et causis subterraneorum* and *De natura fossilium*, published in 1546.

In the former, Agricola rejected from the outset any sort of influence from the stars on the growth of stones. Certainly the forces determining such growth would be exterior to the stones themselves, but those forces would do no more than convey the action of a *succus lapidescens*, the Latin version of Palissy's "congealing water." The evaporation of the waters contained within this sap during heating, or their congealing during cooling, was responsible for the generation of minerals. Thus, in *De natura fossilium*, Agricola treated somewhat contemptuously the magical connotations attached to minerals. The considerable innovation he introduced in this book is of a different nature: he rejected the traditional, alphabetical classification and attempted to make use of all the senses in defining objective criteria for the identification of minerals. He considered color, transparency, brilliance, odor, flavor (on contact with the tongue), warmth or coolness, and dryness or humidity; he also included other, less direct qualities, such as hardness or reactions to water or fire. In this way, Agricola prefigured the modern criteria for chemical composition and crystalline structure, which were at the basis of his groups of properties.

In the ensuing classification, a first level contained the earths (clays, for example), the congealed fluids (salt, alum, even rust), the stones (gems, rocks, marbles), the metals, and the composite minerals. Somewhat behind Palissy on this point, Agricola made no distinction between fossils and stones, not even in the lowest levels of his classifications. While making these classifications qualifies Agricola as the father of mineralogy, his interpretations of the growth of stones were not widely accepted. Through the action of water, which was recognized by Agricola and Palissy, it might seem to us today that it would have been easy to account for the ever-increasing diversity of fossils, which were described with exemplary precision shortly after the Renaissance began. It was nothing of the sort. That Leonardo da Vinci's mysterious, unpublished notebooks had exercised no influence ought not to be a surprise, but that the master potter Palissy and the learned Agricola were also both ignored was not simply a result of resistance to change. New observations had arrived with the Renaissance: for another two hundred years, against the backdrop of Neoplatonism and intense cosmic harmonies, the boundaries between the realms of the living and the mineral were to remain confused.

Sports of Nature

Even though the strangest of speculations were being put forth regarding fossils, taking advantage of the new printing resources for their widespread diffusion, this activity was not simply some sort of salon diversion, in which amusements with words took the place of real substance. As had been done for ancient texts found in libraries, actual objects that had been unearthed were being examined with new attention. Following the example of Palissy, people created the first museums and curiosity rooms, in which the most original productions of Creation were exhibited in a pleasing order. Voluminous works were also printed, in which fossils were depicted, adorned with their most beautiful attire, and with the refinement of detail that the new etching techniques permitted. And stones were cataloged, although without necessarily making the distinctions between fossils and minerals any clearer. For example, Agricola had noted that the basalt of Mycenae resembled wood that had been straightened out, that native silver and asbestos both resembled hair, that certain gems resembled the lobes of ears, that the *leucophtalmos* resembled a human eyeball, and that the *enorchis* resembled testicles. Some extinct species, such as the belemnites or the aptly named horns of Ammon (the ammonites), had evoked images of arrows and horns ever since ancient times, and so there had never been any sort of investigation into an animal origin for them. The analogies between stones and living beings, in fact, went much further: on the one hand, the growth of stones evoked the concept of life, and their alteration, death. On the other hand, spontaneous generation illustrated how the simplest of organisms could appear unexpectedly from the basis of inanimate matter.

"The metallic materials, the metals, and the stones are alive, because materials that have maturity, acerbity, and age also have a life," claimed Girolamo Cardano (1501–1576), another notable figure of the Renaissance,

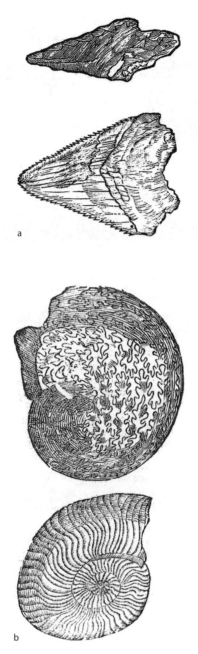

Figure 4.1 The first fossils depicted: *a*: glossopetrae, *b*: ammonites, *c*: belemnites, and *d*: enorchis. From Aldrovandi, *Museum metallicum*, 604–21 (*a–c*); Gesner, *De rerum fossilium*, 13 (*d*).

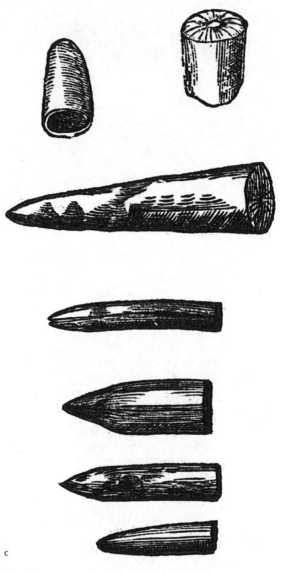

c

Figure 4.1 *(continued)*

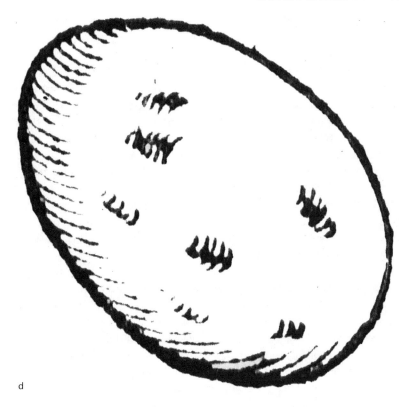

d

Figure 4.1 *(continued)*

in 1551. Cardano was a Milanese mathematician and doctor who also invented the Cardan joint, or universal joint. His analogy was more than formal. As he noted in *On Subtlety*, an attentive examination showed that gems presented tiny defects: pores, cavities, tubules, or inclusions, and also many unfinished organs to provide, as it were, nourishment and growth. "In addition to veins and instruments for nutrition, there are meatuses and small channels, just as we see in stones, and by means of these we can know that they are nourished in the same manner as are plants and the bones of animals: because if they were augmented by access and addition, they would have no need of veins." And just as "the parts of plants, the legs of crayfish, and the tails of lizards or grass snakes are repaired," "stones that are cut will grow, because they are living," with the important difference, of course, that "the generation of stones is effected over the course of a long time, while that of plants and animals is brief." But in the end, as Cardano suggested, "if you contemplate with diligence, you will recognize that stones become pale, or obfuscated, sapped of their appropriate virtues, and they suffer rotting and worm-eating."

Subsequently, one saw stones not only grow—a belief that even today has not totally disappeared from the countryside—but also multiply; and reliable witnesses were even available to describe the sexuality of stones. As depicted in F. D. Adams's description of such prodigies, these analogies between the animal and plant kingdoms illustrated the unity of nature, demonstrating that the cosmos was a place resounding with the great harmonies that were so dear to the Neoplatonists. Although better informed about meteorology and astrology than he was about fossils, Kepler was not uninfluenced by the movement: it was perhaps he who drew the most beautiful of analogies between microcosm and macrocosm: "For as the body puts out hairs on the surface of its skin," he wrote in 1619 in his *Harmony of the World,*

so the Earth puts out plants and trees; and lice are born on them in the former case, and caterpillars, cicadas, and various insects and sea monsters in the latter. And as the body displays tears, mucus, and earwax, and also in places lymph from pustules on the face, so the Earth displays amber and bitumen; as the bladder pours out urine, so the mountains pour out rivers; as the body produces excrement of sulphurous odors, and farts which can even be set on fire, so the Earth produces sulphur, subterranean fires, thunder, and lightning; and as blood is generated in the veins of an animate being, and with it sweat, which is thrust outside the body, so in the veins of the Earth are generated metals and fossils, and rainy vapor.

Beyond the analogies, the manner in which fossils were formed still remained an enigma. Certain fossils were easy to identify, as M. Rudwick noted, in the sense that they were well preserved, they resembled recognizable living species, and they were found near the sea among the shifting sediments. The correct interpretation of their origin did not pose any major difficulties. This was the case for the majority of the fossils described by ancient authors or by Leonardo da Vinci (who extended his analysis to a remarkable degree). Marking a strong contrast to those fossils were other, difficult fossils that had been described progressively, following the example of Palissy's first specimens. These were in a poor state; they did not resemble any known species, or they appeared to represent species of the tropical lands. With his doctrine of final causes, Aristotle declared, "It is plain then that nature is a cause, a cause that operates for a purpose." Accordingly, it had long been accepted that "nature does nothing in vain." But while the general resemblance between fossils and living beings lent credence to the possibility of an organic origin, the idea that there could be extinct species led to the conclusion—difficult to accept—that the Creation had been imperfect. In attempting to find a way out of this philosophical impasse, John Ray (1627–1705), the author of an excellent work describing the flora and fauna of England, supposed that the unknown species lived at the bottoms of seas that had not as yet been explored. But how their debris had come to be included within the hard rocks that made up the highest mountain peaks was still not explained. Even the solution of the Flood had its own difficulties from this point of view.

Within such a framework, the idea persisted that fossils did not have an organic origin, but rather represented what were called *lusus naturae*, or sports of nature. Someone who has never wondered at the perfect cubes formed by metallic sulfides side by side with the sparkling, golden spirals of ammonites that have been transformed into pyrites would undoubtedly not understand how for so long people could have attributed the power of giving form to a plastic force, the *vis plastica*, which was considered capable of shaping not only living beings and crystals, but also the *lusus naturae*, within the depths of the earth. Invoking this force elegantly avoided many thorny issues, such as how fossils of extinct or tropical species came to be within rocks, and how mountains were uplifted. Several variations were attempted upon this grand theme.

For example, according to Father Athanasius Kircher (1602–1680), whose encyclopedic *Mundus subterraneus* was published in 1665, it was through the means of a *spiritus architectonicus* or a *spiritus plasticus* that the petrifactions provided by the classic *vis lapidifica* took their form. But that assumption did not keep this influential and imaginative Jesuit from appealing to his contemporary chemistry for other causes: he remarked that when a kind of slimy earth is "mixed with a sort of parget, which earth, meeting with a nitrous solution in the chinks of mountains," it petrifies with time and "resembles very much a bone in whiteness, being white, porous and brittle." When it is deposited in a cavity, this earth "produces a round ball, which, being broken very much resembles a skull," while "if the mold in which it is cast has the form of a human thighbone (or that of another animal) or a rib, or any other bone, the marl that is contained in it, having the nitrous liquor added to it, will resemble the human os femoris." In the same manner, Kircher could explain how Nature was able to counterfeit within her underground depths the highly embarrassing bones of giants, known since antiquity, such as those of the Cyclopes of Etna (revealed much later to be the bones of dwarf elephants).

The Ancient Coins of an Inventive Physicist

The organic origin of fossils was debated thus on into the eighteenth century. While Kircher was still discoursing on the various sorts of *spiritus*, Hooke—the physicist scorned by Newton—succeeded in delivering some serious blows to the sports-of-nature idea. He had been led onto this terrain by the astronomer and architect Christopher Wren. During the time Wren was preparing a compilation of observations made using a new type of microscope (one with a wide field), to be presented to King Charles II, Wren had become absorbed in other tasks, and he asked his friend Hooke to take his place. As curator of experiments, the latter was required to present a new experiment each week to the Royal Society of London, set to work in September of 1661. In the following year, his observations of the arborescent figures produced by frozen urine and water, as well as of snowflakes, were judged by the society to be conclusive. At the beginning of April 1663, Hooke was even asked to present a new microscopic observation at each meeting. On the fifteenth of the same month, the immense profit to be drawn from microscopy was demonstrated when Hooke introduced the term *cell* in order to describe the microscopic structure of cork. But it was the account he presented on May 27 that caused him to embark on an entirely different course.

On that day, Hooke exhibited his images of thin sections of petrified wood, of wood coals, and of rotting wood. On a microscopic scale it was obvious that they all had pores that were similar. In the petrified wood, "yet did they keep the exact figure and order of the pore of Coals and of rotten Wood, which last also were much of the same size." Hooke hastened to make the leap between observation and interpretation, and he did so all the more boldly because the microscope had also revealed the same similarity of organization between recent and fossilized shells. The conclusion was that all of these bodies, as well as others having strange forms, were shells, which "either by some Deluge, Inundation, Earthquake, or some such other means, came to be thrown to that place, and there to be fill'd with some other kind of Mudd or Clay, or *petrifying* Water, or some other substance, which in tract of time has been settled together and hardened in those shelly moulds into those shaped substances we now find them."

Even though Palissy had already been so explicit, and Leonardo had not needed the microscope to study the anatomical details of shells, this proposal was still new, both in substance and in form. Hooke had shown that figured stones could be divided into two categories: in the first, the form is characteristic of an organism (fossils), while in the second,

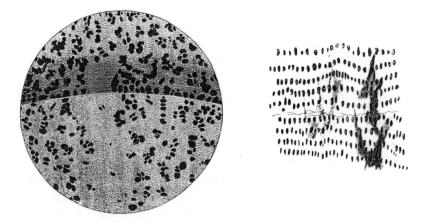

Figure 4.2 Petrified wood (*left*) in its resemblance to wooden coals (*right*) under Hooke's microscope. From Hooke, *Micrographia* (repr. 1961), 106.

it is characteristic of a substance (crystals, whose variations in form he correctly attributed to differences in the stacking of microscopic material spheres). Hooke did not keep his conclusions for his fellow academicians, but he published them in 1665 in *Micrographia*, a 250-page work that included dozens of plates that he himself drew quite adroitly. These were an immediate success: for the first time the single eye of a fly, all by itself, and a louse gripping onto a human hair could be seen enlarged to twenty inches, on a full plate. The persuasive power of the image had its effect on a public that was curious to discover and to accept the world of the minuscule, so much so that one contemporary remarked that the virtuosi would henceforth not be able to admire anything "except fleas, lice, and themselves." But science was also well represented in *Micrographia*: each observation was accompanied by a long discussion, developing, for instance, notions of the fundamental properties of heat or light. Fossilized shells and petrified woods had become a point of departure that was affected neither by the great plague of 1665, nor by the fire that devastated London the following year. Hooke supervised the reconstruction of the city as the right-hand man of Wren, who directed it. The occupation was lucrative but time-consuming, and it also had the side effect that Hooke became a part-time architect for bridges, roadways, churches, theaters, and palaces.

If fossils were indeed the remains of animals, Hooke reasoned, then their distribution about the surface of the globe still remained to be explained. This difficult question became the subject of numerous lectures that Hooke gave between 1667 and 1700 before the Royal Society of London; they were published posthumously in 1705. In the course of the lectures, in which redundancies and even contradictions can be found, Hooke resorted to proposing "earthquakes" as agents capable of provoking the rising or sinking of lands, whether violently or not, and therefore of exhuming the fossiliferous rocks from the sea; earthquakes could also form new deposits on lands that had previously emerged, Hooke said. In 1687 he invoked as the main force behind such changes a continuous variation of the earth's axis of rotation over time: so as to be able to remain normal to this axis, the equatorial bulge would have followed the rotation through its course, provoking vast displacements of the seas. Because the existence of fossils at the peaks of mountains or the presence of tropical species in England no longer presented any particular difficulties, Hooke proclaimed that he was not going to hesitate to turn "the World upside down for the sake of a Shell"! Even though he had left the Flood aside in providing a dynamic and global vision of the earth, his speculations did not meet with much success. In contrast with the very slow displacements of lands suggested by Buridan, he had placed his polar drift within too narrow a temporal framework. But the polar drift theory was inconsistent with the low levels of seismic activity that were detected. It also contradicted astronomical observations, such as those made in Nuremberg by Johann Philipp Von Wurtzelbaur (1656–1713), who concluded that the obliquity of the ecliptic had not varied between 1487 and 1686.

The productive analogy between fossils and remnants from antiquity was better received, and it recurred numerous times throughout Hooke's *Discourses*; this comparison was all the more interesting because it came forth at a time when archaeology did not even exist and ancient coins were just beginning to be cataloged. For Hooke, "these Shells and other Bodies" are "the greatest and most lasting Monuments of Antiquity, which, in all probability, will far antedate all the most ancient Monuments of the World, even the very Pyramids, Obelisks, Mummys, Hieroglyphicks, and Coins, and will afford more information in Natural History, than those other put altogether will in Civil [history]." Certainly, Hooke recognized thereafter, "it is difficult to read them, and to raise a Chronology out of them, and to state the intervals of the Times wherein such or such Catastrophies or Mutations have happened; yet 'tis not impossible, but that, by the help of those joined to other means and instances of Information, much may be done even in that part of Information also." Even though he was evoking times much longer than those of human history, Hooke did not envisage a rupture with the Mosaic chronology. At the end of his life, he even attempted to reconcile his theses with the account of the Creation. And although he was familiar with the works of Steno (discussed below), Hooke did not recognize the importance that the sedimentary layers were going to take on in terms of his own views. Being, above all, a laboratory man did not prevent him from recognizing clearly that species could become extinct, although his knowledge of geology may still have been too fragmentary to allow him to associate fossils and strata within a chronological context. The fact that it took a good hundred years before that link could be established suggests rather that, at the time, this achievement was quite simply impossible.

The Call of a Shark

It was an anatomist who unexpectedly provided the foundations for this new science. Known by the name Nicolaus Steno, Niels Stensen (1638–1686) was born in Copenhagen to a family with deep Lutheran roots. His father was a goldsmith in the service of the court, and of his mother we know nothing other than that she married four times. His childhood was marked by fragile health and a desire to learn from adults, by an appreciation for mathematics, the sciences, and scientific instruments, and by his work in the family's goldsmith shop. He began medical studies in 1656 at the University of Copenhagen, where his reading of the important authors, from Paracelsus to Galileo and Descartes, did not exclude a deep piety. In 1660, Steno spent several months in Amsterdam; there, during his first dissection, he had the good fortune of discovering a salivary duct. That was the first of a long series of discoveries responsible for his now being considered one of the best anatomists of the seventeenth century. To complete his studies, he departed for Leiden, where he struck up a friendship with Spinoza; he also demonstrated while there that the heart is only a muscle, and not the seat of the mind, or of the various faculties that had been ascribed to it since antiquity.

Because he failed to obtain a position in Copenhagen, Steno went to Paris in 1664, where he performed profitable research on the brain. In the preface of his remarkable discourse on the anatomy of this organ, he confided that by contrast with some people, who are "so quick with their affirmations and who will give you the history of the brain as well as the disposition of its parts with the same assurance as though they had been present for the composition of this marvelous machine, and as though they had penetrated into all of the designs of its great Architect, I will make you a sincere and public confession here, that I know nothing about it." Steno also discussed theology with Bishop Bossuet in Paris. He was highly impressed by an experiment performed in his presence by the chemist Pierre Borel (1620–1689): by simply adding two liquids together, one could produce a solid immediately; this demonstration raised questions about whether water that had been blended with various vapors emitted from the earth could produce the same effect in nature.

After a short stay in Montpellier, Steno went to Florence and became physician to the grand duke, Ferdinand II, the following year. A historic accident occurred in October 1666, when a giant shark was beached near Livorno. Upon orders of the grand duke, the shark's head was taken to Steno. Through the course of a pleasantly distracting study on the functioning of the muscles, Steno was struck by the resemblance between the animal's teeth and glossopetrae, the fossils shaped like pointy little tongues, which were found in abundance at Malta and were legendary for their curative properties. After examining them, Steno concluded that glossopetrae were actually petrified shark teeth. Even though he certainly was not the first to declare this, Steno was one of those alert and curious observers, like Hooke, who are ready to identify the important issue behind an observation of seemingly little importance. If glossopetrae really are the remains of animals, could the same prove true for the other figured stones?

Steno went out into the field to seek an answer to the question, armed with a reasoned doubt worthy of the severest of Cartesian precepts to sift through his observations. For example, the same species are found within soils that are very different from one another; fossils do not appear to grow in the soil: on the contrary, they are seen deteriorating there; one never sees any deformities of the kind observed in the roots of plants, which, when growing, conform to the spaces they seek out, the cracks between the rocks; and as Leonardo da Vinci had noticed, fossils reproduce the anatomical details of actual sea animals with remarkable accuracy. In a simple appendix to his anatomical description of the shark published in 1667, Steno concluded that fossils were not sports of nature. He could have stopped at that point, but he added a summary involving geological views of a much vaster scale: he defined sediments and strata and described the manner in which they are deposited. Based on such notions, the regularity of the layers, which was already well known among miners, finally made some sense.

As a great scientist, for whom one observation elicited another and for whom experience alone could be the source of his inductions, Steno had meanwhile gone back to terranes that had been surveyed earlier by Leonardo. To understand how the tooth of a shark could become petrified was merely a single aspect of the more general problem: "given a substance possessed of a certain figure, and produced according to the laws of nature, to find in the substance itself evidences disclosing the place and manner of its production." Tuscany's marked relief, with its deep ravines and high cliffs, certainly offered a much more suitable environment than Denmark or the Netherlands for such reflections. Through the course of his numerous excursions, Steno found plenty of outcrops and mines in which to observe the arrangement of fossils within rocks, the filling in by sediments of what had been cavities, and on a larger scale, the superposition of strata and the accidents that affected them.

The king of Denmark was pressing him to come back to Copenhagen, so Steno hastened to publish his conclusions in 1669 in the *Prodromus* [forerunner] *of Nicolaus Steno's Dissertation concerning a Solid Body* [fossil or other body] *Enclosed by Process of Nature within a Solid* [a rock]. But even though he made repeated field trips over several years in preparing for it, Steno never wrote the dissertation in which he would have expounded authoritatively his observations and the principles established. At the age of thirty-one, the pious scientist had nearly completed his scientific career. Steno converted to Catholicism at the same time as he founded the science of strata, and he remained only two years as royal anatomist in Denmark. Pressed by religious disagreements, he returned to Florence, where he was ordained a priest in 1675. Two years later he was named bishop of Titiopolis, *in partibus infidelium* (in an infidel country) and apostolic vicar of the Catholic populations dispersed across northern Europe. This man who was familiar with princes and philosophers conducted himself in accordance with the edifying life of the ascetic, which led him to the grave in less than a decade. Recalling the fate of the philosopher and mathematician Blaise Pascal, his elder by fifteen years, this end could soon make him the patron saint of geologists: Steno was beatified in 1988, and his canonization is under examination still— apparently a very long process in his case!

The Origin of Strata

In its succinct seventy-six pages, the *Prodromus* expressed the bases of stratigraphy that today are so familiar. The concepts derive for the most part from the fundamental principal that "the strata of the earth are due to the deposits of a fluid." The strata are deposited one atop another, horizontally, on a support that has previously solidified. Their order of succession therefore reflects the order in which they were deposited. The highest layers are the most recent; a layer that encloses within itself the remnants of another must necessarily be subsequent to that one: and so we owe to Steno the paramount conclusion that a layer represents a unit of time. The layers, furthermore, demonstrate a horizontal continuity, which allows one to follow them for vast distances, and they maintain a memory of the conditions in which they were deposited. According to the nature of the objects that are imprisoned within a layer—remnants of marine fauna, of boats, tree branches, pine cones, wood coals, or other burned remains—it is possible for us to recognize whether the objects were deposited under the sea, at the bank of a river, or near a fire. And it was Steno in the end who finally established the science of tectonics, by his assertion that any inclination of a layer reflected some movement of that layer within the past, which would have caused it to lose its initial position; indeed, such movement could even be anterior to the depositing of new, horizontal layers, marking what we now call an *unconformity*.

In Steno's time there were, of course, no means of correctly describing the causes that perturb the initial horizontal disposition of the strata and thus create mountains. Nonetheless, Steno advanced the proposition that mountains could have resulted from underground fires, erosion, earthquakes, or subsidence. On the smaller scale, he established the first notions of crystallography, as Hooke did at the same time, based on reflections regarding the nature and the growth of crystalline facets. And he concluded the Prodromus by presenting the very first series of geological sections using his new concepts, retracing the evolution of the Tuscan formations in a schema that was animated, this time on a small scale, by the subsidences previously noted by Descartes.

The great lesson was that it was possible to deduce history from the observation of sedimentary deposits. Steno was well aware of this, and he did not neglect to mention the time required to deposit the strata: "There are those to whom the great length of time seems to destroy the force of the remaining arguments, since the recollection of no age affirms that floods rose to the place where many marine objects are found to-day, if you exclude the universal deluge, four thousand years, more or less, before our time." As an authentic Cartesian, Steno had no fear of going against preconceived ideas, though here he was unfaithful to his own methods, for once, since he did not test his reasoning by evaluating the rate of sedimentation. He preferred instead to limit himself strictly to the framework of Mosaic chronologies. Would he have been more critical in his proposed dissertation, if he had completed it? As such a question touched upon the very essence of his convictions, we may be doubtful.

While Steno was carrying out his ministry at Hanover, Steno frequented Leibniz, and Leibniz pressed him to abandon his theology, which he judged to be mediocre, and to take up science again. Steno then shared with his friend the deep impression he felt in 1663 during his dissection of the heart that had allowed him to disprove Descartes's physiological speculations. "God saved me from all the subtlety of dangerous philosophers and all the finesse of politic lovers of the same sort of philosophy," he confided in 1677. Recalling the experimental collaboration of a young assistant at an earlier time, Steno continued: "If these gentlemen, who are admired by nearly all scientists, have held as infallible proofs something that I can have prepared in an hour's time by a ten-year old boy, in such a manner that the sight of it alone, without a single word, would cast to the ground the most ingenious systems of these great minds, what assurance can I have about the other subtleties of which they boast?" The conclusion was clear on its own merit: "What I mean is that if they have deceived themselves to such a degree in material things, which are exposed to the senses, what assurance can they give me that they have not also deceived me when they speak of God or the soul?" For Steno, the revelation he experienced on All Saints' Day in 1667, the day of his conversion, was certainly worth more than any experiment. He had already recognized in the preamble to his first brief memoir, "We learn from Holy Scripture that all things, both when Creation began and at the time of the Flood, have been covered with waters." The production of strata over a period of four thousand years presented no difficulty for him. Like so many others of his time, Steno was incapable of imagining longer periods. Even the remains of elephants in Tuscany, which had been unearthed in 1663, did not trouble him: these were merely an indication that Hannibal's Carthaginian troops had passed through the region. Steno's remarkable clear-sightedness was confined within the limits imposed by his faith.

The Wisdom of the Creator

For certain other attentive observers, such as Ray and his friend Edward Lhwyd (1660–1709), the Welsh naturalist, the systems proposed by Hooke and Steno unfortunately revealed an ignorance of even more complex realities of the field. Although Ray considered the glossopetrae to be authentically organized bodies, he denied that the same was true of the belemnites and reserved judgment about the ammonites. Lhwyd, in turn, was astonished to find that certain species of sea urchins were peculiar to the chalk of England, while others were found only in that of northeastern Ireland, and even more so to learn that one finds groups of the same species at different sites: by observing two or three from such a group, one could predict the additional presence of a dozen or two other species. The sense of these curious associations as yet escaped explanation. Furthermore, because fossils are made of the same substance as the rocks in which they are found, often a given species fossilized within one rock would vary considerably from the same species found in another rock. Lhwyd saw this as an argument against the organic origin of figured stones, though here he was in disagreement with Steno. All that remained was to give an account of the prodigious quantities of fossils accumulated within the breadths of the sediments. Toward this end, a stopgap explanation was submitted to Ray by Lhwyd in 1698, proposing that a sort of milt could have been produced by marine animals, the *seminium*, which might have been disseminated afar by spindrifts, winds, and rains; it could then have infiltrated through the rocks, and there, within the depths of the earth, grown to form the figured stones.

From observations reminiscent of those made by Cardano a century and half earlier, the same idea was proposed by the botanist Joseph Pitton de Tournefort (1656–1708). For him, the distinguished naturalist who created the concept of the genus, there was nothing strange in assuming that rocks had semen. Since the Author of Nature had put the germ of the twenty-foot-long narwhal into a tiny egg, why "could he not have been able to enclose a bed of rock within a germ as small as a speck of sand?" And to account for rock stratification, Tournefort added that, by "compressing the Earth's surface equally," it was the air pressure (discovered fifty years earlier) that caused "the growing germs always to flatten horizontally."

In spite of the difficulties raised by such naturalists, however, the organic theory continued to gain ground. The vague analogies between microcosm and macrocosm, within which the sports of nature implicitly drew their substance, were eliminated little by little in accordance with the physics of Descartes and Newton. There was indeed a whole current behind Hooke and Steno that thus began taking form. In England especially at that time, it appeared necessary to link fossils closely with the Flood and within the framework of short chronologies.

In 1681, before Halley's theory on comets, Thomas Burnet (1635–1715), a jurist and theologian, wrote *The Sacred Theory of the Earth*, representing the Flood as a major episode. Within a universe that was supposedly ancient, the initial terrestrial chaos had dissipated when the elements formed concentric envelopes of increasing density: earth, water, oil, and air. As they deposited upon the oil, particles of dust from the air subsequently constituted the smooth, unctuous, fertile soil of Paradise; and the earth, whose axis of rotation was normal to the ecliptic, enjoyed a uniformly pleasant climate. The sun's heat and the actions of subterranean waters had caused this crust to become fragile, and it broke abruptly, and so the Flood came at just the right moment to chastise the sins of humanity, leaving the reliefs created by the crustal deformations after the receding of the deep waters. This was the point at which the earth's axis took on its inclination under the effects of the disequilibrium that had been induced in the mass, and the time when alternating seasons began to replace the consistently mild climate. From that point on the earth began preparing slowly, at its depths as well as on the surface, for the universal conflagration. A period of peace and tranquility would follow before the earth was transformed into a star, and everything would end, finally being consumed as predicted by the Scriptures.

This *Theory* did, in fact, provide something new, which justified the success evidenced by its various Latin and English versions that continued appearing until 1702. Burnet had incorporated into a schema of the earth's evolution, which was highly inspired by that of Descartes, a history that pivoted on the Flood, the most ancient of disasters, whose actuality was asserted by a multitude of traditions. What resulted was an account that was both moral and physical, a tale of a twofold decadence, that of the earth and that of humanity, and one that was consistent with sacred teachings. Burnet was not a naturalist, and he knew little about the question of fossils; but certainly he was incapable of attributing an enormously vast timescale to the earth—but not ageless eternity, which he denied absolutely. "So if there be evidence," he asserted, "either in Reason or in History, that is because it is not very many Ages since Nature was in her minority, as appears by all those instances we have given above, some whereof trace her down to her very infancy." This account of an earth that changed its reliefs did not contradict Mosaic timescales. In the opinion of the guardians of orthodoxy, however, Burnet's ideas depended too much upon allegory, and they did not leave enough initiative to God the Creator. This criticism cost Burnet his positions as chaplain and secretary to the cabinet of King William III.

Following Burnet, another Englishman, the doctor and naturalist John Woodward (1665–1728), gave the Flood an even more important role, in his *Natural History of the Earth and Terrestrial Bodies*, published in 1695. This time, fossils held the central position. Having flattered himself that he had written from facts of experience alone, Woodward commenced the work with a long and influential chapter proving that fossils had assuredly been "the real Spoils of once living Animals." The major difficulty presented by the Flood was that its forty days were far too short a duration for the depositing of the ensemble of the strata. Woodward resolved this problem by supposing that the waves that rose from the abysses of the earth had had the effect of completely dissolving some of the rocks: "the whole Terrestrial Globe was taken all to pieces and dissolved in the Deluge, and the particles of Stone, Marble, and all other solid Fossils dissevered, taken up into the Water, and there sustained together with Sea-Shells and other Animal and Vegetable bodies." The earth was thus formed by the depositing of "that promiscuous Mass of Sand, Earth, Shells" in strata, of which the *specific gravity* was greater in the depths than at the surface. Even though there was no regular increase of density to be detected in descending into these strata, the synthesis produced by Woodward from his own observations and the recent theories of Newton and Steno proved, just the same, to be a lasting success. Fifty years later, Buffon, as one example, felt obliged to contest Woodward's thesis, pointing out his surprise that this great dissolving that had engulfed all the rocks had somehow been able to spare the shells.

As noted earlier, the various theories that were proposed to account for fossils were precarious, and necessarily so, because the timescales that were considered acceptable were simply too brief. For a long time the choice to be made was which kind of crutches one was willing to use. Neither the Flood nor the sports of nature—the two great terms of the alternative—could give a satisfying explanation within a framework of Mosaic chronology for the omnipresence and the diversity of the fossils. But natural history did not have its La Peyrère: there was nobody who perceived this dilemma, or at least nobody who had the audacity to mount a public protest regarding it. A meticulous observer such as Ray might have noticed that the accepted brevity of time was excessive. In examining the bottom of a well in Amsterdam, he recognized that the bed of sand and shells represented the ancient bottom of the sea. He noted in 1673 in his *Observations Topographical, Moral, & Physiological* that the floods of the rivers had accumulated nearly a hundred feet of sediments over the top of that bed, "which yet is a strange thing, considering the novity of the World, the age whereof, according to the usual Account, is not yet 5600 years," but he did not continue any further with his reflection.

The attention of naturalists such as Ray was attracted to entirely other horizons. This prodigious force of life, which was manifested through the profusion of fossils buried deep within the hearts of mountains, appeared particularly splendorous in view of the multitude of animal and plant species. Such a multitude could, at first glance, be overwhelming. "I imagine a man looking around him at the infinite number of things: plants, animals, metals. I do not know where to have him begin his experiment," exclaimed the philosopher Michel de Montaigne (1533–1592) with perplexity in regard to the remedies offered man by Nature. A century and a half later, the myriad species that had already been described, of domestic or exotic plants, of insects, birds, fish, and beasts, and the number, doubtlessly even larger, of fossils and other inanimate bodies—all this elicited nothing but wonder for the great classifier Ray. "What can we infer from all this?" he asked in *The Wisdom of God Manifested in the Works of Creation*. "If the number of Creatures be so exceedingly great, how great, nay, immense, must need be the Power and Wisdom of him who form'd them all?" The multitude of beings organized within a perfect nature corresponded to the immensity of the space that so fascinated Newton. Aristotle hadn't lost everything: in spite of Descartes's efforts to exclude it from natural philosophy, Aristotelian teleology remained present indeed. Through the planets and the stars, through plant and animal life, a youthful Nature celebrated the grandeur of a Creator to whom, alone, eternity had been reserved.

FIVE

The March of the Comets

THE GREAT VIEWS OF COSMOLOGY   An active debate throughout the eighteenth century was whether fire or water was the principal agent acting through time upon the earth. It was in this context that a cosmological and geological pathway was sought by Buffon, the distinguished Newtonian, to prove the great antiquity of the earth. On the basis of experiments conducted on balls of various materials and sizes, heated to high temperatures and then cooled, Buffon claimed in his *Epochs of Nature* that the earth was seventy-five thousand years old; he wrote in his private notebooks, however, that its actual age of more than 10 million years simply could not be grasped.

Boudoir Science

During the solar eclipse on 12 August 1654, "all of Europe was overcome with fear—I am speaking of the ignorant people—to the point, even, of believing that it was the end of the world, and of preparing themselves for death, or of hiding in cellars to avoid the evil influences," complained Pierre Petit (1594–1677), the king's geographer and administrator of fortifications in France. But even more terrorizing were the comets, he added, because of the wars, plagues, famines, floods, earthquakes, and fires that were attributed to their passage. Not only the ignorant were among the ranks of people who viewed the stars as "the heralds of arms who come to declare war against mankind on behalf of God," according to critic Pierre Bayle (1647–1706); consider Tycho or Kepler, for example, two illustrious observers of the skies who harbored these fears as well. After the passage of the comet of 1607, Kepler recalled: "It is in accordance with history that, with the appearance of comets, there commonly arise long-lasting troubles, which together with the death of a great many people also bring fear and grief to the survivors."

Meanwhile, new attitudes appeared later on in the century. When the comet of December 1664 appeared in the firmament, Louis XIV assigned Petit the task of dissipating his subjects' anxieties. The geographer adroitly depicted comets as stars having regular, elliptical movements, which cleaned the skies of exhalations released by planetary atmospheres. Shortly before, when courtiers thought proper to tell the dying prime minister Jules, cardinal de Mazarin, that it was fit "to honor his agony with a wonder," Mazarin had quipped that "the comet had done him too much honor." Madame de Sévigné (1626–1696), who reported the anecdote, added, for her part, that "human pride does itself too much honor in believing that there are grand affairs about the stars when oneself must die." The great comet of 1680 had just passed; it had been less a source of apprehension. In fact, it inspired the youthful Fontenelle to write a comedy produced the following year, and it similarly inspired Bayle's classic *Various Thoughts on the Occasion of a Comet*, in which an attack against superstition was used to screen a plea for liberty of conscience.

Although the fear of comets did not disappear, at least it had been assuaged. It had not been long, it is true, since the audiences of the new academies in London and Paris had demonstrated that interest in science extended well beyond the circles of specialists. After his scathing 1659 *Les précieuses ridicules* (*The Pretentious Young Ladies*), Molière, in *The Clever Women*, took mocking aim at pedantic women's academic pretensions thirteen years later. Even if he had evoked, prior to Halley, the formidable effects of the possible impact of a comet, it was obviously not chance that caused him to have a pedantic schemer say:

While we were sound asleep, we had a near escape:
Another alien world passed close to earth last night
It fell right through our vortex in its hurtling flight;
It could have bumped right into us as it did pass—
It would have smashed us into pieces just like glass!

The fashion in salons was not merely to have at hand a telescope, such as would allow any honest person to look at stars or comets, but also to be occupied with arduous problems by which social life did not necessarily gain anything. In the 4 March 1686 issue of *Journal des Sçavants*, an article deplored that ever since the Mathematicians found the secret of introducing themselves down even into the boudoirs, and of introducing the terms relating to a science as solid and serious as mathematics into the ladies' cabinet, it is being said that the empire of gallantry is going to ruin—that nobody talks there about anything other than problems, corollaries, theorems, right angles, obtuse angles, rhomboids, etc., and that it has been found that, not long ago, two young misses of Paris had had their brains muddled to such an extent by this sort of knowledge that one of them did not want to hear any sort of talk at all about proposal of marriage unless the person seeking her had learned the art of making telescopes..., and that the other had rejected a perfectly honest man because he had not been able, within the timeframe prescribed by her, to produce any further advancement regarding the squaring of the circle.

In contrast with these learned, light-hearted banterings, which perhaps hinted something about the Parisian mind-set, the speculations of the severe minds of England, such as Burnet or Woodward, were dedicated to serving science with the latest advances of exegesis. Comets and their majestic tails were integrated more easily and naturally into vast perspectives than were fossils, whose implications were still vague but whose importance could be sensed. Because they did not represent anything more than meteorological perturbations, as Aristotle had supposed, comets had first of all been expelled from the corruptible world by Tycho. Their transformation into aborted stars by Descartes subsequently offered a way for them to lose their evil influences. Since comets were of the same material as the stars and the planets, they would not be long in taking their places among the theories of the earth. That this occurred first in England ought not to be a surprise.

Of the Proper Usage of Comets

In England, where a vivacious Puritanism contributed to the Glorious Revolution in 1688 and to the Stuarts' fall, comets were no pretexts for comedy. After Newton's and Halley's demonstrations that comets' movements were governed by the laws of gravitation, they lost their status as erratic bodies, which had been one of the sources of belief in their occult qualities. While discussing the effects of a comet's passing near the sun, Newton had noted, among other things, that a comet ought to cool very slowly, "for a globe of iron of an inch in diameter, exposed red-hot to the open air, will scarcely lose all its heat in an hour's time." Since the ratio between a body's mass and its surface (through which the heat is dissipated) diminishes with the size of the comet, the rate of cooling ought to diminish more rapidly than the diameter increases. "Therefore," added Newton, "a globe of red-hot iron equal to our earth, that is, about 40,000,000 feet in diameter, would scarcely cool in an equal number of days, or in more than 50,000 years." Possibly in order to avoid the problem of cooling too slowly, Newton suspected, though incorrectly, that "the duration of heat may, on account of some latent causes, increase in a ratio yet less than that of the diameter." And he concluded, "I should be glad that the true ratio was investigated by experiments," without pursuing it, however, any further himself. As Buffon, one of his most ardent disciples, would show, the question had a considerable interest indeed.

Rather than remaining the portent of evil omens, one special comet had meanwhile been transformed by Halley into the agent of the Flood. Whereas comets previously had only served as forewarnings of celestial chastisement, one of them now had become the very instrument of divine wrath. The germs sown involuntarily by the great skeptic Halley were soon to find fertile ground in Whiston. His *New Theory of the Earth, from its Original to the Consummation of all Things* was an exhaustive application of Newtonian philosophy; its success was so great that six editions were published between 1696 and 1752. At the source of this account—duly punctuated with lemmas, corollaries, and scholia—which began with a reminder of the principle of inertia, was the observation made by Halley that the great comet of 1680 had a period of 575 years. It was the only one that had grazed the earth, remarked Whiston, and what was more, it had appeared around the years 2344 BC and 2919 BC, two of the dates proposed by chronologists for the Flood. If, as was suggested in Genesis, "the Ancient chaos, the Origin of our Earth, was the Atmosphere of a Comet" having an extremely eccentric orbit, it followed that God must have decided, one fine day, to put an end to this chaotic youth by dispatching our planet into a circular orbit. All through the period during which plant and animal life prospered, the earth did not rotate, until the moment in which Divine Will caused another comet to appear (the one that returned in 1680). The comet was attracted by the earth and pierced the earth's surface crust with its dense head, and thus it was that the Flood was set off, by masses of water liberated from great abysses within the earth, which blended with the water present in the comet's tail, on 28 November 2349 BC or 2 December 2926 BC, at two o'clock in the morning, Beijing time, where Noah must have been living at that time. Among other consequences, the earth all at once took on its bulging figure and its elliptical orbit, inclined the axis of its poles, acquired its rotational movement, and lost its population of at least half a trillion multiple-centuries-old inhabitants.

Because this *Theory* exploited the precision of astronomy in a marvelous way and reconciled the Scriptures elegantly with the new physics, it is almost superfluous to note that it was "well approved" by Newton, to whom it was dedicated, by an author who was about to succeed him at Cambridge. It also pleased the great empiricist John Locke (1632–1704), who judged, in a letter to one of his mathematician friends, that Whiston was "one of those sort of Writers, that I have always fancied should be most esteem'd and encouraged." Locke added that, by explaining "so many wonderful and, before, inexplicable Things in the Great Changes of this Globe," Whiston had become one of those builders who promote the advance of knowledge, while the "Finders of Faults, the Confuters and Pullers-down do not only erect a barren and useless Triumph upon human Ignorance, but advance us nothing in the Acquisition of Truth."

To reveal the sense of the most obscure passages of the Bible, Whiston completed this audacious work in 1717 with his *Astronomical Principles of Religion, Natural and Reveal'd.* Hell, for example, had been described as a "State of Darkness," of "outward Darkness," of "blackness of Darkness," of "Torment and Punishment for Ages," or "for Ages of Ages," "by Flame," or "by Fire," or "by Fire and Brimstone, with Weeping and Gnashing of Teeth," where "the Smoak of the Ungodly's Torment ascends up for ever and ever." Referring to such portrayals, Whiston remarked, "This Description does in every Circumstance, so exactly agree with the Nature of a Comet, ascending from the Hot regions near the Sun, and going into the Cold Regions beyond Saturn, with its long smoaking Tail arising up from it," that it was difficult not to think "that the Surface or Atmosphere of such a Comet [is] that Place of Torment so terribly described in Scripture." Traveling through the skies, where the righteous enjoyed eternal happiness, comets thus offered "a most useful Spectacle to the rest of God's rational Creatures," which would "admonish them above all Things to preserve their Innocence and Obedience." Locke never was able to appreciate these audacious inferences, since they were published thirteen years after his death, but the spectacle of comets in 1742 inspired Maupertuis, who was still a good Christian, to quite different sorts of reflections: "These stars, after having been the terror of the world for so long," he related, "have fallen into such a degree of discredit that nobody believes any longer that they could cause anything more than a cold!" What would Whiston, who claimed to be the period's best apologist, have thought about that?

Water or Fire?

Participating in a glorious period marked by the seals of Hooke, Newton, and Halley, the English theories of the earth dominated the first half of the eighteenth century. Descartes's evolutionary schema persisted like a watermark behind the accounts of Burnet, Woodward, and Whiston, but it was intermingled with the Mosaic account and therefore subject to untimely divine interventions. These theories also spurred some caustic remarks, such as those of Voltaire, who was, notwithstanding, an Anglophile: "It would take more time than the Flood even lasted to be able to read all the Authors who have made their wonderful systems out of that," he quipped. "Each of them destroys and remakes the earth in his own manner, just as Descartes formed it. Because the greater part of philosophers have put themselves, naturally, into the seat of God, they think they can create the universe with the word." In England, water remained the principal agent of the earth's history, conforming to the Mosaic account, according to which, right after the separation of light from darkness, Elohim had commanded: "Let there be a vault through the middle of the waters to divide the waters in two." The two essential moments of this history remained the Creation and the Flood, and these were coupled in a certain way with the history of humanity, which itself was given orientation by the Creation and the Passion of Christ. Fire was thus reduced to a subordinate role within the inaccessible depths of the earth. Paradoxically, on the Continent it was a theory aimed at attacking Descartes's system that placed fire at the forefront. Its author was Gottfried Wilhelm Leibniz (1646–1716), the man who was on the verge of becoming Newton's most celebrated adversary.

Already renowned for his vast erudition and his memoirs on law and physics, Leibniz had spent five years in Paris with the aim of entering its Academy of Sciences. He returned to Hanover with his hopes dashed. There he was engaged in 1677 by the duke of Brunswick-Lüneburg as adviser, librarian, and supervisor of the mines of Harz and made responsible for the modernization of the mines. When he was asked by the duke to write the history and genealogy of his house, Leibniz departed to search in Italy and Germany from 1687 to 1690 for archives, antiquities, and natural curiosities. This memoir, which was never finished, helped the duke to become elector to the empire; and Leibniz, in turn, who had studied Agricola and was familiar with Steno, gained the opportunity to demonstrate that he, too, was a sensible observer. One fact struck him particularly, that the ensemble of sedimentary rocks had vitrifiable material for its basis, the great granite masses. This fact started him thinking that the sediments had been deposited upon the residues of an immense, primitive crystallization. In 1691, upon returning from his excursions, Leibniz mentioned in a letter his suspicion "that the globe of the earth could have been in a state of fusion at the beginning, when the light separated from the darkness." As a result, "the base of this great body (*ossamenta ejus*) would be a type of vitrification," and the sea was a sort of deliquescent solution, because "after the cooling of the melted mass, the vapors had condensed and washed the surface of this mass, producing a water charged with fixed salts."

This important thesis was developed in *Protogæa*, a work published in Latin thirty-three years after Leibniz's death; nonetheless it was known somewhat earlier, owing to a two-page abstract of the work published in 1693. The thesis was part of a systematic critique of Descartes and his *Principles of Philosophy*. After calling into question his physics, Leibniz attacked his cosmogony, which he found far too inconsistent with sacred history and experience. Demonstrating his orthodoxy in relation to the Scriptures, he proposed that the formation of this "volcano" we inhabit was "owing to the divorce of light from darkness signaled by Moses"; and it was observation, not speculation, that permitted him to assert, moreover, that "the globe of the earth was, first of all, regular in form, and, from the liquid that it was, it became solid." Thus he could identify the subtle burning matter of Descartes as common rocks in a state of fusion. Since he could not have learned of Newton's remark about a comet's loss of heat (it was not published until after Leibniz's death), Leibniz did not account for the extreme slowness of cooling of an immense melted mass, although he had recognized that, by "the accumulation and the condensation of material at the surface, the heat had been concentrated, and the cooled crust had been strengthened." Notwithstanding its subtitle, *On the Primitive Aspect of Nature, and Traces of a Very Ancient History, Comprising the Very Monuments of Nature, Protogæa* dealt in reality with a Mosaic history. Because "even a slight notion of great things has its price," as Leibniz indicated in his preamble, if one "wants to go back to the remotest origins of our country, we need to say a few words about the earliest configuration of the earth, of the nature of the soil, and of that which it contains." This desire to include a picture of the primitive earth within the family history of Brunswick-Lüneburg implicitly demonstrates the narrow chronological limits of Leibniz's framework.

Slightly preceding *Protogæa*, which was intended for a knowledgeable public, a curious book that described water, instead, as the first principle also had a certain impact. It was written between 1692 and 1708 by Benoît de Maillet (1656–1738), a diplomat, and it had the advantage of numerous reviews (Fontenelle provided one even before it was published). Ten years after the death of its author, it finally appeared in the classic form of dialogues, entitled *Telliamed* [de Maillet written backward] *or Conversations between an Indian Philosopher and a French Missionary on the Diminution of the Sea, the Origin of Man, etc.* Its success in the bookshops was assured in part by its daring transformist speculations, in which the blunt rejections of Mosaic chronologies pronounced by the Indian philosopher, Telliamed, must certainly have played a role. The naturalist Dézallier d'Argenville was not the only one to be scandalized. "Only a kind of godlessness could invent such dreams," he ranted, after inveighing: "What a folly in this author to substitute Telliamed for Moses, to bring man out of the depths of the sea, and, for fear that we should descend from Adam, to give us marine monsters for ancestors!"

Maillet became consul general of France in Egypt and later inspector of French establishments in the Levant and on the Barbary Coast. He was clever as a diplomat, renowned for his courage, on familiar terms with beys and pashas, and well read in the great Arabic authors. According to his biographer, "he knew very much, without acting like a scientist," and "he loved flattery, which is a common defect of witty people—but he liked it fine, and delicate." Maillet proved his finesse through his own studies of the action of the sea on the shores and on sedimentation. He remarked, for example, that by contrast with the fossils that are found among the surface layers of the earth, those present at the bottom of the sediments belong to species that now are extinct. It should not surprise us that this well-informed observer of the Mediterranean shores reintroduced, based on traces left by the Greeks, the eternity of ancient time.

At the foundation of the system expounded through the intermediary Telliamed was the allegation that the sea level had diminished continually since the beginning of historical times. Maillet thought evaporation had been the cause, because the variation had become faster over time, following the very slow change in the distance between earth and the sun and the change in solar luminosity, both effects being due to the displacements of stars within the Cartesian vortices. Evaporation had progressively brought forth the primitive mountains and valleys, which the interactions of marine currents had modeled at the bottom of the sea by the accumulation of sediments and shells. Subsequently erosion of the shores and sedimentation produced other mountains at the edges of the continents, while the seeds of plant and animal life present in the universe found a favorable environment for their development near the coastal waters. From the sea, then, life had been able to reach the air and the land. "The transformation of a silkworm or of a caterpillar into a butterfly would be a thousand times more difficult to believe than that of fishes into birds, if this metamorphosis did not occur daily before our own eyes," added Telliamed, after describing how the transformation of fins into wings had given rise to the first, beautiful flights. As for humanity, the fossils of human bones and the fragments of petrified ships (which, in reality, were the remains of ancient vertebrates and trunks that had been converted into silica) attested to the fact that it was at least a good million years old.

Although it was immense, such an age became negligible before the great cycles of some 5 billion years marking the time for the activity of the earth as it proceeded in its trajectory through space. Maillet added his sentiment that "before the invincible proofs that we have had a diminishing sea for over two billion years," the stars were consumed through the course of durations of time that also could be counted by the billions of years. But these durations figuring in the manuscript of *Telliamed* really shocked people too much. They disappeared completely from the first printed edition, and in the second, three or four zeros were removed from the figures of the billions and millions of years. And in order to foil the reasoning of Telliamed, who lashed out at his interlocutor, the French Missionary, that his "system of a beginning of matter and of movement through time repels reason, and has no foundation whatsoever," Maillet's publisher printed "Divine Providence" in all the many places where the word "chance" was used in the manuscript. As was witnessed by Dézallier d'Argenville, this editor's censure did not really disarm the critics. But like Georges-Louis Leclerc (1707–1788), better known by the name Buffon, the readers of *Telliamed* were not all disgusted. The first volumes of Buffon's monumental *Natural History* appeared just one year after *Telliamed*, at the same time as *Protogæa*. And it was at the end of his life of incessant labor that Buffon delivered, in his *Epochs of Nature*, the "great views" that contradicted both the brevity of Mosaic times and the eternity of the Greeks to which Maillet had returned.

A Young Man in a Hurry

Georges-Louis Leclerc, the future count of Buffon, lord of Montbard, marquis of Rougemont, viscount of Quincy, and vidame of Tonnerre, had the good fortune of being born well off. He was the eldest of five children and the son of an attorney at the parliament of Burgundy, who had acquired the duty of collecting the salt tax. His godfather was his mother's uncle, the rich director general of the farms of the duke of

Savoy, king of Sicily, who, opportunely, had no progeny of his own. Georges-Louis's path seemed to have been well marked out: after three of his sisters and brothers eventually entered into religious orders, it was presumed that he would pursue the social rise undertaken by his ancestors. A mediocre pupil of the Jesuits at Dijon, where his parents had settled, while still an adolescent he discovered a passion new to his family: mathematics. But despite a pronounced inclination toward Euclidean geometry and integral calculus, around the age of sixteen he turned in the direction of law. Three years later, armed with his diploma, he decided to defy his father and dedicate himself to science. After he had left for Angers to study, a bloody duel obliged him to return hastily to Dijon. There was only enough time for him to become the traveling companion of the duke of Kingston, an adventurous English aristocrat slightly younger than he, and an original fellow who accompanied them, as tutor and doctor to the duke.

Through the long course of a year, the three travelers discovered the towns of the South of France and Italy, their peoples, and their pleasures. During their peregrinations, opera, local gastronomy, and simple, leisurely activities did not cause Leclerc de Buffon, as he had begun calling himself, to lose all sight of mathematics. In correspondence with a mathematician from Geneva, he formulated subtle analyses based on the calculation of probabilities. Nature, though, remained a stranger to him, even in Italy. Upon returning to Dijon in the spring of 1732, he settled his inheritance from his mother, against the will of his father, then repurchased the Buffon estate before going to Paris to take up his abode there. His rank and fortune permitted him, at the age twenty-five, to devote his time fully to science. Buffon, as he signed his name from 1735 on, then adopted a pattern that he chose to follow until his death: autumn and winter in Paris, and the rest of the year in his Burgundy estates.

Buffon preferred "to let the water run, and to plant hops," rather than appearing out in the world or at court. At Montbard he found an environment that gave free rein to his love of concrete: he razed an old feudal donjon and constructed a "chateau" that was limited in luxury and surrounded by gardens and terraces. He installed outbuildings, which became his workplaces, and he enlarged and exploited his domain, closely supervising the revenues from his seignorial rights. After he reached age sixty, Buffon invested heavily in forges, which he insisted be exemplary in quality. But like his other enterprises, they were far from profitable, because science scarcely left him the time to manage them carefully. Buffon worked everywhere, in his forests as well as in his study, which was an isolated room at the top of a rampart; it was devoid of books and furnished with one simple table of black wood and a portrait of Newton. Moreover, he loved to ride about and inspect his domain. And he entertained, he went to mass every Sunday, and he knew how to act with generosity, often discreetly. In a word, the future count lived the part of his rank.

In Paris, Buffon was approaching closer to the Royal Academy of Sciences. He was interested in the calculation of probabilities, such as determining the chance that a coin thrown onto a tiled floor would land directly on a tile or, instead, on the dividing line between two tiles, and so forth. The solutions to this new type of problem, which Buffon provided using a combination of probability calculation and integral calculus, brought him admission to the Academy as assistant teacher in the mechanics class. At that time, the Academy was under the care of the count of Maurepas, who was also minister of the navy. Under his twofold title, Maurepas had asked the Academy about methods for improving the quality of the lumber to be used in producing ships. Buffon applied himself to this question, which the Academy had no means of studying. With his forests, he had the best advantages available for investigating the influences of soil on the growth and quality of wood, or the mechanical properties of wood, or a beam's resistance to breakage as a function of its length and section. To obtain Maurepas's good graces, in view of a promotion within the Academy, would undoubtedly fit in well with his plans. But the forested property-owner Buffon did not foresee that by addressing himself to such questions, which were fundamentally botanical, he would soon bring about a dramatic change in his own destiny.

So Buffon distracted himself from mathematics and became a specialist in the physics of trees. He passed from the Academy's mechanics class to the botany class, where he advanced to the level of associate in June 1739. One of the first convinced Newtonians in France, he also had excellent relations in England: shortly after this time he was admitted to the Royal Society of London, and his translation of Newton's *The Method of Fluxions and Infinite Series* appeared the following year. But a crucial event occurred this same year of 1739 when smallpox claimed the life of the efficient keeper of the Royal Botanical Garden and Cabinet, Charles-François Du Fay, a scholar whose interests ranged among all sorts of specialties. The succession was coveted. While discreetly remaining at Montbard, Buffon presented his case with tact. "I would request my friends to speak for me, to speak openly that I am suitable for this position; it is the only reasonable thing I have to do at present," he wrote to an Academy friend, also specifying: "The directorship of the Royal Botanical Garden requires a young, active man who can tolerate

Figure 5.1 Buffon at the age of fifty-four. Painting by F. H. Drouais.

the sun, who knows plants, and who knows the ways to make them multiply, who is a bit of a connoisseur in all areas where it is demanded of him, and moreover, one who understands buildings." Well thought of by Maurepas and recommended by Du Fay before his death, Buffon was heard, to the surprise of many, by the king himself: on 26 July he was named keeper, receiving an enviable dwelling place and a remuneration of three thousand pounds a year.

From One Description to the Next

As enterprising in Paris as he was in Burgundy, Buffon spent forty-eight years as keeper of the Royal Botanical Garden. He doubled its size, provided it with new buildings, and enriched its collections considerably. He surrounded himself, above all, with competent collaborators and maintained a vast network of correspondents, whom he asked to send animals, plants, and minerals to Paris, not for the purpose of flattering the eye, but to depict Nature in a way detailed enough that its outline could be grasped. While this work was going on in Paris, another adventure was being conducted in the calm of an office study at Montbard. The first task set by Maurepas for the new keeper had been a description of the king's collections. According to the formula of the nineteenth-century critic Charles Sainte-Beuve, Buffon was "until then, a genius in waiting, whose object was yet lacking"; he abandoned to the anatomist Louis Daubenton the development of that description, while he initiated the project of an encyclopedia of nature, from the mineral realm to man, such as never had been compiled since Aristotle or Pliny. It was as he wrote his *Natural History, General and Particular*, volume after volume, that Buffon developed into a naturalist. He gradually took possession of his subject, uniting the "detailed attention of a laborious instinct applying itself to a single point" with the "great views of an ardent genius, embracing all with a single glimpse." He renounced his own theories where necessary and without any dogmatism invited his successors to resume his work, which, in the end, treated neither plants nor the lower animals.

At Montbard, everything was subordinated to the needs of this great work, from the domain transformed into a menagerie to the daily life that was ordered with a monastic precision. "The soul of the great philosopher was pregnant with natural history; it had to be brought forth," related Chevalier d'Aude, Buffon's last secretary, adding that "one will not draw the absurd consequence that the count of Buffon was insensitive to glory (it was his favorite mistress) or to delicate praise (he loved it to a fault); but he merited those, and he breathed their incense with a voluptuous pleasure, which might have seemed prideful, except for that those were not at all the motive force of his potent faculties: if he wrote and created, it was because he was tormented by this twofold need." Buffon did nothing to endanger the intellectual independence that was assured by his fortune. Endowed with an exceptional liberty of tone and conventions, he kept his distance from clans, coteries, and polemics, always judging offenses to be unworthy of responses. He spent less and less time in Paris and ended up neglecting the Academy of Sciences, just as he had lost interest in the Académie Française, which had admitted him as a member in 1753. But withdrawing thus did not impede him from addressing himself to the empress Catherine II of Russia, for example, "from sovereign to sovereign" or from seeing his Buffon estates raised by Louis XV to the status of an earldom.

Behind this famous man whose vanity has often been misunderstood, another Buffon was concealed, one whom his correspondence allows us to glimpse. He was bawdy, for example: "Daubenton is beautiful, and certainly has the loveliest boobs in the world"; a moralist: "true happiness is found in tranquility"; and on one occasion simply a father who has suffered loss: "I lost a child who was just beginning to make himself understood, that is to say, to love." Married late to a young lady he loved, then widowed early, Buffon applied himself to a life that was private and free, though untarnished by scandals, based on utilitarian views that also influenced the beginnings of his work as a naturalist.[1] To "dispel the superfluous," an expression employed in a letter of his youth, "he was the delicate flatterer of women whose conquest affected him but little." That is to say, Aude continued, "that the difficult pleasures would have cost him too much time, and that he abandoned them to the amateurs. Young girls and simple women were quite a bit to his taste, provided they were also pretty." But nothing was allowed to distract his thoughts or to trespass upon the some dozen hours reserved for his daily study. Of this indefatigable activity of Buffon's, there is no shortage of witnesses, such as Marie-Jean Hérault de Séchelles, a young, ambitious attorney and the future temporary president of the National Convention who drafted the first constitution during the Revolution. He was invited to Montbard in 1785, and there he recorded: "To someone who was astonished at his fame, [Buffon] said, 'I have spent fifty years in my study.'" And thirty years after he had proclaimed that "Style is the man himself," Buffon confided to his guest, "I am learning every day to write," and in a similar vein, "Genius is nothing more than having a greater aptitude for patience."

The History of the Earth and of Time

Because "the general history of the earth ought to precede the particular history of its productions," Buffon began his *Natural History* with a volume titled *History and Theory of the Earth*, which he published in 1749. He was adventuring into unknown territory since, with the exception of youthful journeys into Italy and England, he had rarely traveled far from Paris or Montbard. Because he had made few observations in the field, even in Burgundy, his descriptions were based mostly on his readings, of Herodotus, Strabo, Descartes, Steno, Halley, Leibniz, Maillet, the English authors of theories of the earth, and naturalists and travelers of his period, such as Nicolas-Antoine Boulanger (1722–1759), the engineer

---

1. In his first discourse, Buffon wrote, using a phrase that has remained a classic, "Wouldn't it be better," in a classification, "to have the horse, which is solidungulate, followed by the dog, which is many-toed, and which does, in effect, have the custom of following it, than by the zebra, which is little known to us and which has, perhaps, no relation with the horse other than that it, also, is solidungulate?" *Histoire naturelle, premier discours*, in *Œuvres philosophiques*, 18. But he subsequently adopted points of view in his Natural History that were clearly less like caricatures.

of bridges and roadways who was close to the Encyclopedists and whose most important geological work remains yet unpublished. Buffon's originality came therefore from his arrangement of the facts more than from their novelty, and from the audacious conclusions he drew from them, which proved much more suitable for the greater public than for specialists.

The first of his principles was that of actual causes, which had already been expounded by Strabo, for example, but in Buffon it was distinguished by his clarity of expression. In order to understand the earth, one must exclude the extraordinary: "The shock or the approach of a comet, the absence of the moon, the presence of a new planet" were causes that could "produce anything one could have wanted, and from just one of such hypotheses as these, one could draw up a thousand physical romances, which their authors then call a Theory of the Earth." The trio Burnet, Whiston, and Woodward, as one might have guessed, was particularly brought to mind by this comment, and Buffon concluded his criticism with a solemn statement: "As historians, we refuse certain vain simulations; they proceed toward possibilities that suppose an upheaval of the universe in order for them to come about, in which our globe, as a spot of abandoned material, escapes the eyes and is no longer considered an object worthy of our attention." In order to understand the past, declared Buffon, one must begin from the present. "From effects that happen every day, from movements that succeed one another and that renew themselves ceaselessly, from constant operations that are reiterated every day—these are our causes and our reasons."

As an essential consequence, natural history ceased being merely descriptive and began to involve a historical account. But Buffon was jumping the gun, because his *Theory of the Earth* had not yet been able to break free from a rudimentary history based on two moments, the Creation and the Flood, in the style of the Mosaic chronology. And paradoxically, in order to explain the origin of the earth, Buffon resorted again to his account of the impact of a comet. "Cannot one imagine with some likelihood," he advanced, "for a comet to have fallen upon the sun's surface and to have displaced that star, and that it would thereby have separated some of its parts from itself, to which it would have communicated an impulse of movement in the same direction and by the same shock, in the way that comets would have become integrated into the body of the sun in the past, and that they would have been detached from that by an impulsive force common to all, which they conserve even today?"

By contrast with the divine interventions postulated by Whiston, this hypothesis was sustained by veritable arguments: there was only one chance in 64 that the planets would fortuitously turn in the same direction, and there was only one chance in 7,692,624 that their orbits would be inclined as they are in relation to the ecliptic, declared Buffon. Based on that point, the current movement of all the planets could only have resulted from the impact of a comet, an event that certainly is exceedingly rare, but which, at the same time, is not dependent upon a supernatural will; furthermore, owing to the abundance of comets, it does not require qualification by any sort of extraordinary character. Then, practically without transition, Buffon noted that the fire of the material uprooted from the sun had been extinguished and that water had become the principal agent on the surface of the earth, endlessly modeling and remodeling the reliefs that globally remained unchanged. The omnipresence of shells, the horizontal stratification, and even the forms of the mountains, for Buffon, as for Maillet, were all proofs that the sea had submerged everything and arranged everything before putting the reliefs into form by means of its currents. But the reasons for the ebbs and flows of the sea escaped him, because a properly historical vision was still lacking.

Buffon had counseled in his work: "Each time one wants to be so rash as to try to explain theological truths by means of physical reasons, or one allows oneself to interpret the divine text of sacred books in terms of purely human views, or one would reason upon the will of the Almighty and upon the execution of his decrees, one necessarily falls into darkness and chaos." But this humility did not shield him at the beginning of 1751 from the censure of the deputies and the syndic of the Faculty of Theology at Paris, then under Jansenist influence. Fourteen of Buffon's propositions were judged contrary to the dogma, among them: "the waters of the sea were responsible for producing the mountains, the valleys, and the lands," and "the sun will probably extinguish itself for lack of combustible material." Only four of these theses concerned the system itself, while ten concerned philosophical questions that essentially touched the soul and truth. It is not clear whether the danger to faith would have come more from this "truth," which could be attained through observation and reason, or from the concrete role of the waters and the possible extinction of the sun.

Buffon was preoccupied. He preferred to keep his distance at Montbard, but the censure did not trigger any sort of anxiety or metaphysical torments for him. Affirmations such as "The first religion is for each one to keep his own, and the greatest happiness of all is to believe the best," could have called into question the reality or depth of his faith, but Buffon did not have to proclaim, as he did: "The deeper I penetrate into the breast of Nature, the more I admire and respect her Author." Buffon, an assiduous parishioner who often invited his chaplain to dinner, in any case manifested a very modern attitude regarding the separation of science and religion. Having much better things to do than to get bogged down in a controversy with the deputies and the syndic, he responded in ten brief points. Only the first concerned the part of his *History* that was actually scientific: "I have abandoned those ideas in my book regarding the formation of the earth and in general everything that could be contrary to the account of Moses, and I have only presented my hypotheses on the formation of the planets as though a purely philosophical supposition." The about-face was not new; the faculty was contented with it. At that time pamphlets circulated in secret, such as *The Machine Man*, by Julien de la Mettrie (1709–1751), a manifesto of materialism and atheism. And the time was not far off when Louis XVI would declare, in order to refuse the archdiocese of Paris to Monseigneur Loménie de Brienne (1727–1794), whom he later made his own minister: "The archbishop of Paris, at the least, ought to believe in God!" Contending against the diffusion of pernicious ideas, the historian P. Hazard summarized: "The period was controlled by inconsistencies, because it was controlled by facilities. One resisted, and one ceded to a general feeling that flattered the sweetness of life."

And so, the *Theory of the Earth* did not risk the flames. Buffon resumed his work, and it was three decades later, after his *History of Quadrupeds* and his *History of Birds*, that an authentically historical perspective was established in his *Epochs of Nature*. In relation to the *Theory of the Earth*, where "the unit of time of the Creation" so dear to Newton was still being mentioned as a "very essential thing," a considerable change had been achieved. "A philosophical will to reach as far as possible in the search for natural causes, rather than purely scientific reasons, pushed Buffon to refuse Newtonian Creationism and to introduce history as if by accident," J. Roger summarized, adding: "Buffon, setting out to write a natural history, ended up by devoting himself to the history of nature." As testimony of this shift, we read in the famous first page of *Epochs*:

Just as on civil history, we consult titles, we research medals, we decipher ancient inscriptions in order to determine the periods of human revolutions and certify the dates of the moral events. In the same way, in natural history, we need to search through the archives of the world, draw out the old monuments from the bowels of the earth, gather their remains, and reassemble into one body of proofs all the clues about physical changes that permit us to go back to the different ages of nature. This is the only way to fix some points within the immensity of space and to place a certain number of milestones on the route of eternal time. The past is like the distance; our sight diminishes within it, and it, itself, would become lost, had not history and chronology placed some lanterns, some torches at the darkest points.

The Key to Time

Buffon himself placed some lanterns along the way of the succession of "various changes" that gave the earth, beginning from its earliest moments, a form that is very different from the one it currently has. "These are the changes that we call its epochs," he wrote, associating with each of them not only an essential event, but an age as well.

First epoch: when the earth and the planets took their form [zero time]; *Second epoch*: when the material, in the process of consolidating, formed the inner rock of the globe, as well as the great vitrifiable masses that are at its surface [2,936 years]; *Third epoch*: when the waters covered the continents [35,000 years]; *Fourth epoch*: When the waters receded and volcanoes began to be active [50 to 55,000 years]; *Fifth Epoch*: When the elephants and the other animals of the South came to inhabit the northern lands [60,000 years]; *Sixth Epoch*: When the continents separated from one another [65,000 years]; *Seventh Epoch*: When the power of man seconded that of nature [75,000 years].

Even though they were seven in number, these epochs did not owe anything to Moses. "It has been upon the results of my experiments that I have based all my reasonings," proclaimed Buffon. The paradox was only apparent: "It is indeed on the condition of its being firmly anchored in experience that speculation can expand" and that "even the most conjectural ideas, or those that could appear to be too risky" can obtain, by means of relations "that will not be missed by those who know how to evaluate the force of inductions and appreciate the value of analogies." In identifying his seven epochs, Buffon referred to *facts* and *monuments*, the latter of which are more numerous and apparent in proportion as the past in question is more recent. The reconstitution of the first moments of the earth, of which we recognize hardly any remains, rests therefore upon facts alone, and of the five that Buffon highlighted (listed below), four are important from this point of view:

- First fact: "The earth is elevated at the equator and lowered at the poles, in the proportion that the laws of gravity and centrifugal force require." As was calculated by Newton and verified by Maupertuis and his colleagues, the figure of the earth is flattened and could be thought to represent a remnant of its original fluidity.
- Second fact: "The terrestrial globe has an interior heat that is its own and that is independent of that which the sun's rays communicate to it." This permanent exhalation of heat is evidenced by the heating that can be felt in mines, previously mentioned by Kircher in his *Mundus Subterraneus* (*The Underground World*), which was measured for the first time in 1744 at the request of Mairan in the salt mines of Giromagny, in eastern France, by Antoine de Gensanne (?–1780), the longtime director of mines in Franche-Comté and Alsace.
- Third fact: "The heat emitted to the earth by the sun is rather small in comparison with the terrestrial globe's own heat; and this heat emitted by the sun would not by itself be sufficient to maintain nature in a living state." This conclusion, which was quickly invalidated, was drawn by Mairan from terrestrial heat budgets, especially in his *Dissertation sur la glace* (*Dissertation on Ice*) of 1749.
- Fourth fact: "The materials that make up the earth's globe are, in general, of the nature of glass and can all be reduced to glass." Regarding this point, Buffon was content to follow Leibniz. In agreement with the many experiments made by glass and ceramic industries of the period, the chemists' practice demonstrated, in effect, that rocks, such as granite, melt when heated to quite high temperatures and that they solidify into glass when they cool again.
- Fifth fact: "We find all over the surface of the earth and even in the mountains up to 1500 and 2000 fathoms in height an immense quantity of shells and other remnants of sea life." Even if not new or pertinent for the primitive earth, this observation, certainly, remained capital.

Buffon asserted, in confirmation of the views of Descartes and Leibniz, that "the primitive liquefaction by means of fire of the entire mass of the earth is therefore proved, in all the rigor that the strictest logic demands." To proceed from this observation to its chronological implications, Buffon put the experimental method into operation. For a forge master, passing from conjecture to an experimental study in cooling did not present any considerable difficulties. Over a period of five years, Buffon directed numerous experiments on steel balls ranging in diameter from 0.5 to 6 inches, which were heated until they were white-hot. The amounts of time required to cool them, first until they could be touched with the fingertips and then to room temperature, were measured. These results were reported in 1774 and complemented later by new measurements involving balls constructed of different mineral substances. By mentioning the discussion of the cooling of the comets given in

Newton's *Principles*, Buffon finally acknowledged his debt: "A passage of Newton's gave rise to these experiments," he recognized.

It was in extrapolating these results to bodies of greater dimensions that Buffon determined the durations of the most important stages of the evolution of the earth. He found, after correcting the results of these effects by factors such as the heat received from the sun, that 2,936 years had been necessary to consolidate the earth's molten crust. In order to arrive at the current temperature conditions, 72,000 more years had had to elapse. From the achievement of current temperatures, 93,000 years would remain for the existence of life. After that, in accordance with the irreversibility of evolution, temperatures would have become too low to sustain life on earth. Following the same method, Buffon also calculated the cooling times of the other planets and their satellites, establishing, for each one, the time intervals during which temperatures would be compatible with life. If "organized nature such as we recognize it has not yet been born on Jupiter, whose heat is still too great to be able to touch the surface," it is extinct on the smaller satellites, including the Moon, and in "full existence" on the majority of the planetary bodies.

"With the view of performing a great deed," Buffon wished "to reconcile natural science once and for all with theology." Because of such immense ages, or a complete neglect of the Flood, or an end of the world caused by cold, and not by fire, as had been announced in the Scriptures, the Faculty of Theology was not at all convinced by the interpretation of the first verses of Genesis that had been inserted into the *Epochs*. But because Buffon enjoyed so many protections and considerations, the faculty preferred not to censure him. Rather, they ascribed "this new attack against faith" to "the rantings of his old age." Still, the durations published in the *Epochs* had intentionally been reduced to a comprehensible minimum. "Instead of going back too far in the limits of the duration," Buffon had admitted, "I brought them in as close as possible without obviously contradicting the facts delivered in the archives of Nature." In his manuscript notes, we read that "the obstacles that oppose the emission of heat" vary less with the size of the body than he had supposed, and "that alone would have augmented our scale by ten times more and would have given me ten million years in place of 600,000 for the duration of our epoch." The rupture with Moses, then, was complete and total.

In support of these extremely long times, Buffon estimated the rates of sedimentation, considering the Normandy coasts as examples. "Are we not forced to admit," he wondered, "not only to centuries, but hundreds

COMMENCEMENT, FIN & DURÉE de l'existence de la NATURE ORGANISÉE dans chaque PLANÈTE.			
COMMENCEMENT.	FIN.	DURÉE absolue.	DURÉE à dater de ce jour.
de la format.	de la format.	ans.	ans.
V.Satel. de Sat. 5161 des Plan.	47558 des Plan.	42389...	6 ...
LA LUNE. 7890	72514	64624...	0 ...
MARS...13685	60326	56641...	0 ...
IV.Satel. de Sat.18399	76525	58126..	1603 ...
IV.Satel. de Jup.23730	98696	74966..	23864 ...
MERCURE... 26053	187705	161712..	112033 ...
LA TERRE.35983	168123	132140...	93291 ...
III.Satel. de Sat.37672	156658	118986..	81826 ...
II.Satel. de Sat.40373	167928	127655..	93096 ...
I.Satel. de Sat.42021	174784	132763..	99952 ...
VÉNUS. ...44067	228540	184473...	153708 ...
An. de Sat....56396	177568	121172...	102736 ...
III.Satel. de Jup.59483	247401	187918 ..	172569 ...
SATURNE...62936	262020	199114...	187188 ...
II.Satel. de Jup.64496	271098	206602..	196266 ...
I.Satel. de Jup.74724	311973	237249...	237141 ...
JUPITER.... 115023	483121	367498...	

Figure 5.2 The beginning, end, and duration of the existence of organized nature in each planet. A reproduction of Buffon's original in *Recherches*.

of centuries, in order for these marine productions to have been, not only reduced to powder, but transported and deposited by the waters in such a manner as to form the chalks, the marls, the marbles, and the limestones?" The newly discovered immensity of time, revealed by an examination of the earth as an entirety and by the observation of rocks that formed at the bottoms of the seas, caused people's heads to spin. "Although it is very true that the more we expand time, the closer we approach the truth and reality of the use that nature has made of it," Buffon emphasized in his notebooks, "nonetheless, we must constrict it as much as is possible, in order to conform to the limited powers of our intelligence." In terms of the scale of a human life, or of the Mosaic chronology, a hundred thousand years did not differ much from 1 million or 10 million years. Buffon's caution was not at all due to religious preoccupations; nor were the barriers that stopped him those of nature; they were the limits of understanding of most of his contemporaries.

The Benefit of the Doubt

Buffon died a glorious death. Before a large public, he received the last sacraments from the church on 15 April 1788 and passed away the following day. After his grandiose funeral in Paris, his mortal remains were transported to Montbard to be buried in the chapel he had had built thirty years earlier. "Make this place solid; I will be here longer than anywhere else," he had ordered the masons. The Revolution, which Buffon had seen coming, commenced the following year. Although violated under the Convention, his burial escaped profanation, reflecting the ambivalent attitude that even then surrounded the man and his work. People contented themselves with carrying away the lead that garnished his coffin, to make balls destined "to strike down the barbaric hordes." Shortly afterward, his illustrious name did not spare his only son, who was accused of conspiracy, from the guillotine.

Already, while Buffon was still alive, the weak points of his *Natural History* had been pointed out. In a letter following the publication of *Epochs of Nature*, the fine naturalist Guettard, whom we will meet again later, asked: "Still more of these tricks of Buffon. My dear count, how long will you play the role of Cyrano de Bergerac?" And he continued: "That the lesser planets are not habitable because they have lost their primitive heat, that the earth will no longer be habitable in some thousands of years, that Saturn and Jupiter will be so in several other thousands, who will believe this of you, my dear Count?" His own immediate response was "Weaklings, and more weaklings!" To the great Buffon, who had been "made to bring Light into the most sublime of Minds," Guettard threw back: "You are just a superficial physicist!" In 1792 the naturalist and archaeologist Aubin-Louis Millin (1759–1818) admitted that Buffon was responsible for the improvements of the Royal Botanical Garden, as well as "interesting and little-known facts." But he hastened to add that "in rendering justice to the literary talent of this great man, one cannot deny that he has retarded the progress of true knowledge in natural history because of the scorn he has shown and inspired for systems and methods, without which this science would offer nothing but confusion and would be nothing but an inextricable labyrinth." Millin also noted in passing: "As for the cosmogonical romances of Buffon, they have been destroyed by others who will soon suffer the same destiny."

Starting with its cosmogonical part, the *Theory of the Earth* had quickly been attacked on purely physical bases. It was shown first that, because of the laws of gravitation, any matter that had become detached from the sun after an impact would quickly return to its source. It appeared afterward that the comets, rather than being dense objects, were, at any rate, much too light to be able to produce "solar splatterings," which Guettard mocked, in a similar manner. But Buffon did not abandon his system in the *Epochs of Nature*; he contented himself with justifying it awkwardly. Nevertheless, he was not disconcerted when the physicist Jean-Baptiste-Louis Romé de l'Isle (1736–1790) rightly contradicted it, declaring that "the action of this [primitive] heat is negligible at the surface; and consequently, that which we experience cannot come from anywhere else other than the sun's action on our atmosphere and upon all the sublunary bodies."

By comparison with the English theoreticians permeated with the Bible, or with the naturalists obsessed by the Flood, Buffon had to judge these critics as accessory. To go back to the origins of the earth, the mechanism of formation was less important than two essential attributes, the initial fluidity and the very slow cooling, from which arose the seven epochs and the complete rupture with Mosaic chronologies. Without perceiving the perspectives that he had created, his adversaries did not think so much about those attributes, regardless of whether they themselves were devout or of scientific mind. By contrast with his contemporaries who had also perceived the immensity of time, Buffon was not content to make it only a vague incident within a treatise. Rather, he considered it a subject worthy of methodical study, for which all expedients of natural philosophy were to be summoned. Struggling "against 'faculties' and final causes, as well as against God, who would explain everything, and chance, which explained nothing," according to J. Roger's formulation, Buffon had allied grand views with a twofold experience, that of the laboratory and that of nature. As a great Newtonian, he had completed an essentially Cartesian cosmogony, impressing upon it the twofold mark of history and time.

The Meaning of Nebulae

Because of the subordinate role that he accorded to God, Buffon distinguished himself from the cosmogonists of his times. Emmanuel Swedenborg (1688–1772), a prolific scientist of Sweden, had been nourished as much by mythology as by Ovid, Lucretius, Descartes, or Burnet. He thought that "the constitution of the visible heavens never can be understood without first understanding the constitution of the invisible." In his *First Principles of Natural Things*, published in 1734, he wrote, alongside some metallurgical and crystallographic considerations, that material that had been ejected from the solar vortex could proceed farther and farther from its source and then cool and condense to give rise to the planets. Ten years later, following a mystical crisis that caused him, first off, to renounce women, and then to see appearances of Christ, Swedenborg became the prophet of a new cult, still practiced nowadays, the Church of the New Jerusalem. Even if his influence on Buffon remains uncertain, Baudelaire and Balzac, certainly, were affected by his *Conjugal Love: Delights of Wisdom Relating to Conjugal Love*, which was followed by *Pleasures of Insanity Relating to Licentious Love*.

Equally explicit were the religious backgrounds appearing in the manuscripts and the *Original Theory or New Hypothesis of the Universe* by Thomas Wright (1711–1786). Wright was an attentive reader of Whiston and an self-taught scientist who was appreciated by the English aristocracy. For him, faith alone would allow one to discover the majestic organization of the universe; observation, by itself, could provide only an image that would prove too fragmentary. To Wright the universe appeared ordered about a "sacred throne of omnipotence," upon which God and the angels were seated, forming the center of a sphere, the "Gulfe of Time or Region of Mortality." This sphere included the stars and the planets and was itself limited by a second sphere, that of the "Desolate Regions of [the] Damn." Since the region of mortality could only be spherical— otherwise the sky would have been covered with stars in all directions— Wright concluded that it was shaped like a disk, the Milky Way. And in this system in which the sun was but a peripheral star, the stars were prevented from collapsing under the influence of gravity by means of their rotational movement about the Divine Center, which "cannot be made in less than a Million of years." Even at the speed of a cannonball, the immense distances of "near 1,000,000,000,000,000 Miles" would be traversed on a timescale approaching hundreds of millions of years, as Wright remarked. But he did not draw from this observation any implications about the age of the universe. In his spiritual perspective, it was not the speed of light that was important, but rather that of souls, as they traveled across space. From displacements that necessarily must be rapid over such distances, Wright advanced, we may truly infer that "the Soul must be immaterial, and that in all Probability there may be states in the Universe so much more longer lived than ours, that, compared with the Age of man, the Age of such Beings may be almost as Eternity, or rather, as that of the human Species to that of a Sun-born Insect."

A young German philosopher, Immanuel Kant (1724–1804), had not read anything of Wright except an incomplete summary of *An Original*

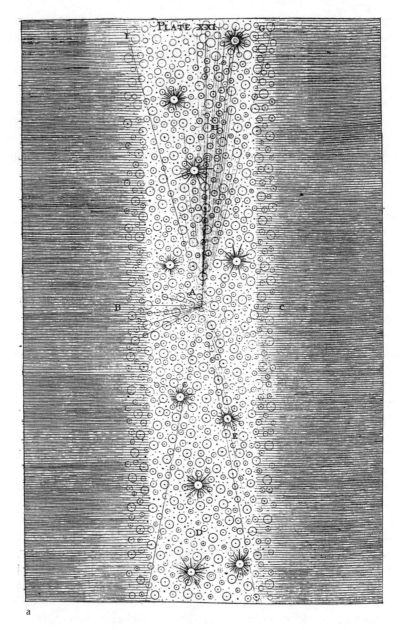

a

Figure 5.3 The Milky Way (*a*), represented by Wright, with its "finite view of infinity." A partial view (*b*) of the infinite number of sidereal systems and planetary worlds similar to the known universe. From Wright, *Original Theory* (1971), 65 (*a*) and 83 (*b*).

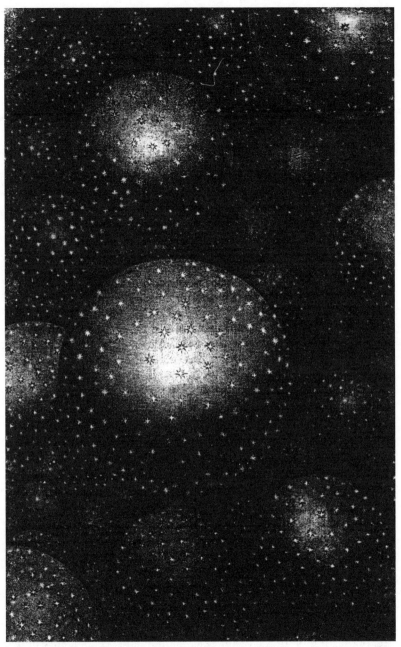

Figure 5.3 *(continued)*

Theory, so he could only guess that this universe had been peopled by celestial volcanoes (the stars) and that it had been ordered about a Divine Center, which itself was devoid of stars. Following Kant, others have sometimes hastened to recognize Wright as the author of the first correct description of the galaxies. Even more important was the fact that Kant examined *An Original Theory* in light of the Newtonian theory of gravitation, and of observations demonstrating that nebulae, previously considered to be clouds of dust, had elliptical forms. Why had not this gravitational force, which determined the movements of planets, incorporated the dusts of the nebulae to form planets that gravitated locally about themselves and, on a larger scale, systems analogous to the Milky Way? By contrast with Wright, who was blocked by the uniqueness of the Divine Center, Kant, in his *Universal Natural History and Theory of the Heavens*, published in German in 1755, was free to multiply his worlds on out to infinity.

This infinity of the universe nonetheless remained of divine origin, and it extended through time as well as through space. "Millions and entire mountains [bundles] of millions of centuries will flow by, within which always new worlds and world-orders form themselves, one after another, in those reaches [so] distant from the center of nature, and reach perfection," Kant affirmed. But this eternity retained more of a philosophical sense than a physical sense: "Hundreds or perhaps thousands of years are needed before a great celestial body obtains a firm state of its material," he noted in one of the rare passages where a concrete duration was mentioned. In spite of its subtitle, *Newtonian Science of the World*, the *Universal Natural History* was qualitative. It also had the fault of including speculations—censured by admirers of the philosopher at the time—regarding the plurality of worlds, such as the idea that the creatures of the other worlds "become more excellent and perfect in proportion to the distances of their habitats from the sun," this relationship having "a measure of credibility which is not far from demonstrated certainty."

All through the eighteenth century, it was a French school led by Alexis Clairaut and Jean Le Rond d'Alembert that improved celestial mechanics upon the foundations cast by Newton. The figure of the earth, the complex movement of the moon, the planetary orbits, the precession of the equinoxes—nothing escaped the influence of gravitation. Pierre-Simon de Laplace (1749–1827), the son of a Norman farmer, was a professor at the military academy in Paris before temporarily becoming minister of the interior for Napoleon, then marquis and peer of France under the Restoration. He is the one who assured the crowning of this system of mechanics. In his *System of the World*, he demonstrated the intrinsic stability of the solar system. For example, "From the sole circumstance that the motions of the planets and satellites are performed in orbits nearly circular," remarked Laplace, "in the same direction, and in planes which are inconsiderably inclined to each other, the system will always oscillate about a mean state, from which it will deviate but by very small quantities."

The planets did not risk falling in onto the sun; not only was it superfluous to require Newton's God to maintain them in their orbits, but a strict determinism was imposed upon all of nature. We ought "to consider the present state of the universe as the effect of its previous state and as the cause of that which is to follow," Laplace had stated as early as 1776, in his *Philosophical Essay on Probabilities*. "An intelligence that, at a given instant, could comprehend all the forces by which nature is animated and the respective situation of the beings that make it up, if moreover it were vast enough to submit these data to analysis, would encompass in the same formula the movements of the greatest bodies of the universe and those of the lightest atoms. For such an intelligence nothing would be uncertain and the future, like the past, would be open to its eyes." Although Buffon could in this way obtain the good graces of Laplace, his cosmogony accounted only for the movements of the planets in the same direction, and not for the rotation of the planets and their satellites upon themselves or for the slight eccentricities of their orbits. Like Kant, Laplace thought, in 1796, that the condensation of a nebula represented the "veritable cause" of the solar system. In cooling, an incandescent fluid mass of the size of this system had contracted and differentiated itself, and—an important improvement in relation to Kant's schema—the rotation of the ensemble of the nebula, which originally was slow, became more and more rapid in proportion as it reduced further in size.

By contrast with the speculations of Wright or Kant, who had awakened but few echoes, this theory of the protosolar nebula endured because it quickly received confirmation based on observations made by William Herschel (1738–1822). A flutist who had emigrated from Germany to England and a self-taught astronomer, Herschel dedicated his time to discovering the limits of the universe using the great telescopes he had constructed with his own hands—including even the polishing of the mirrors. At the heart of Herschel's "construction of the heavens," were the nebulae. He not only discovered them by the thousands but also studied the successive stages of their condensation. The nebular material was scattered among clusters and condensed about a nucleus, which could be more or less brilliant, and enveloped, or not, by an atmosphere. As summarized by Laplace, "the nebulae, classed in a philosophic manner, indicate, with a great degree of probability, their future transformation into stars, and the anterior state of nebulosity of existing stars." At that point, astronomical distances imposed their repercussions upon time. "Certain stars may be able to be seen several million years after they have been annihilated, because the light emitted from them requires several million years to cross the space separating them from the earth," the Swiss mathematician Leonhard Euler (1707–1783) revealed. But it was Herschel who integrated the immense durations in a real and positive way: "The state into which the incessant action of clustering power has brought the Milky Way at present is a kind of chronometer that may be used to measure the time of its past and future existence," he wrote in 1814. He had just confided to the poet Thomas Campbell about his observation of the lights from ancient worlds: "I have looked farther into space than ever human being did before me. I have observed stars from which the light, it can be proved, must take two million years to reach this earth." The road had been long. A century and a half after Römer had measured the speed of light, the old science of astronomy was only just beginning to probe the depths of time.

SIX

Heroic Age, Relative Time

THE BEGINNINGS OF GEOLOGY   Numerous controversies beset the newborn geology at the turn of the nineteenth century. But these did not prevent reaching a consensus about the very old age of the earth in less than fifty years during the "heroic age" of the discipline. Based on proof of species extinctions, fossils began to be used at that time for relative dating, and the term *geologic durations* came into use for describing immense periods of time.

Geognosy, Geologists

Most especially, beginning in the second half of the eighteenth century, the new generations of naturalists went out to explore the plains and the mountains, in order to observe the strata and rocks from up close. For these field-oriented scientists, the age had well passed when one could search back for the origins of the world without leaving one's desk. Seated upon improbable hypotheses, the great syntheses that had been successful in the salons shattered when confronted with the astonishing complexity that was revealing itself to these new scientists. For example, Cuvier, in spite of a sincere respect for his predecessors, reviewed the theories of the earth in 1812, with quill dipped in irony. Whiston, in particular, was subjected to his taunting humor: he had "created the earth with the atmosphere of one comet, and deluged it through the tail of another; the heat which remained to it from its first origin excited all mankind to sin; thus they were all drowned except the fishes, which apparently had passions less unruly." About such an astronomical-theological tour de force, this was all that Cuvier, the convinced Protestant, could find to say.

The time for audacious inductions had passed also. What was important now was to go beyond speculation, to discover in the field what were the right questions to ask. This exploration was contemporary with the beginning of the Industrial Revolution. All across Europe, the search for coal led to prospecting in zones that had remained distinct from the traditional mining districts; their exploitation led to searching through underground excavations with feverish ardor; and the examination of the coal led to the discovery of surprising fossilized flora. But doubtlessly even more important was a pure curiosity in relation to the rocks, which until then had been viewed from an essentially utilitarian perspective, as is evidenced by the term *limestone*. In the spirit of the Renaissance anatomists, the naturalists applied themselves to discovering the interior of a great object: the earth. A multitude of new observations now had to be organized; a nomenclature remained to be created; a science was putting itself into place: *geology*. First appearing in 1344, then later mentioned episodically in the seventeenth century, this term began to take on its current sense around 1760. From its heroic age (1775–1825), the uncertain steps of the new discipline were reflected in the hesitations of its vocabulary. At one point, different terms were variously accepted, *geology*, *geognosy*, *geogeny*, and *geopony*, in company with the gracious *oryctognosy* (descriptive mineralogy), before the *geologists* generically supplanted the *geognosts* and other *geotechnists*.

For these geologists, the organic origin of fossils had been progressively accepted. Voltaire, with his scorn for the "people in whose minds a petrified pike on Mont Cenis and a turbot found in the land of Hesse have more power than all the reasonings of sound physics," belonged, from this point of view, to the old guard. His irony was quickly found to have been exercised at his own expense when he quipped in 1768: "There are shells named *Conchae Veneris*, shells of Venus, because they have an oblong slit gently rounded at the two ends. The gallant imagination of the physicists gave them a fine title, but this denomination does not prove that these shells are the remains of ladies."

Like gems and fossils previously, minerals and rocks were gathered in great collections. The notion of crystal and the differences between the minerals that constituted the rocks (quartz, talc, or mica, for example) and between rocks themselves (granite or basalt) continued, at any rate, to be poorly understood. Like the hexagonal prisms of basalt columns, the geometrical forms in which certain rocks were found perpetuated the misunderstandings. "Sometimes one finds the *Basaltic Crystals* solitary and detached, sometimes grouped rather regularly among themselves, from the last degree of smallness to enormous sizes," noted Romé de l'Isle in 1772. "Most often they are mixed in a confused way among other stony bodies, with which they are closely united. Those to which they most ordinarily belong are the quartz and rock crystals, granites, petrosilex, talcs, micas, etc." When he announced in 1783 the observation made by his assistant, Arnould Carangeot (1742–1816), that the angles formed by the faces of a crystal were characteristic of a species, the rigorous identification of rather large specimens became possible, in principle. But the small size or the irregular aspect of minerals too often masked their crystalline nature. That minerals, in general, were crystals was not recognized until the middle of the nineteenth century, with the invention of the polarizing microscope and its use to examine rocks in thin, translucent sections. During this period it was also understood that rocks such as chalk are composed of minuscule fossils. The confusion to which even an expert such as Romé de l'Isle attested was thus to linger on.

While *crystallography*, a term coined by Romé de l'Isle, was being constituted, quite soon finding a master at the Museum National d'Histoire Naturelle in the person of René-Just Haüy (1743–1822), *petrography* (the science of rocks) labored to emancipate itself from mineralogy. Complementing the paths traced by Agricola, chemical analysis, meanwhile, provided important criteria for the characterization of rocks. With their great mining traditions, the Germans and the Swedes were the first among these analysts. Even before the notion of a chemical element was specified and order was put into the system of nomenclature under the direction of Antoine-Laurent de Lavoisier (1743–1794), minerals had aroused the attention of chemists. This interest was all the greater because, with the exception of the gases of the air, minerals constituted the only source of new principles, which had been avidly sought since the middle of the century. During the ensuing period, the best analyst was the German apothecary Martin Klaproth (1743–1817), who distinguished himself by new procedures: he crushed and dissolved the minerals (in molten carbonates, for example) and implemented an attentiveness aimed at reducing the errors of his analyses, the results of which he had developed the habit of publishing. Within a chemistry that did not yet depart clearly from that based on the air, water, fire, and earth of the ancients, systematic analyses of minerals had thus led to distinguishing numerous types of earths, metallic oxides for the most part, such as lime or magnesia.

As the mass of observations regarding the most varied of terranes grew, a need began to be sensed for graphical representations. Jean-Etienne Guettard (1715–1786), a chemist who also served as botanical doctor to the duke of Orléans, applied himself to this exercise in a systematic manner. A great traveler, and an adversary of systems and of Buffon, Guettard presented a "preliminary mineralogical map" of France to the Academy of Sciences in 1746. Symbols distributed over a geographical map localized the most notable works of nature, while a wide oval "belt" of marl, centered on Chartres, separated an inner, "sandy" belt from a vast external "schistose or metallic" belt, distinguished according to the mineral substances found there. This vision, which was essentially horizontal, had no sort of chronological connotations, but it permitted Guettard to make some interesting correlations, such as the continuity of the marly and schistose bands between the Paris and London basins. The public utility of these representations was quickly recognized. With his pupil Lavoisier, who was the son of one of his friends, and whom he encouraged to study chemistry, Guettard was charged by the government in 1766 with describing all of France in two hundred maps. The project was too ambitious and was never properly completed, but it brought about a lasting geological prospecting effort on the part of Guettard, Lavoisier, and their colleagues. Such cartography developed rapidly in other countries as well, especially in Germany and in central Europe, with their rich mining districts. The maps, which at the beginning were figurative, became interpretative little by little, and the use of color was introduced to identify the terranes.

Buffon was concerned only indirectly by these works; at the intersection of two epochs, that of the grand visions and that of the observation of detail, he did not really need them in order to go deeply into his reflection. He found himself somewhat out of fashion, but in accord with his ideas regarding this point, the need to invoke long timescales had become pressing. The Age of Enlightenment was over. Having brought man closer to the universe by means of the laws of mechanics, science had opened the way to a reasoned interpretation of nature. No one did a better job than Fontenelle, the great Cartesian and "scientific novelist," of persuading the public that the truth, by now, was to be found in science and that science could serve to emancipate conscience. With his celebrated parable of the roses that are astonished, generation after generation, to see an identical gardener bending over them, he had illustrated in 1686 the insignificance of human life in relation to the very slow evolution of celestial bodies.

A mistrust of tradition and authority began to develop, caused by the progress of the sciences, and a belief in constant progress for man and society began to take shape. Paul Henry Thiry, the baron of Holbach (1723–1789), author of hundreds of articles in Diderot's *Encyclopédie* (on earth sciences, in particular), illustrated the evolving views in a striking manner; in his *System of Nature*, which, however, was accused of the "crime of lese majesty, human and divine," he proclaimed that "reason guided by experience ought, in the end, to attack the source of the prejudices of which the human species was so often victim." Another change was coming to a conclusion: reason was taking the place of the Scriptures as the essential driving force for history, and it pushed people (though not without opposition) to abandon the most literal of the interpretations of Genesis. The skepticism that affected all Mosaic chronologies, and even the impossibility of any sort of chronology—as some went so far as to assert—were, of course, a part of this mind-set. Voltaire, the most illustrious representative of the eighteenth century, was better inspired than he had been on the subject of fossils when he wrote in 1768: "I still confess that it is demonstrated to the eyes that it must have required a prodigious multitude of centuries to perform all the revolutions that have occurred to this globe, regarding which we have incontrovertible testimony."

Three Abbés in Quest of Time

Well before Herschel had flattered himself by saying he could see light arriving from millions of years earlier, philosophers and geologists had already become accustomed to discerning traces of much earlier ages within nature. Certainly, people were far from being unanimously convinced, especially in England, but by this point the discussions were confronting observable facts, rather than entertaining essentially philosophical arguments. Among the pioneers of the long duration of time were three influential abbés. The first was the Englishman John Turberville Needham (1713–1781), who was ordained in France, where he completed his studies. His renown was due to various experiments, such as those he performed with Buffon regarding the juice of boiled mutton, which allowed him to account for spontaneous generation by the formation of animalcules. Needham desired to reconcile faith and practice to the point of being accused of materialism. He finally settled in Brussels in 1768, where he founded an academy of sciences. He did not incur censure from the church even though, in the following year, he asserted in his *Nouvelles recherches physiques et métaphysiques sur la Nature et la Religion* that "The chronology established by Moses is by no means that of the earth, nor that of our system, and even less that of the entire universe, which might, according to innumerable and successive systems, penetrate well forward into eternity. To speak rigorously, it is the chronology of the human species alone, expressed only in terms of the generations that have succeeded one another, and calculated on their duration." And so, being ecclesiastical did not prohibit the adopting of new ideas, "provided that reason, supported by experience, requires it." And since the "days" of Genesis simply referred to long periods, Needham felt no compunction in stating, on the basis of the rates of advancement of the areas of land into the seas, that "the future destruction of the two current continents of Asia and Africa, which will form new continents elsewhere, will require about three million years, as is easy to verify by geographical measurements."

The lay abbé Pierre-Bernard Palassou (1745–1830), who was a friend of Guettard and Lavoisier, had explored the Pyrenees in his youth, with the encouragement of Buffon, within the project of producing the mineralogical map of France. He was particularly struck by the devastating action of torrents that were capable of toppling entire boulders, as well as of gradually demolishing a mountain chain. Rejecting any past action of the sea, which had often been invoked on this subject, and focusing on the future, Palassou announced in 1781: "One day, posterity will be able to say: *There are no more Pyrenees.*" But he added, making reference to the conclusions drawn by Gensanne in 1777: "We can conceive of how distant this epoch is from us. Gensanne found, by means of observations that he claims are unequivocal, that the surfaces of the mountains reduce by about ten inches per century. So, in supposing them to be only fifteen hundred fathoms above sea level, and always susceptible to the same degree of reducing, a million years will pass before their total destruction."

The remarkable self-taught naturalist Jean-Louis Giraud Soulavie (1752–1813) was another abbé who saw erosion as the only agent capable of modeling reliefs. He was the son of a prosecutor and served for a time as vicar in the Vivarais, in southern France, before settling in Paris. As soon as he was ordained, he began the task of writing his vast *Histoire naturelle de la France méridionale* (*Natural History of Southern France*). This work treated the past rather than, like Palassou and Needham, the future. He claimed that one could determine the ages of the existing valleys because "since the time within which a destruction occurred was determinate, one only needed to inquire how much time remained to destroy a given mass that filled a whole valley." His conclusion, which was based equally upon Gensanne's observations, was that "millions of years of flowing water had been necessary to dig a valley of our simple Vivarais in the heart of a plateau of lava, then through the lower, fundamental granite mass, to raise subsequently at the bottom of this valley a new volcano, which itself is so ancient that its flows and appendices are falling into ruin." But, facing the open hostility of an influential Jesuit who was far too attached to the Mosaic account, Soulavie, a changeable person, decided not to publish his views. After producing seven volumes between 1780 and 1784, he interrupted his *Histoire naturelle*. History itself became his preoccupation instead. Soulavie served in the Revolution as a Jacobin diplomat who sought "the friends of France, and the enemies of our enemies." He was later a controversial man of letters, a collector of archives and engravings, and an informer of the scandals of the ancien régime. He came to the end of his life in a pious and leisurely manner as a counterrevolutionary who had, meanwhile, through the vicissitudes of the times, married the same woman four times.

As a complement to erosion, which so attracted the attention of Needham, Palassou, and Soulavie, the slow process of sedimentation had been discussed previously by Buffon. When Lavoisier declared in 1789 that "the sea had a very slow oscillating movement of ebbs and flows, and these movements were cadenced over a period of several hundreds of thousands of years, and this has already been repeated a certain number of times," his originality was not due to the durations to which he appealed, but rather to his conclusion that "it must result from this that by taking a section of the horizontal layers between the sea and the great mountains, this section ought to present an alternation of littoral layers and pelagic layers [those deposited in deep waters]." Long durations of time had thus been established, and this identification of the conditions of sedimentation led to the notion of facies and to the history of the advancing and receding of the seas.

Naturalists in Germany, England, and elsewhere in Europe were affected by this movement, but few attempted to quantify the immense durations that were beginning to appear before their eyes. Buffon, who recognized the slow cooling that entrained an inexorable evolution, was opposed by an English doctor, George Hoggart Toulmin (1754–1817), who was convinced that no traces of any sort of origins could be discerned in those terranes that had been deposited since the earliest ages. Toulmin published his *Antiquity and Duration of the World* in 1780; here he stated, in particular, that the sedimentation at the bottom of the sea would provide, in time, for the formation of new lands and would compensate for the destruction by erosion of the continental masses. "In the circle of existence, in vain do we seek for the beginning of things," he stated, and likewise, that it is "needless to multiply facts any farther, in proof of a succession of events of an amazing duration." He was an ardent partisan of an extreme actualism: "Nature is invariably the same, her laws, eternal and immutable," he concluded, thus undeniably reflecting

Aristotelian accents. But above all, he asserted that man as an integral part of nature participated in the same eternity. Perhaps this heresy, as R. Porter has suggested, was the cause for the discredit, and even the censure, that condemned his system.

Primitive and Secondary

When Toulmin declared that it was fruitless to search for traces of a beginning of time, he was actually proceeding countercurrent. The majority of geologists, by contrast, saw an unambiguous order in deposition of rocks. As Leibniz expounded in his *Protogæa*, "One must not believe that stones all come from fusion. It is primarily for the first masses and the base of the earth that I admit this mode of formation." He stated that it was upon this base of vitrifiable matter that the actions of the waters had subsequently accumulated strata: "I do not doubt that the waters that later spread over the earth's surface, after the calm had settled again, had deposited a great quantity of matter detached from the surfaces of the bodies, and of which one part has formed different sorts of terranes, while another has hardened into rocks, and that the various layers of these superposed upon one another call out the different orders of precipitations that occurred in different times."

The order of the formation of strata had been discerned at several times before the publication of *Protogæa*, testifying to the vitality of the emerging discipline. In spite of the distances that separated them, the terranes observed from one end of Europe to the other led to similar conclusions. Guillaume-François Rouelle (1703–1770), a chemistry professor at the Royal Botanical Garden, was impressed by the general symmetry in the Paris basin, from Brittany to the Vosges, and by the regular succession of its strata. He identified, in 1740, an "ancient land," which was a granite base, not stratified, upon which had been deposited a "new land" formed by the slow accumulation of fossiliferous sediments. A teacher characterized by a rare eloquence, he had Lavoisier among his pupils. Among his auditors were Turgot, Rousseau, and Diderot, of whom the latter left twelve hundred pages of notes from his courses. Rouelle was well heard, even though he published hardly anything. At the same time the abbé Anton Lazzaro Moro (1687–1764), from Friuli in Italy, produced his theory of the upheaval of mountains by the action of a subterranean fire. Within the Alps, he distinguished the primitive mountains, which were massive, and the secondary, which were stratified.

Fifteen years later in Germany, the well-known chemist and doctor Johann Gottlob Lehmann (1719–1767) pushed the analysis even further. In the Harz and the Erzgebirge (the metalliferous mountains between Saxony and Bohemia), he differentiated primitive mountains, formed of veined rocks, dating from the earth's earliest moments; secondary mountains, in layers, witnesses to a revolution that had been felt throughout the globe; and tertiary mountains, which were formed by local revolutions that had deposited sediments at the bottom of the sea. Shortly after Lehmann came a reserved doctor and theologian, Georg Christian Füchsel (1722–1773), whose work did not have quite the audience that Lehmann's had. His 1761 book, however, included a remarkable geognostic map of Thuringia. Furthermore, he defined for the first time the notion of a *formation*, which was soon to take on great importance: "Each deposit forms a stratum, and a succession of strata of the same composition constitutes a formation, or an epoch, of the history of the globe." On top of this "fundamental soil" thirteen such formations were recognized between Harz and the Thuringian Forest. Without specifying the durations of its periods, Füchsel declared, moreover, in 1773 in another work: "If there were plants and animals in the ancient periods, then necessarily there would have been men, but these creations were not unique. The ancient continent was peopled throughout as many centers of creation as there were native tongues." Even though he did not model himself upon the Mosaic framework of Lehmann, Füchsel did not distinguish clearly between natural and human history.

Giovanni Arduino (1714–1795) of Italy was another great figure concerned with the beginnings of geology. He was born into a humble family and worked from an early age in the mines of the Upper Adige, before the Republic of Venice, of which he was a citizen, entrusted him with important agricultural and mining responsibilities. From the Alps to the Po plain, "the series of these strata that form the visible crust of the earth appear to me to be classed into four general and successive orders," he wrote in 1760. "These four orders could be conceived as four great strata (as they are, in fact) of such a manner that, in whatever place they are exposed, they are disposed one on top of another, always in the same fashion." These similarities in the stacking of strata had long before been recognized by coal miners, but this generality contributed by Arduino gave them a fundamental character. Arduino designated the four orders as "primitive, or primary, secondary, tertiary, and of the fourth order," and he recognized that each of them was made up of innumerable minor strata, formed at different times and under different conditions.

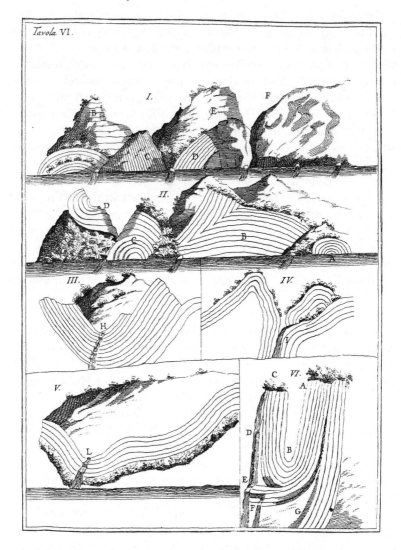

Figure 6.1 Folds of strata. From Moro, *De Crostacei*, (plate 6), reproducing a 1715 plate published by Antonio Vallisnieri in his *Lezione accademica intorno all'origine delle fontane*.

The German Simon Pallas (1741–1811) explored even as far as the Urals, and everywhere he went, the hearts of the mountains seemed to enclose the most ancient testimony of the earth's past. In these mountains the inclined disposition, or even the folding of the strata, attested to the violence of phenomena that had occurred since the original horizontal deposition. In the tectonics that was gradually emerging from the works of Moro and of the Swiss naturalists Johann Scheuchzer and Horace-Bénédict de Saussure, the concept of geological sections became indispensable for following the strata all along the upheavals that they had undergone. But the principle of the superposition of layers, established by Steno, was to be applied only to deposits effected in water, enveloping "solid bodies." Beginning with the granite of the mountains or the vast solid basement that one noticed at the edges of the sedimentary masses, there were numerous terranes that appeared patently devoid of fossils. What was their connection with the sedimentary rocks? Had they been produced by the action of fire, as Leibniz and Buffon had supposed, like the lava expelled from volcanoes? Or had they been formed in the water, before life existed? That origin would explain their lack of fossils, whence their name of *primitive* terrane, in accordance with an old opinion that was soon to be upheld forcefully by a famous German school.

Vulcan Regained

Volcanoes, in addition to stirring fascination and terror, have always stimulated the imagination. For Father Kircher, they were the superficial parts of an immense subterranean network that linked "centers of conflagration," an idea that suffered from the discredit that had rapidly begun tainting Kircher's speculations. In a world that was dominated by water, one preferred to see in volcanoes the actions of more modest forces. In order to explain "earthquakes," Hooke referred to the eruptions caused by the combustion of vapors produced by the action of seawater upon sulfurous minerals. And when Telliamed asked his French missionary, "From whence do you say that volcanoes originate, then, if not from the oils and the fats of all these different bodies inserted into the substance of these mountains?" he expressed only ideas that had already been well received. Nicolas Lémery (1645–1715), an apothecary and chemist, gave a solider foundation to this sort of explanation in 1700, when he demonstrated that the spontaneous combustion produced by a mixture of iron filings and sulfur powder appeared to be "quite capable of explaining the manner in which the fermentations, turmoils, and blazings occur in the bowels of the earth, as happens on Mount Vesuvius, on Mount Etna, and in several other places."

Fontenelle drew from this experiment a lesson that was to have a much more general impact. "The best way to explain nature, if one could often put it to use," he advised, "would be to counterfeit it, to provide representations of it, so to say, causing those to produce the same effects, by using causes we recognize and that we have already put into action. Then we would no longer be guessing, we would be seeing with our own eyes, and we would be sure that the natural phenomena have the same causes as the artificial ones, or at least, causes that are quite similar." In certain of their impetuous aspects at least, geological phenomena seemed to be a matter for laboratory experiments. Nevertheless, the difficulties were great, owing to the characteristically elevated temperatures or pressures; and besides, nature was formidably complex. In spite of the skepticism often displayed, however, the effects of fire on minerals were studied with an ardor that was equally sustained by research that was undertaken all through the eighteenth century for the purpose of competing with the Chinese production of porcelains.

Even if the volcanic scoriae visible in nature was attributed to such subterranean fires, it was only the active volcanoes that were generally recognized as such. An excursion by Guettard in connection with his mapping projects gave rise to a profound change in point of view. In 1751, while he was going through the town of Moulins, to the north of the Central Massif, his attention was attracted by some curious black stones. Although he had never seen a volcano, he recognized that the stones were lava. When he asked from which quarry it had come, he was told that it was from Volvic. According to Condorcet, Guettard then exclaimed "Volvic! Vulcani vicus [the village of Vulcan]!" Whether the tale is apocryphal or not, the fact remains that Guettard quickly resumed his route heading toward Auvergne. "I recognize a volcano, this is what Vesuvius looks like! or Etna, or the peak of Tenerife—I've seen engravings of them!" he exclaimed at the heights of the peak dominating Volvic; and "the craters, the lavas, the inclined, parallel layers—they must have been formed by molten material!" No legend had retained the memory of the last eruptions of the Auvergne volcanoes (which we know now were only sixty-seven hundred years ago). Even though unmindful of this possible Latin etymology of Volvic, popular wisdom apparently remembered the volcanoes of the Chaîne des Puys as piles of ancient metallic scoria. Guettard may have aroused some consternation when he announced that "this realm has had volcanoes in distant centuries at least as terrible" as those of Naples or the West Indies, "which perhaps would only require the slightest of movements or the least of causes for them to become enflamed anew," but the discovery did not lead naturalists to search elsewhere for other witnesses of such violent activities of the past.

This renaissance of the Auvergne volcanoes included a second episode, in which Nicolas Desmarest (1725–1815), from Champagne, was

Figure 6.2 The Giant's Causeway, Northern Ireland. Engraving from Diderot and d'Alembert, *Encyclopédie*, s.v. "pavé des géants" (Giants' Causeway).

the principal protagonist. His precocious naturalist talents had attracted the attention of d'Alembert, before the encyclopedist Jacques Turgot, intendant of Limousin, charged Desmarest in 1762 with inspection of manufacturing in his province. During his travels the following year to inspect the Auvergne paper mills, Desmarest made a detour through the Puy de Dôme in order to see the volcanoes described by Guettard. There he was struck by the prismatic columns of basalt and their similarity to the columns of the Giant's Causeway, on the north coast of Ireland, which had recently become famous, though remaining mysterious. Because of their association with lava, the Auvergne basaltic formations caused him to suspect a volcanic origin for them as well, and Desmarest thus returned in the following years to map their outcrops around the Monts Dore. His presentiment proved justified. These columns represented the ends of long black trails, or flows, which had descended from the mountain, bordered by scoriae and pumice stones, and whose cooling had produced the characteristic prismatic pillars. The Cimmerian rocks that Pliny and Agricola had named *basalts* had thus issued from volcanoes and formed plateaus; erosion had subsequently cut valleys in them, and the depths of the valleys were measures of the ages of the basalts.

As an illustration of how we often do not see something unless we are looking for it, after Guettard's discovery, and especially after that of Desmarest, extinct volcanoes began emerging from one end to the other of the old tracts of Europe. They were found in Languedoc and

Vivarais, where Soulavie had studied them after Guettard; in Hesse, on the shores of the Rhine; in Venetia; and in Catalonia. But their great number did not call back into question the nature of the sources of the heat they were associated with. For Guettard volcanic action was due to the subterranean fires of bitumen, a material that was seen oozing at Clermont-Ferrand, and for Desmarest it was caused by the combustion of coal, as was suggested by the presence of the scoriae analogous to what was produced in blast furnaces. No connection had been established between central fire and volcanism. For Desmarest, as for the majority of naturalists, water remained the principal geological agent.

Neptune at the Apogee

There was nobody who sustained this cause, which is called *Neptunism*, with greater force of persuasion than Abraham Gottlob Werner (1749–1817). As the son of a forge owner, he discovered the wonder of minerals early, for his father gave him specimens to play with from early childhood. While he was still a student in the mining school in Freiberg in Saxony, at age twenty-five, Werner became known for a book he wrote, *On the External Classification of Minerals*, in which he combined the criteria of exterior form and chemical composition to classify the two hundred kinds of minerals known at the time. Shortly afterward, he was engaged by his school to teach mining and be the curator of the mineral collections, and he began to teach geognosy and oryctognosy. In 1786 his twenty-eight-page work *A Short Classification and Description of the Various Rocks* opened the path for clearer distinctions among rocks and minerals. Students began to flock to Freiberg from all over Europe. Generous, methodical, punctual almost to a fault, and an indefatigable speaker, Werner was capable of discoursing intelligently on military art, on ancient and oriental languages, and even on the determinism exercised upon those by the nature of soil. And so he was revered during his own lifetime like few before him. According to Cuvier, this "great oracle of geology" had the misfortune of being so overcome by paralysis at the mere sight of a pen that, in his praiseworthy concern for consistency, he even reached the point of no longer opening his mail. Nevertheless, Werner left us with several books and voluminous reports, and his ideas were spread throughout the world by his numerous disciples who "went out to ask questions of Nature, in his name," according to an apt remark by one of them, the French Jean-François d'Aubuisson (1769–1841).

Within this system that dominated the heroic age of geology, Werner resumed Füchsel's concept of formation. The alternations of strata, layers of chalk and silex, for example, or of coal, marl, and schist, have the quality that although they can be observed in locations quite distant from one another, they are always similar in aspect. But still, one can generally only see a limited thickness of strata peering through the surface, and it is rare to see several formations in the same location. This observation raised the question of the possible spatial-temporal relationships linking the same types of formations seen protruding through the surface at different locations. According to Werner, each formation would represent a given temporal unit, a capital notion that one of his pupils illustrated by comparing the stacking of the strata to a street, where it is the numbering of the houses that is important, rather than the various materials with which each of the houses is constructed. In boldly supposing that the formations would have vast geographical extensions, Werner concluded that the geologist's aim should be to reconstitute an ideal stacking, ordered in time, of the formations that had been observed out in the field in a fragmentary manner. With this stacking, which was called the *stratigraphic column*, geology attained its full historical dimension.

Werner's historical view of geology became obvious so quickly that its author was forgotten; only a caricature remains of Werner's theses, the picture of the "earth-onion" with its uniform, concentric layers. If Werner was denigrated during the golden age of geology (1825–75) as much as he was admired while alive, it was because only the causal part of his geology remained by that time, and it proved to be much less productive. Even so, Werner had rejected the empirical classifications of the miners and mineralogists, and he had excluded floods and catastrophes from his system. And in his courses, he mentioned the extremely ancient periods when water covered the earth, beyond any sort of Mosaic framework. Unfortunately, the sedimentary rocks were not the only ones that needed to be integrated into his Neptunism. Whereas Leibniz and Buffon had seen the vitrifiability of granite as a proof of its igneous origin, Werner concluded that the formation of a glass after fusion—rather than quartz and feldspars, the characteristic minerals of this rock—indicated quite a different origin. Chemistry had already demonstrated that water dissolves silicates better at high temperatures. Werner found it possible to sustain the idea that granite had precipitated chemically in a primordial, hot ocean that was heavily charged with matter. As for the basalts of Harz and Erzgebirge, which were well known to Werner, these were perched on plateaus, and they seemed to constitute the final stage of the formation of sands, clays, and clayey stones, far from any volcanic remains. In accord with the vitrifiability demonstrated by basalts, Werner maintained with renewed enthusiasm the old idea that these rocks had precipitated in water. When this conclusion was contradicted by Desmarest, he did not relent. The principle of the superposition of layers was applicable to these aqueous basalts; accepting a volcanic origin would have called back into question the sequence of the depositing of the principal types of rocks that he had identified, from the primitive granite solum to the recent alluvial earths and volcanic mountains, passing through the stratified and secondary terranes.

The Great Cycles of Fire

In defending the aqueous theory of basalts, Werner found himself taken up in one of the great controversies that have punctuated the history of geology. It was raging because a Scottish natural philosopher, James Hutton (1726–1797) was meditating, by contrast, on the action of fire in the depths of the earth. Hutton and Werner had more than one point in common: inveterate deists and aged bachelors, they both had studied law and published parsimoniously throughout their long careers. But Werner had scarcely surveyed beyond the old sedimentary terranes of Saxony and Hesse, whereas Hutton knew, above all, the omnipresent granites of the Isle of Arran, in Galloway, and in the Highlands of his own country. Their references had nothing in common, though, and in fact, few pushed the criticism of Neptunism as far as Hutton did. If all rocks came from the water, he argued, why did one find no traces of water within the pores of the granites or the basalts? And how could water percolate through solid rocks? And how could it have dissolved such rocks if, as was more reasonable to believe, the primordial waters had not been so very different from the current oceans? Fire, by contrast, could pass through anything, it could extinguish itself without leaving any traces, and it could even obliterate, through fusion, any traces of earlier states. And fire is endowed with a tremendous mechanical force. Could it even raise mountains?

The path that led Hutton to these reflections was still far from being direct. Hutton was born into a well-to-do family and succeeded in becoming an attorney for a brief period, before resuming his medical studies and then traveling. He spent two years in Paris, where he studied chemistry and anatomy and probably took the courses of Rouelle. He returned to Scotland in 1750 and dedicated himself to agriculture, using methods that were intended to further his scientific aims. Soils and rocks began to interest him so much that, from 1753 on, the earth became one of his primary subjects of study. He settled in Edinburgh fifteen years later, during the period of Scottish Enlightenment. He was a very active member of the local scientific societies, through which he experimented with chemistry; and he became friends with the economist Adam Smith, the engineer James Watt, and the physicist Joseph Black; the latter two were authorities on the subject of heat principles. Except for his medical thesis, which he produced at Leiden in 1749, Hutton had published very little up to this time. According to the mathematician John Playfair (1748–1819), he was "one of those who are much more delighted with the contemplation of truth, than with the praise of having discovered it." At fifty-nine, an age when careers for some amount to nothing more than the managing of academic honors, his first important text appeared, the *Abstract of a dissertation read in the Royal Society of Edinburgh, concerning the system of the earth, its duration, its stability*. This ambitious subject was taken up again three years later, in the same society's journal, and was finally developed in 1795, two years before Hutton's death, in his long *Theory of the Earth*, drafted at the insistence of his friends, who feared that his illness would end his life too quickly. Although one could criticize the obscurity of its prose, the *Theory of the Earth* was read immediately. It was even translated into German shortly thereafter; and in 1802 Hutton's audience broadened when his friend Playfair expounded his theses in *Illustrations of the Huttonian Theory of the Earth*, where he corrected the most controversial points of the theory, presenting it in the form of a deductive model, expurgated of Hutton's profoundly deist teleology.

Hutton had, in effect, founded his theory upon deist presuppositions, which he foregrounded by affirming that "the globe of this earth is evidently made for man" and that its structure was "erected in wisdom, to obtain a purpose worthy of the power that is apparent in the production of it." The question, then, as described by R. Laudan, became how, without any other divine intervention, the conditions at the surface of the earth had remained favorable to life ever since the dark beginnings of time. In his attempts to demonstrate the mechanisms that provided for the permanence of these conditions, within a world that is apparently disordered, Hutton had to posit a strict actualism, or uniformitarianism, which supposes that the forces of nature have always been the same, and a postulate that the world necessarily continues to be reconstructed at the same rate as it is being destroyed. If erosion diminishes the reliefs and if the particles transported by rivers are deposited at the bottom of the sea, it follows that the sediments will consolidate and be uplifted subsequently to form new lands, which in their turn will also erode.

The difficulty was that, by contrast with the episodes of destruction, the manner of the reconstruction remains elusive to our observation. It was through the aid of philosophy, Hutton insisted, that reconstruction could be described, and it was through the action of fire that it could be explained. At the high pressures that predominated within the depths of the earth, fire exerted a primary action by consolidating all sediments, including limestone. At even greater depths, fire would subsequently melt them, whether partially, to produce gneiss, or totally, to give rise to granite. And Hutton asserted, furthermore, as one who was familiar with the mechanical force produced by steam-powered machines, that it was the expansive force of this fire that provoked the upheavals of new mountains, with the volcanoes acting as pressure-releasing valves, thereby preventing excessive upheavals and limiting the earthquakes. Now if the granites resulted from the action of fire, and not water, then they would have intruded into the surrounding rocks, rather than being deposited upon the underlying rocks. Hutton went out to the field to prove this conclusion. In a certain valley of the Grampian Mountains, at the heart of the Highlands, one could clearly see red granite veins intercrossing with black schist mica, stratified and fossiliferous. Before these "objects which verified at once so many important conclusions in his system," according to his friend Playfair, "the guides who accompanied him were convinced that it must be nothing less than the discovery of a vein of silver or gold, that could call forth such strong marks of joy and exultation!" And Hutton realized, moreover, that the deposition of new sediments upon an ancient substratum that had been deformed, folded, or straightened before being leveled by erosion must necessarily determine an irregular interface. Such unconformities could also be found in the field, and these were not secondary accidents, as Werner, for example, had claimed, but rather the indubitable record of the emergence of new worlds.

In breaking with the idea of strata that were continually stacked upon a primordial granite, ever since the earliest beginnings of the earth, Hutton discovered the traces of a "succession of worlds," in which periods of tranquil sedimentation had alternated with violent upheavals. The great idea of the geological cycle was formulated through the pen of Playfair, who substituted "heat" for Hutton's term "fire." For Hutton, a mechanical analogy was apparent: this world was "to be considered thus merely as a machine," which is "destroyed in one part, but is renewed

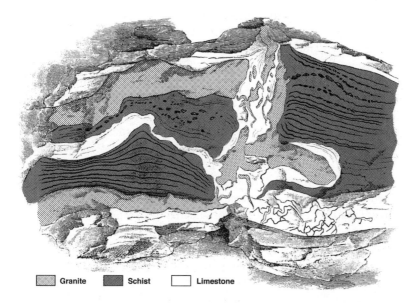

Granite   Schist   Limestone

Figure 6.3 Hutton's observation, in the valley of Glen Tilt, of the "total confusion maintained by substances found at the juncture between granite and stratified rocks [schists], which crumbles practically with the pressure of a pencil." From MacCulloch, "Geological Description of Glen Tilt," plate 18.

in another." Hutton simply concluded from these indefinitely repeating cycles: "The result, therefore, of our present enquiry is, that we find no vestige of a beginning,—no prospect for an end." In truth, Werner's excesses were met by those of Hutton: on the one hand was a history that would prove to be too linear, while on the other was an eternity, from which history was excluded. And where Werner saw nothing more than the marks of the waters, Hutton saw the marks of fire and fusion in every sort of rock—including shell-bearing limestones, chalk flint, and rock salt. But it had always been known that when limestone is heated to high temperatures, it decomposes into lime; Hutton consequently supposed that the elevated pressures resulting from the burial of sediments impeded limestone's loss of carbon dioxide.

At first the chemist James Hall (1761–1832), a friend of Hutton, found such ideas to be repugnant, but later he decided to test them experimentally. Because Hutton was opposed to the idea, blaming those who "judge of the great operations of the mineral kingdom, from having kindled a fire and looked at the bottom of a little crucible," it was only after Hutton's death that Hall began his work. He chose, for a vessel that would be capable of withstanding elevated pressures and temperatures, a "common gun-barrel, cut off at the touch-hole, and welded very strongly at the breech by means of a plug of iron." Limestone powder was placed into a porcelain crucible lodged at the base of the breech, which had previously been sealed, while pounded clay was used to fill the rest of the cannon, whose muzzle was finally welded with a more fusible metal. The breech was then heated in a common muffle, whereas the muzzle was "kept cold by means of wet clothes." These experiments were dangerous, and not all of them succeeded. Nonetheless, Hall was able to confirm in 1804 that the original powder, under pressure, had formed a solid piece of limestone, or even of marble, rather than decomposing. Shortly before, he had discovered that the product resulting from the fusion of basalt depended crucially on the manner of cooling. With a rapid cooling, glass was produced, while if the cooling was slow, crystals would form: vitrifiability was proved to be consistent with an igneous origin. But Werner's followers were long insensitive to these experimental arguments, which were incompatible with their own ideas. In the 1830s, however, the idea of an aqueous origin for these basalts was abandoned. Werner's disciples, who had gone to discover the volcanoes of Auvergne, finally found out what Desmarest meant by "Go and see," the response he gave to those who contradicted him.

Time of Life, Time of Fire: One Common Immensity?

Jean-Baptiste de Monet (1744–1829), knight of Lamarck, had no objections to the immensity of time that had been proclaimed by Hutton. In a book that he published in the same year as Playfair's, he exclaimed: "Oh, how very ancient the earth is! And how ridiculously small the ideas of those who consider the earth's age to be 6,000 odd years!" This work, *Hydrogeology*, was published at the author's own expense, illustrating the difficulties from which Lamarck never seemed able to extricate himself. Widowed three times and the father of a large family, Lamarck appeared to have been plagued by money troubles all through his life, although he did receive some convenient royalties and stipends. It has been said that he did not even leave enough to provide for the expenses of his burial. He was the eleventh child of a noble though penniless family, and while still adolescent he engaged himself heroically to fight in the Seven Years' War. When peace had been restored, he began studying botany to liven up his dismal life among the garrisons. When he quit the army, he did not abandon natural history; instead he broadened his perspective into medicine and chemistry. He published a flora of France in 1779, which helped gain access for him, through the help of Buffon, to the Academy of Sciences and then to the Royal Botanical Garden. In 1793 the National Museum of Natural History was created, and he was entrusted with the chair of insects, worms, and microscopic animals, which certainly nobody would have cherished. Lamarck was not annoyed. In spite of the attention given to fossils and shells, these groups remained little known. He gave them the name *invertebrates* and began to classify them methodically.

This work did not distract Lamarck from his reflections concerning the foundations of science; he believed there were close connections among physics, chemistry, and natural history and that there must be some influence of the moon on the atmosphere, for example, or of the climate on organisms. Terrestrial physics could be subdivided into meteorology, hydrogeology, and biology, the latter two terms coined by him. But Lamarck was hampered by his materialist (and vaguely deist) ideas and by the unorthodox and outdated versions of chemistry and meteorology that he held to. His *Hydrogeology* was, as its name suggests, the history of the earth as dominated by water. Lamarck was familiar only with a bit of the geology around the Paris basin, with its fossiliferous, horizontal strata, marked by their absence of granite and basalt. Fire did not have any role in the work, but still Lamarck shared Hutton's firm conviction regarding the immensity of time. Even though the eighteenth century's most eminent geometers had already demonstrated that the axis of the earth's rotation could vary only to a negligible extent, if at all, Lamarck hypothesized that the great cycles appropriate to the extremely slow displacements of the seas resulted from the fact that the earth's center of gravity does not coincide with its center of volume.

Within such a framework, which would not have disoriented Aristotle or Buridan, Lamarck advanced some radically new ideas. He asserted the fixity of species in 1794, but in 1800 he declared their mutability after his efforts in the classification of invertebrates. He did not consider the alterations to be evolution, but rather the natural process of producing the whole range of organisms through progressive changes effected over long periods of time. His *Zoological Philosophy*, which appeared in 1809, expounded at length that the notion of a species is subjective, that extinctions are but one point of view, and that catastrophes are nonexistent. His system also proposed that as a response to external changes, an increasingly complex continuum of plants and animals had been produced, beginning with the simplest of beings. Such extremely slow processes would require immense timescales. Lamarck had already announced in *Hydrogeology*: "The great age of the earth will appear greater to man when he understands the origin of living organisms and the reasons for the gradual development and improvement of their organization. This antiquity will appear even greater when he realizes the length of time and the particular conditions which were necessary to bring all the living species into existence. This is particularly true since man is the latest result and present climax of this development, the ultimate limit of which, if it is ever reached, cannot be known." With Lamarck, time had burst in upon biology.

The Order of Fossils

Fossils played scarcely any role in Lamarck's thesis of the gradual perfecting of life. During the course of the eighteenth century, it became apparent to some that the deeper one penetrated along the stratigraphic column, the less the fossils resembled currently living beings. From the extinct species noted by Palissy to the curious associations of fossils discovered in many places by Lhwyd, reports converged to establish this observation. To mention but two of the previously noted authors, Arduino remarked that the fossils were "worn and imperfect" in the secondary terranes, but in the tertiary they were "quite perfect, and completely similar to those we see in modern seas." Shortly afterward, in 1780, Soulavie was more precise when he distinguished and defined four ages as follows: "First Age: reign of the shells whose analogies do not exist today at all; Second Age: later part of the preceding reign, including some shells similar to those living today; Third Age: exclusive reign of the shells living today in our seas; Fourth Age: reign of the known fishes and plants, of our days."

Although these ideas had begun hazily, at the very beginning of the nineteenth century they started to crystallize, suddenly giving fossils a central role in the burgeoning geology. At the time when historical geology was being constituted, under the impulse of Werner's school, it became essential to compare and correlate formations. Nevertheless, the identification of rocks from criteria that were essentially mineralogical offered little aid in establishing their relative ages. For example, it was not easily possible to distinguish between two different samples of limestone, or of sandstone, or of schist, from this point of view. And accidents, such the lateral variations of facies (from sandstone to clayey schists, then to limestone, for example), and unconformities made the exercise more difficult. According to an analogy suggested by authors of the time, the historians of the earth found themselves confronted with the same problems that historians of humanity had already encountered. Folds, sedimentary gaps, faults, upheavals, unconformities, and other sorts of accidents corresponded to various types of ravages that had occurred to the ancient texts. Just the same, chronology had formed the framework of the universal history, from Diodorus to Eusebius and from Scaliger to Newton: in order to establish it, one had to harmonize, through laborious cross-checks, the dates and calendars gathered from the archives of different societies. For the history of the earth, it was the fossils that were going to render this formidable task possible.

Along with the German Johann Friedrich Blumenbach (1752–1840), one of the first naturalists to make reference to this role of fossils was Jean-André De Luc (1727–1817), a Genevan friend of Rousseau. He was a great traveler, but business misfortunes had forced him to settle in England, where he became a reader to Queen Charlotte in 1773. He was a Calvinist and the last of the renowned naturalists of the French language who remained impregnated with Mosaic literalism. De Luc noted in 1791: "The layers of this *clay* contain a great variety of *organized bodies*. And however they may change in their different locations, we can still recognize that they are of the same *period*, in that the same characteristic bodies of this *period* can be found in numerous places." But his still rudimentary reflections, surrounded by a mass of less interesting observations, were rapidly eclipsed by those of more cautious observers.

Among the first of these was the Englishman William Smith (1769–1839), a surveyor of the environs of Bath who announced in 1796 that strata of the same nature could be distinguished by the fossils they enclosed. He had drawn this conclusion from observations made during the three years he worked on the excavation for the canal that was to serve for transporting coal in Somerset. In 1799 he went into business for himself, offering services as a surveyor, a civil engineer, a mineralogist, or a geologist. His business prospered, and his numerous journeys permitted him to extend his observations and increase his fossil collections. In 1802 he developed, with the approval of the Royal Society of London, the project of producing a geological map of England and Wales. By the time his work was completed in 1815, he had revealed the regular succession of the secondary terranes there and said proudly of his map: "Alone I did it!" But this grand enterprise and others, in the final accounting, were financial setbacks: before the end of his days, when he was honored by geologists and had been given a pension by the king, Smith had to sell his goods and collections, and even, in 1819, spent two months in debtors' prison. Throughout his career he was a practical man, dedicating himself to the service of farmers and landowners. As he explained in 1816 in his *Strata Identified by Organized Fossils*, the fossils "furnish the best of all clues to a knowledge of the Soil and Substrata." By contrast with chemistry or botany, and the extensive knowledge required to identify the soils and the methods for improving them, "the organized Fossils (which might be called the antiquities of Nature) and their localities also, may be understood by all, even the most illiterate." However, biology, the extinctions of species, and even chronology, remained beyond the scope of Smith's preoccupations.

The life and career of Georges Cuvier (1769–1832), though they were marked just as notably by the theme of fossils, shone with a completely different splendor. Cuvier was born in Montbéliard, which was then part of the Germanic Empire, to a family who came from a village also named Cuvier in the Jura mountains. Owing to his passion for reading, kindled by his mother, he had already discovered Buffon and had made his beginnings as a collector and a naturalist by the age of twelve. Cuvier had a phenomenal memory and pursued his studies brilliantly at Stuttgart, before going to Normandy in 1788 to become the tutor to a noble family of Protestants. As he spent his winters at Caen and his summers near Fécamp, the land and the sea gave him ample material for observation and dissection. Cuvier had already demonstrated his taste for positive knowledge and his disdain for grand theories, and those opinions were soon to make him the fierce enemy of Lamarck and of Geoffroy Saint-Hilaire. And his skills as a naturalist astonished a friend of his host family, an agronomist of the Academy of Sciences, who accordingly introduced him to the scientific circles of the capital. That was how, in 1795, Cuvier became an assistant professor at the Museum in Paris, teaching comparative anatomy during the following year. Undisturbed by France's movement from Empire to Restoration, his career continued to ascend toward wealth, glory, and the highest scientific and administrative functions. His life as a member of the Academy of Sciences and of the Académie Française, a chancellor of the university, a state councillor, a baron, and finally as a peer of France, even so, was darkened by the loss of his four children. Although he was often depicted as a courtier—"He praised me just the way I like," Napoleon said once after receiving him— this reputation bears qualifying. It was Cuvier's freedom of speech that impressed the English geologist Charles Lyell, who said in 1823, "He is more liberal and independent than I believe most Frenchmen are," after hearing Cuvier publicly praise the officially shamed Napoleon.

Isolated in his Norman retreat, Cuvier formed in 1792 the conviction that was to guide his entire science: anatomy ought to be rational and not merely descriptive, and it ought to be applied to the fossil species just as well as to the living. With their sparse, fragmentary, and mutilated remains, sometimes even mixed in with others, the quadrupeds offered a practically virgin field of study from this point of view. Cuvier dedicated himself quickly to that area of research after his first efforts on worms. In order to reconstitute fossilized quadrupeds, he utilized his two great principles of the subordination of organs and the correlation of forms. He announced in 1812 that an organism constitutes a whole whose parts are so well accommodated to one another that "the least prominence of the bone and the smallest apophysis have a determined character relative to the class, the order, the genus, and even the species to which they belong, so that whenever we have only the extremity of a well-preserved bone, we may, by scrutinizing it and applying analogical skill and close comparison, determine all these things as certainly as if we had the whole animal."

These principles had allowed Cuvier to establish, in 1796, the existence of four kinds of "elephant" fossils, in contrast to the single species recognized by Buffon; two of the four appeared to have become extinct. Because such species probably would not have escaped notice, even if they lived at the far end of the world, and because the quadrupeds could not have escaped the disastrous advancing or receding of the seas, as aquatic shell animals could, it would be the disappearance of quadrupeds that could demonstrate the reality of the extinction of species. Catastrophes could therefore be shown to have punctuated the history of the earth: the proof would be the resulting disappearance of entire groups of fauna. Cuvier had little to say, however, regarding how new groups of flora and fauna came into being after a catastrophe. The important point for him was that global revolutions could be revealed only through comparative anatomy. "A science that at first would not appear to have such close ties with anatomy, that which considers the structure of the earth, which gathers the monuments of the physical history of the globe, and which attempts to paint with bold hand the picture of the revolutions that that has undergone—geology, in a word," he stated, "cannot establish with certainty several of the facts that it uses as a basis, except through the aid of anatomy."

The Paris basin was found to be especially plentiful in species of all sorts; Guettard drew attention to the wealth of fossilized birds and quadrupeds in the gypsum of Montmartre, and Lamarck himself described more than six hundred species of invertebrates, mainly in the layers of coarse-grained limestone. It became clear that these fossils belonged to distinct periods, but what were their respective ages? In order to complete the vast history of the extinct quadrupeds of the Paris basin that he had undertaken, Cuvier enlisted the collaboration of Alexandre Brongniart (1770–1847), a mining engineer, to whom he consistently attributed the merits of their common labors. Brongniart was a son of the renowned architect who built the Paris Stock Exchange. He became director of the Sèvres porcelain factory and succeeded Haüy at the Museum. Of the same age as Cuvier, he had, until Cuvier enlisted him, occupied himself with mineralogy and zoology. Criss-crossing the Paris basin in company with their assistants, Cuvier and Brongniart applied themselves to deciphering lateral variations of facies caused by the incessant movements of the seas that Lavoisier had previously noted. The simple solution to such complicated situations was summarized in 1808 by the two in their "Essay on the Mineralogical Geography of the Surroundings of Paris": "The method we employed for recognizing, out of so many limestone beds, one that has been observed previously in a very distant location, is drawn from the nature of the fossils enclosed within each layer. These fossils are generally always the same within the corresponding layers, and they present quite notable differences of species from one system of layers to another. This is a sign for recognition that has, to the present, not deceived us."

Where Werner had seen nothing but a mass of undifferentiated terranes, Brongniart and Cuvier were able to define a new era, the Tertiary. Through correlations of fossils, taken to an unimaginable degree of refinement, they introduced order into the complexity of the entire Paris basin. The capacity to be so specific with fossils stemmed, above all, from the extinctions whose reality had just been demonstrated. By contrast with the classifications of Lehmann or Werner, the mineralogical constitution had only a causal interest, since it characterized the conditions of the formations of the rocks, but not their ages. Lamarck had become the founder, by accident, of invertebrate paleontology, and his labors of classification had made stratigraphy possible. This use of fossils appeared so well founded to Brongniart that, in 1821, he could finally guarantee: "I consider the characters of the epoch of the formation as drawn from the analogy of organized bodies, as primary in value for geognosy, and as more important than all the other sorts of differences, however great those may appear." With this affirmation, then, fossils prevailed over everything, even over the principle of the superposition of layers: if an older fossil was placed atop another, more recent one, for example, one had to deduce that an upheaval had inverted the strata. And in thus correlating the chalks of the Paris basin with the limestone perched at two thousand meters of altitude in the Alps, Brongniart was the first to establish that rocks of the same age could have very different histories.

The Final Splendor of Teleology

While sea levels climbed or fell and mountains rose up or eroded to plain, there were certain species, though relatively rare, that seemed to defy time. They traversed the eras without any sort of change, a fact that bestowed on them a certain immortality. It was the ones with a transient existence that were more useful from the point of view of stratigraphy. The presence of specimens such as these within the sediments would authorize not only the dating, but also the correlation of terranes that were distant from one another, even separated by oceans. Patient study of outcrops and correlations led closer and closer to the establishment of a relative scale for geological formations, which was ceaselessly being revised and reworked to refine its generality and precision.

Even before fossils began to play this key role, correlations among terranes had extended beyond the limits of Europe and into the New World. Based on observations made during his great expedition to South America, from 1798 to 1804, Humboldt advanced a vast synthesis in 1823 of the geognosy of the two hemispheres. From one side to the other of the Atlantic, he saw in the successions of the formations a regularity that merited a *pasigraphic* method. He assigned to each rock a Greek letter (α = granite, β = gneiss, etc.), to which he added an accent to indicate the presence of shells, stating,

In order to unite the principal phenomena of the positions of rocks in the primitive, intermediary, secondary, and tertiary formations, I propose the following series:

α, $\alpha\beta$, $\beta + \pi$, $\beta\gamma$, $\gamma + \tau$, α, γ, δ, α, β, δ, $\bar{o}$ $\|$ κ^g, τ', δ_τ', δ', δ', $+ \pi$, γ, τ', $\sigma\pi$, $\sigma + \alpha$, $\sigma\pi$, $\bar{o}$ $\|$ $\pi\kappa^a + \xi$, $\tau^a + \theta$, κ^n, τ^m, κ^q, τ°, χ^1, τ^c, $\|$ κ^{2^l}, τ^p . . .

[and he added] As the notation I present here may be variously graduated by the manner in which the characters are accentuated, in uniting them as coefficients in complex formations, or in adding exponents, I doubt whether the names of the rocks arranged in series beside each other would address itself as forcibly to the eye as the algorithmic notation.

Humboldt, however, had hardly any success with his new scheme, which, indeed, was far too general to be correct.

It even appeared possible to reconstitute the positions of the ancient lands and seas and their climates on the basis of the nature of the rocks and the fossils. On the instigation of William Buckland (1784–1856), one of the most notable geologists of the first half of the nineteenth century, the prefix *paleo-* was finding its way into new words, such as *paleogeography* and *paleoclimatology*, which followed *paleontology* and preceded

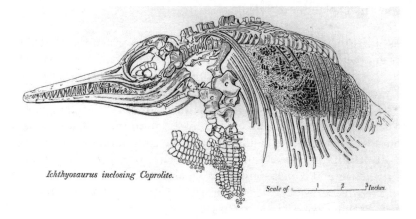

Ichthyosaurus inclosing Coprolite.

Scale of |____1____2____³ Inches.

Figure 6.4 Ichthyosaur with its coprolites, unearthed by Buckland. From his *Geology and Mineralogy*, vol. 2, plate 13.

paleoecology. Highly in demand by his listeners because of his enthusiasm, Buckland was a professor at Oxford for a long time before becoming the dean at Westminster. In 1824 he published the first correct description of a dinosaur, the megalosaur. When he discovered *coprolites*, or fossilized excrement, one of his conclusions was that carnivorous species, such as the ichthyosaurs, had existed ever since the dawn of time. To his horror, he realized that the conviction that death had been a consequence of Original Sin would be contradicted if carnivorous species had existed before the creation of man. Buckland responded that this anomaly was just another proof of divine benevolence, since the only function of carnivorous species was to reduce the suffering of sick or aged individuals. Buckland defended such theses, with his customary talent, in *Geology and Mineralogy Considered with Reference to Natural Theology*. He published this treatise in 1837, through the support of a bequest from the earl of Bridgewater, who was an eccentric humanist aristocrat known for his strong propensity for ancient texts, dogs, and illegitimate children and whose deep concern was to demonstrate "the Power, Wisdom, and Goodness of God, as manifested in the Creation."

As nature continued to be examined through the prism of theology, it was in England, long frozen by fears of the French Revolution, that adherents of biblical literalism continued to be prominent the longest, and it was the English who had the most difficulty in allowing such derisive fragments unearthed from the soil to prevail over the divine word of the Scriptures. For a paleontologist such as Buckland, "a science which unfolds such abundant evidence of the Being and attributes of God," could not reasonably be viewed in any other light than as "the efficient Auxiliary and Handmaid of Religion." His teleological argumentation, which did not differ much from that of John Ray, 150 years earlier, was not always well received on the Continent. Aubuisson, a follower of Werner, confided in 1830 to Charles Lyell (1797–1875), who had just recently published his *Principles of Geology*: "We *Catholic* geologists flatter ourselves that we have kept clear of the mixing of things sacred and profane, but the three great Protestants, De Luc, Cuvier, and Buckland, have not done so. . . . Have they done any good to science and religion?" Anxious that his criticism of the "Mosaico-geological system" might hamper the sales of his own *Principles*, Lyell had to respond "No." But since teleology did not in any way exclude seeing the Bible as merely a history of mankind, Buckland could subscribe to extremely long durations: "Millions of millions of years may have occupied the indefinite interval between the beginning in which God created the heaven and the earth and the evening or commencement of the first day of the Mosaic narrative," he wrote, thus following the example of Pope Pius VII, who had officially admitted at the turn of the century, according to Jean-Jacques Nicolas Huot, the geographer, that the "days" of Genesis were merely indeterminate periods of time. In the old debate between faith and reason, as C. C. Gillispie emphasized, the delimiting line was not between religion and science, but rather between the old and the modern of each. Even if that line was a sinuous one—a matter that for geologists such as Buckland depended on the nature of the problem at hand—the immensity of geological time in reality was no longer being contested among scientists. By the middle of the century, it was only religious or general literature that maintained the last vestiges of Mosaic times.

New Principles

Inaugurated by the creation of the term *dinosaur*, the blossoming of paleontological discoveries at the beginning of the nineteenth century contributed, more than anything else, to the success of geology in the eyes of the public. But the profound causes for the apparent upheavals experienced by the earth's crust remained confused. Although the radius of the terrestrial globe is sixty-four hundred kilometers, only a thin film of the surface is available for observation, so geologists are limited in the same way as the doctor who could only diagnose based on the patient's skin. Nevertheless, Werner had argued that searching for the reasons of the movements of the sea was meaningless, since even the slightest change occurring at the impenetrable depths of the earth could translate into considerable effects at the surface.

In describing this situation, fertile ground for controversy, the old-fashioned believers in the water theory (Neptunists) were often contrasted with the modern followers of the fire theory (Plutonists). Likewise, progressives, who believed in a globally unchanging world (uniformitarians), were contrasted with conservatives, who saw in the global revolutions the mark of recurring upheavals (catastrophists). Charles Lyell, the herald of uniformitarianism, was no stranger to these reductive oppositions. He came from a well-off Scottish family and was interested in entomology from childhood. In 1817 at Oxford, he discovered mineralogy and geology with Buckland, during a period when his law studies did not prevent him from traveling on the Continent. He practiced geology between 1825 and 1827 while also engaged in the profession of law, which problems with his eyesight later forced him to abandon. The year 1827 was also when Lyell discovered Lamarck's *Zoological Philosophy*. "His theories delighted me more than any novel I ever read," he wrote to a friend, "but though I admire even his flights, and feel none of the *odium theologicum* which some modern writers in this country have visited him with, I confess I read him rather as I hear an advocate on the wrong side, to know what can be made of the case in good hands." The extremely long durations of time postulated by Lamarck aroused his particular enthusiasm: "That the earth is quite as old as he supposes, has long been my creed," he continued, "and I will try before six months are over to convert the readers of the *Quarterly* to that heterodox opinion." But Lyell journeyed for nearly a year before making that attempt; his travels took him from the Paris basin to the Central Massif and to Sicily. He became familiar with fossil fauna through the company of Gérard Deshayes (1797–1875) of Paris, another of the great classifiers, from whom he borrowed the idea that the proportion of extant species should be a criterion of the main stratigraphical divisions. Using this concept as a basis, he subdivided the units of the Tertiary in his *Principles of Geology*, published in three volumes between 1830 and 1833. By 1875 twelve editions appeared, evidence of the audience and the consideration that Lyell rapidly enjoyed. He was appointed professor at King's College in London in 1831, knighted in 1848, and promoted to baronet in 1864.

Putting the talents of an attorney to good use, Lyell expounded the methods and results of his discipline from the very first chapters of *Principles*. Lyell practically wiped clean the slate of the past, since such a title, which was inspired by Descartes and Newton, seemed to imply that geology as a science was born with his treatise. The subtitle of the work, *Being an Attempt to Explain the Former Changes of the Earth's Surface, by Reference to Causes Now in Operation*, indicates the nature of the actual causes being invoked: for as far back as one can go into the past, the laws of nature have remained the same, not varying in degree or kind. Lyell's credo was thus a rigorous uniformitarianism, nourished by Hutton's great ideas. Although Lyell recognized that Lamarck "has been courageous enough and logical enough to admit that his argument, if pushed as far as it must go, if worth anything, would prove that men may have come from the Orang-Outang," Lyell still completely dismissed the idea of affiliations between species. He did not even point to any degree of improvement within living beings since earliest times. Certainly, there were innumerable species that had disappeared, but the classes had not been disrupted. It was difficult to find a reason for the presence of fossilized specimens of tropical species in the cold of Europe, if it was not due to a slow cooling of the globe through the course of geological times. Because he limited such a cooling, in principle, to the very earliest moments of the earth, Lyell took inspiration from Humboldt to assert that climates are determined by the distribution of the continental masses. Historical geology demonstrated that this had varied constantly, so there was no need to deny uniformitarianism to explain the repeated changes in climate that fossils revealed.

With time, Lyell's work came to be considered a beacon of the new discipline. Although fire gradually triumphed over water, it still posed new problems. And as we shall see, it was not an easy task to reconcile uniformitarianism with the heat of the depths, which was required by Plutonism and by Hutton's vast cycles. Nevertheless, the quarrels over doctrines had few effects on the activities out in the field, where innumerable slices of the earth's past were being revealed. Aubuisson, for example, advised at the beginning of his *Traité de géognosie* (*Treatise on Geognosy*): "This method [after Werner] of assembling facts has so little place in my work, that the geologist who would substitute an entirely different one, or who would, for example adopt Hutton's system . . . would have no more than a few pages to rewrite in this treatise: the observed facts and their consequences would still remain the same." From the recent Quaternary formations to the oldest fossiliferous terranes, a nomenclature was adopted to describe geological history. Following the works of Humboldt or the Englishmen Adam Sedgwick and Roderick Murchison, for example, three great *eras* were defined and subdivided into *periods*, *epochs*, and *ages*, which exhibited progressively less pronounced changes in fossils and rocks. The extremely old terranes, largely

Appearance of the great biological groups throughout geological time

Cenozoic	Quanternary
	Recent (1833)
	Pleistocene (1839) Evolution of man
	Tertiary (Age of mammals and flowering plants)
	Pliocene (1833) Australopithecus
	Miocene (1833) Modern fauna and flora
	Oligocene (1854) Monkeys and gorillas
	Eocene (1833) Placental mammals: carnivores, rodents, elephants, etc.
Mesozoic	Secondary (Age of reptiles)
	Cretaceous (1822) Primates and flowering plants; end of the Dinosauria, of the ammonites, etc.
	Jurassic (1799) flying reptiles, Dinosauria, birds; flowering plants
	Triassic (1834) Ichthyosaurs, marsupials
Paleozoic	Primary (Age of invertebrates, then fishes [Devonian], and finally, amphibians)
	Permian (1841) Major extinctions in marine environments
	Carboniferous (1822) Reptiles, winged insects; conifers, cycads, seed-bearing ferns
	Devonian (1839) Bony fish, amphibians, wingless insects; mosses, ferns.
	Silurian (1835) Armored fish, scorpions; terrestrial vascular plants.
	Ordovician (1879) Corals, urchins, jawless fishes, graptolites.
	Cambrian (1835) Diversified marine fauna: arthropods (trilobites, crustaceans), cnidarians (octopuses, jellyfish), worms, mollusks, foraminifers, etc.
Cryptozoic	Precambrian: Cyanobacteria, algae, stromatolites, soft-bodied metazoans

Note: The dates indicate when the different periods were named.

devoid of fossils, were simply grouped together into one more era, the Precambrian.

Since the most recent terranes were also the best known, unavoidable proximity distortion led researchers to magnify the most recent of the

Figure 6.5 The great ages of the earth, reconstructed: "Ideal view of the earth during the Lias period," from Figuier, *La terre avant le Déluge*, 199.

periods. Although they were aware that the earliest periods ought to be the longest ones, stratigraphers were not equipped to move from relative dating to actual ages. And even though the increasing complexity of the history buried in the depths of the soil meant appealing to longer and longer durations, there was not yet any way to measure them. The background question, whether geological intensity could possibly vary through the course of time, remained unanswered. Some scientists had asserted, counter to Lyell and uniformitarianism, that the abundance of granite in the ancient terranes was evidence of the continuous cooling of the globe. Others came to the same conclusion based on observations of the diminishing volcanic activity over the ages. But certainly none of the arguments could be considered conclusive without rigorous timescales.

SEVEN

The Long History of Two Barons

HEAT AS A CLUE TO TIME Fourier formulated his celebrated heat theory in order to give a quantitative basis to Buffon's cosmological work. Kelvin used it along with the first two laws of thermodynamics to claim that the age of the earth and the sun could not be as great as was asserted by geologists, first setting an upper limit of 100 million years for the sun and an interval ranging from 400 to 20 million years for the earth.

Revolutions

While the expansion of paleontology permitted scientists to reconstitute an entire animal starting from a single jaw-bone, before an astounded public, the physical sciences pursued their conquests at a brisk pace. In France especially, scientists scored success after success: Lagrange, Laplace, and Monge; Biot, Dulong, Petit, Malus, Fresnel, and Gay-Lussac; Berthollet, Chaptal, and Foucroy—all these individuals, now recognized by street names in Paris, testify to the vigor that animated mathematics, physics, and chemistry. In the space of a few decades, Paris became the center of an effervescence that has never since been equaled anywhere.

Neither the expulsion of the Jesuits from France in 1764 nor the beheading of Lavoisier thirty years later hindered the scientific fervor in any way. At the end of the ancien régime, the colleges were taken over by the Benedictines, the Oratorians, and the Minims. After the clergy were ousted for good, the high schools and the first Grandes Écoles were created at the twilight of the Revolution. A new species, the teacher-researcher, emerged at the École Polytechnique and the École Normale Supérieure, to educate and train large numbers of schoolteachers and officers who were to constitute the skeleton of the new Republic. Scientists were no longer maintained by inquiring aristocrats; they had become real professionals, with whom the Voltaires and the Kants of the new century no longer attempted to blend. Modern science was asserting its presence.

Some of the branches of physics, such as optics and mechanics, had already attained a certain maturity. By contrast, the science of heat was at its earliest stages, and it still remained the business of chemists. Among the various kinds of earth that had been distinguished, one, the *inflammable earth*, was thought to escape from burning bodies. In assuming the role that previously had been assigned to fire, it received the more scientific-sounding name of *phlogiston*. The German chemist Georg Stahl (1660–1734) was its most eloquent propagandist. Lavoisier, with the help of his balance, became its executioner: at the end of the eighteenth century, phlogiston was abandoned because its postulated release was found incompatible with the increases of weight observed during combustion or calcination. Lavoisier demonstrated that combustion represented a chemical combination (generally with the oxygen of the air) that liberated heat.

Heat was therefore devoid of mass. For some, it was natural to imagine that it represented the flow of an imponderable fluid, the *caloric*, which was neither created nor destroyed in passing from hot bodies to cold. For others, adopting an idea that had been sustained by Aristotle and other philosophers of antiquity, heat represented the degree of agitation of the particles of matter. The debate was open. The word *energy*, so familiar to us now, was foreign to science of that day, and it bore only its original meaning, "the force of a discourse, of a sentence, of a word." The notion of force, however, had been generalized after Descartes and Newton. It was forces that animated the physical world, though these forces could be quite different in nature. They were connected with gravity, movement, electricity, magnetism, capillarity, and chemical affinity, and while they were studied for themselves, the nature of their interactions remained confused. The nomenclature reflected this confusion, such as the use of the term *vis viva* (living force) to designate what we now call kinetic energy. It was the force of steam that brought about a change of perspective: it captivated the engineers; it was the reason that coal had taken the place of animals, wind, and waterpower. The machine age was establishing itself. The Industrial Revolution had begun.

Machine Power

Through his many inventions produced between 1765 and 1790, the Scottish engineer James Watt contributed more than anyone else to making the "fire engine" the driving force of the nineteenth-century industrial expansion. *Condenser, puppet valve, double action, governor, sun-and-planet wheel*—so many terms for the poetry of mechanics; they testified to its creative genius. The primitive combustion engines pioneered by Newcomen, Papin, and Savery at the turn of the eighteenth century were transformed into powerful apparatuses for exploiting mines or moving ships. "If you were now to deprive England of her steam engines, you would deprive her of both coal and iron; you would cut off all the sources of her wealth, totally destroy her means of prosperity, and reduce this nation of huge power to insignificance. The destruction of her navy, which she regards as the main source of her strength, would probably be less disastrous," Sadi Carnot (1796–1832) proclaimed in 1824.

The economic efficiency of these new machines was certainly a fundamental question. Engineers progressed by trial and error in seeking to take full advantage of steam's force by controlling its expansion. They also brought steam to higher pressures and inquired whether they might better replace it with air or alcohol. In comparison with the empiricism of the English engineers, the French physicists distinguished themselves by their theoretical approaches. It was Sadi Carnot who pushed the analysis the furthest. His father, Lazare Carnot, was not only a renowned scientist but also an eminent statesman. Under the ancien régime, he published *Essay on Machines*, in which he introduced the concept of the work of a force, which was defined as the product of the force and the distance over which it is applied. Although he was celebrated as the "Organizer of Victory" for preparing the plans for fourteen armies during the wars of the Revolution, the elder Carnot abandoned politics at the beginning of the Empire to dedicate himself to science and the education of his children. He resumed service as minister of the interior only during the Hundred Days, when Napoleon escaped from Elba to regain power briefly.

Sadi Carnot had been in good hands when he entered the École Polytechnique at the age of sixteen. Unfortunately, he lived only a bit longer than did his contemporary, Franz Schubert; both left a deep mark and a work incomplete. In the only one of his works to be published during his lifetime, *Reflections on the Motive Power of Fire*, Carnot concentrated on the essentials. He introduced some notions that have since been accepted, such as that of the cycle or of reversibility, and he decomposed the operation of a machine into a series of elementary steps (compressions and expansions) through the use of which a precise evaluation of forces could be established. Just as the work done by a mill increases with the height of its falling water, the power of an engine appears to be determined by the difference in the temperatures of the bodies between which the steam is transferred (the boiler and the condenser). The nature of the fluid has no effect, nor does increasing the pressure or the quantity of heat produced. It is the difference in temperature, alone, that determines the yield of the engine.

With this reasoning, Carnot had put his finger on the *irreversible* nature of the transfer of heat from hot to cold bodies. And so, the growing field of thermodynamics received an outline for its second principle. Nevertheless, Carnot's ideas were so new that it took physicists a good twenty years to appreciate their wealth. The same was true of engineers, who waited a while before advancing no further into the fog than finding out what factors come into play or determining the maximum power their engines could provide.

Carnot also laid the foundations for the first principle, the conservation of energy. But his days were numbered, and cholera had the last word. Fortunately, Carnot's notes on the conversion of heat into mechanical work escaped the flames to which the personal effects of the epidemic's victims were normally consigned. The notes remained unpublished for a long time, though, and in the meantime, the first principle was formulated by others.

During the same period, the mathematician Joseph Fourier (1768–1830) became closely interested in another important aspect of heat, its propagation. Curiously, although both Carnot and Fourier considered heat intensively during the same period, no particular relationship resulted between the two men. And even though Carnot's dissertation was read before the Paris Academy of Sciences, it evoked no sort of reaction in Fourier. Perhaps the lack of special relations was a reflection of the great difference in their horizons: Carnot's inspiration came from engines, while Fourier's came from theories of the earth and from Buffon, whose influence was still significant. "For us, the problem of the earth's temperature has always been one of the greatest objects of cosmological studies," wrote Fourier, "and we have this principally in mind in establishing the mathematical theory of heat." In so proceeding, he developed fundamental theoretical tools that were also useful for many other branches of modern science. "The profound study of nature is the most fertile source of mathematical discoveries," he declared. Through Euler, Poincaré, Legendre, Laplace, and Gauss, celestial mechanics illustrated this principle magnificently. But never again after Fourier would mathematics return to take such productive inspiration from geological problems.

Neither Monk nor Soldier: Baron, Prefect, and Scientist

The convulsions experienced by France during the Revolution in her struggle "to doff the 'swaddling clothes' of routine, superstition, and privilege," as the astronomer Arago put it, had actually embellished many lives with burning piquancy. Fourier was younger than Lavoisier or Condorcet, though he barely escaped their tragic lot before completing his brilliant demonstration that a mathematician's life is not necessarily monotonous. He was born at Auxerre in Burgundy, only twelve leagues away from Montbard, the fief of Buffon. His father was from Lorraine, a tailor who had three children by his first wife and twelve by his second. Joseph was ninth of the second series and had already lost both father and mother by the time he reached the age of eight: the ingredients for a melodrama were in place. But the orphan's vivacity had not escaped the notice of a well-meaning neighbor, who recommended him to the bishop of Auxerre. The latter had him enroll in the local military school directed by the Benedictines. This school was the Burgundian equivalent of the one administered by the Minims at Brienne, which had received at the same time a certain Napoleon Bonaparte, the enterprising second son of an old Corsican family.

To be sure, the young Joseph was a precocious child; by twelve he had already written "several sermons that were highly applauded in Paris, according to the accounts of high church dignitaries," recorded Arago, although the latter did not substantiate his compliment. As the years passed, Joseph's talents asserted themselves, and he developed a furious passion for mathematics, hiding himself during the night to study by the light of bits of candles he managed to put aside during the day. He discovered his first theorems at the age of sixteen. Still sponsored by his bishop, he went to Paris to pursue his studies. He was tempted to enter a military career, but in spite of the support of the illustrious mathematician Adrien Marie Legendre, he was not even admitted to the examination for the artillery. Supposedly the minister had responded "Fourier is not a noble and cannot be allowed to enter the artillery—not even if he were to prove to be a second Newton!" Failing the military, he turned to the church. After a short stay at Auxerre, Fourier arrived at the abbey of Saint-Benoît sur Loire to take on the Benedictine novice's habit in 1787. He noted with a hint of envy on 22 March 1789, "Yesterday I turned twenty-one. By this age, Newton and Pascal had already acquired a good many claims to immortality." Without the upheaval that commenced shortly afterward, what sort of opus magnus would Dom Joseph Fourier have written during a life devoted to religious studies? In autumn, religious congregations were forbidden. The novice never even pronounced his vows. He left the monastery and returned to Auxerre, where he taught mathematics at his former school. And then he submitted his first dissertation to the Academy of Sciences.

Fourier became less and less indifferent to the Revolution. He participated in the local revolutionary bodies and ended up being taken in a skirmish near Orléans in the autumn of 1793. Stripped of his functions, he went to Paris to plead his defense before Robespierre, who ordered him put into the camp of terrorist enemies of the Revolution. Robespierre's fall on the 9th of Thermidor by the Republican calendar (27 July 1794) saved him from the guillotine. Shortly afterward the department of Yonne in Burgundy chose him to be part of the first heterogeneous promotion of the short-lived first École Normale Supérieure. There Professors Joseph Lagrange, Laplace, and Gaspard Monge quickly noticed his talents, and through their influence he obtained a position at the recently created École Polytechnique. But a several-month-old denunciation had followed him, triggering a new arrest in the middle of the night. The poor old woman who was porter at his residence was terrified when the chief henchman growled at her, "You'll be able to come fetch him back yourself; he'll be in two pieces!"

Somehow or other Fourier's situation finally settled down. After teaching analysis for three years at the École Polytechnique, he was sent off to the campaign in Egypt. There he was engaged in numerous missions. For one, he became permanent secretary of the Egyptian Institute, which was founded immediately. He also served as one of the diplomats entrusted by Napoleon with delicate negotiations. He started off in the service of an army that was required to provide for its own needs of steel, arms, powder, and other munitions. In spite of everything, he found time to publish dissertations on various topics: the general resolution of algebraic equations, the aqueduct that took water from the Nile to Cairo, and statistical research on the state of the country. He even wrote a compendium of Egypt's revolutions and customs. He was finally able to return to France, with the very last unit from the expedition, in 1801. Fourier was told to direct the drafting of a record of the Egyptian Institute's work. In the end, he wrote only the *Preliminary Discourse* of the monumental *Description of Egypt*. Describing Napoleon's role in the expedition made it a delicate exercise: the *Discourse* had to wait until 1809 to be finished.

Fourier had already resumed his place at the École Polytechnique and had been there for some months when Napoleon named him prefect of Isère, in the Alps. The isolated location made the assignment a sort of exile, but the orders of the First Consul were not to be discussed. He left for Grenoble, but without any particular haste. The twelfth child of a respectable tailor of Auxerre, he was said to be capable of giving "lessons on theology to a bishop and in courtesy to a parliamentarian born before 1790." He would know how to make himself appreciated. He was urbane and charming, with a clarity of ideas, a good memory, and good sense— his qualities were enchanting. But his scientific contributions were still somewhat meager by the time he had reached thirty-three. Because he was a bachelor, it was not his family but his teaching that had occupied much of his time until that point. Naturally, Fourier had to perform his administrative tasks as prefect at Grenoble, and he also directed the draining of the marshes of Bourgouin and the construction of Alpine roads. He even participated in the local artistic and scientific life. In fact, it was to him that the young Jean François Champollion owed his passion for Egyptology. And it was at Grenoble that Fourier commenced the work through which he developed into a theoretical physicist belonging "to that very select band including Galileo, Newton, Maxwell, Planck, and Einstein, who by the originality, importance, and influence of their work effected revolutions in various branches of the subject," in the words of his biographer J. Herivel.

The Analytical Theory of Heat

While in his prefecture, Fourier found the time to delve more deeply into a problem that he had already treated briefly before, namely, the conduction of heat through a body. The source of interest for this seemingly innocuous problem was the question of the terrestrial "fire" that Buffon had brought back into the foreground. As Fourier explained it,

This heat, which dissipates in the planetary spaces, is proper to the earth, and primitive: it is due to causes that existed at the beginning of this planet. It abandons the internal masses slowly, so that these conserve a highly elevated temperature for an immense amount of time. This hypothesis of internal, central heat has been renewed throughout all the ages of philosophy, because it presents itself to the mind as the natural cause of various great phenomena. What is needed is to examine this opinion through a precise analysis founded on the laws of mathematics and of the propagation of heat.

This problem had already been posed by the Academy of Sciences in 1736, when "the nature and propagation of fire" was the topic of a prize contest. Among some fifteen competitors were Voltaire and the marquise du Châtelet, who each sent in a separate dissertation. Even Voltaire's was appreciated by the Academy. Subsequently, the question of the temperature profile in an iron bar whose extremities were maintained at different temperatures received a preliminary response. Fourier went beyond this special case by posing two complementary problems in the broadest manner: How does the temperature inside a body of any sort of geometry vary when provided a continuous source of heat? and How does the temperature of a body vary with time when the source of heat is discontinued? Possible applications of the problems to the earth's situation are obvious.

Like other great discoveries, Fourier's appeared in retrospect to be stamped with the seal of the obvious. Nevertheless, he needed to dissipate the confusion perpetuated by some of the greatest minds of the time regarding the different modes of heat's action. Fourier ventured no hypotheses regarding the actual nature of heat; for him, to be able to define it was sufficient. He recognized that a mathematical analysis of the propagation of heat must be founded upon the fundamental observations that "different bodies do not in fact possess the same degree of power to *contain* heat, to *receive or transmit* it across their surfaces, nor to *conduct* it through the interior of their masses."

Fourier used a simple heat budget to evaluate the relative importance of these effects. Within an infinitesimally thin slice of matter situated inside a body of varying temperature, a portion of the heat that comes at a given instant from *all* the surroundings must simply pass on through. This portion is a function of a parameter called the *thermal conductivity*, and it is particular to a given body. A metal, for example, will communicate its heat much more quickly than wood will. The remainder of the heat produces a variation in temperature. It is released (upon cooling) or accepted (upon heating) by the slice of matter being considered, and this also depends upon another parameter that is particular to the body, namely, the *specific heat*. Fourier called the first part, which is not absorbed, the *heat flux*, and he finally postulated that this flux is proportional to two factors, the thermal conductivity and the temperature variation at the point of the body under consideration (referred to as the *gradient* in the language of physics). Heat is emitted (or yielded) only

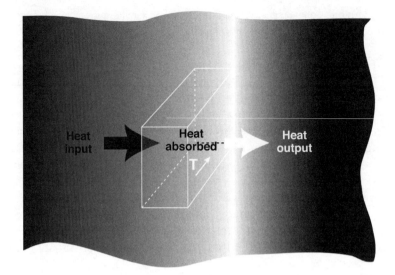

Figure 7.1 Heat transfer within a solid body.

on the surface in the form of radiation, the intensity of which increases considerably with temperature.

In establishing his heat budget, Fourier obtained an equation (with partial derivatives) of grand elegance, and he wrote its definitive form in 1807. Integration of the equation gives the temperature at any point within the body as a function of time and of the initial thermal conditions. But this was a new type of equation, and its solutions obviously depended upon the geometry of the body being considered: a bar would not conduct heat in the same manner as a sphere. In order to integrate such an equation under various limiting conditions, Fourier created a new branch of mathematics. In particular, he demonstrated that any algebraic function could be expressed as the sum of trigonometric functions. This is how the classic series and integrals bearing his name were born.

Contrary to Fourier's expectations, his work was received with mixed feelings. His trigonometric series and his conception of heat flux were disconcerting, and he was accused of a lack of rigor. With embarrassment, the Academy decided to have a new prize contest on this question in 1811. It resulted in the first recognition for Fourier, to whom the prize was, in a sense, offered. He carried the prize off with his head high; and so he should have, for he had submitted the same dissertation as before, with just a few changes. Under the name of *The Analytical Theory of Heat*, this "exquisite mathematical poem," according to one expression that became attached to it, was still not published by the Academy until 1822. Even though cosmological problems had been the author's real source of inspiration, they were mentioned only in the introduction.

A Very Slow Exhalation

As prefect in Grenoble, Fourier had been, by chance, on Napoleon's return route from the Isle of Elba. In a somewhat comical scene, his order for arrest turned into a rally that was sanctioned by a twofold promotion: the baron Fourier became count and was named prefect of the Rhône. He held this position for only two months. Because he had opposed the orders of Lazare Carnot, which were too strict for his taste, he was discharged from his assignment, and he returned to Paris. Unfortunately, after Waterloo, his attitude during the Hundred Days brought him the rancor of Louis XVIII, and thus ended the respectable income that would have permitted him for the first time to devote himself entirely to science. Fourier had never run after fortune; he was generous when he had funds to spare, and yet now he found himself nearing financial straits. His requests for a pension in recompense for his twenty years of loyal service were obstinately refused. Through the aid one of his former pupils who had become prefect of the Seine, he obtained the post of director of statistical services for the department. Naturally, the statistics he compiled became models of their kind.

As time did its work, Louis XVIII had to confirm Fourier's election to the Academy of Sciences in 1817. Before he became its permanent secretary in 1822, Fourier published his first dissertation in which he, at last, applied his theory to the earth. He proposed that the earth's heat had three possible sources: space, the sun, and the earth itself. Fourier quickly found the way to end in Buffon's favor the old debate regarding how much of the elevation of temperatures observed in mines was due to the effects of the sun and how much to the heat of the earth itself: "The analytical solution demonstrates that this increase of temperature could not be the effect of solar heating. It is entirely due to a primitive heat that the earth possessed at its origin and that diminishes over the course of the centuries, dissipating itself at the surface." A second question was how this heat was evacuated. Two practical problems arose. The thermal conductivity of terrestrial materials was unknown: as yet only that of wrought iron, which was doubtlessly much higher, had been measured. And nobody knew what the earth's temperature was right after it was formed. Was it the melting temperature of iron, Fourier wondered, or a temperature of only 500 degrees centigrade, which "would be more than ten times lower"? Regardless of the hypotheses retained, the fact is that heat as a general rule tends to move slowly in solid bodies. And because the distances are immense within the earth, cooling at depth is an endless process.

Mathematically, the temperature at the earth's surface is proportional to the square root of the thermal conductivity and to time, and so the numerical applications are simple. Fourier, who was immersed in antiquity, concluded with certainty that "the lowering of the temperature over a century is less than 1/57,600 of one degree centigrade. From the time of the Greek school in Alexandria until the present day, the dissipation of central heat has not occasioned the lowering of the temperature by even 1/288 of a degree" at the surface. This slow lowering of the temperature is so slight that it cannot be measured. But Fourier was cautious, and he did not press the matter any further. He left it to the reader to discover that it would have required 29 million years to cool the surface of the earth if it had initially measured 500°C. If the original temperature had been 5,000°C, the required time would have been ten times as long. According to Fourier's calculations, some hundreds of millions of years thus became well-founded estimates for the age of the earth, for the daring minds who amused themselves with such rules of three.

The rates of cooling for modest cannonballs, when extrapolated to the earth, had given ages much too brief: half a century after Buffon, the boldest estimates of the great naturalist were reduced to the order of brief episodes. In spite of the uncertainties affecting the thermal parameters of the earth's depths, one question that escaped experience could thus be treated. This was an important step, because it was one of the first times that a mathematical theory opened up totally new perspectives, rather than being limited to giving accounts of phenomena that were possibly complex but still had already been recognized in an empirical manner. Arago pointed out that the primitive heat would not raise the surface temperature by more than one-thirtieth of a degree. Contrary to Buffon's claims, the thermal death of the earth would have little impact on life.

To complement his demonstration, Fourier applied his ideas to the temperature of the heavens. At that time, an absolute zero of temperature had not yet been established, and unwarranted extrapolations of the air's cooling with increasing elevation had yielded negative temperatures of several thousand degrees for the immensity of "planetary space." By contrast, using his theory Fourier arrived upon the more correct conclusion that stellar radiation maintained temperatures in space between –50°C and –60°C. At that point Fourier ceased his cosmological reflections. Surrounded by a circle of young scientists, his faithful domestic Joseph, and some colleagues from the Institut de France, he lived his old days peacefully and was even welcomed into the Académie Française in 1827. He became friends with Cuvier, who in many respects was his opposite. In the salon of this ardent conservative Cuvier, who preferred facts to grandiose theories, Fourier associated with Stendhal, Humboldt, and Geoffroy Saint-Hilaire. His rise in physics, and especially in mathematics, had quite quickly become noteworthy. He had connections with naturalist circles and had justified the foundations of many of Buffon's ideas, which had had their impact. He also exerted a more subtle influence in geology, even though this was a terrain into which he had not wanted to venture. "It is less our object to discuss the special applications of this theory at the surface of the terrestrial globe, whose constitution remains unknown," he wrote, "than it is to establish the mathematical principles of this order of phenomena." By publishing his calculations as a simple illustration of his theory, Fourier gave considerable credit to the hypotheses of the protosolar nebula and primordial heat, but still he did not get involved in the debates that divided the geologists. In England, however, the chronological implications of terrestrial heat were still to be exploited. And meanwhile, the sun took the central place that was proper for it in the discussion.

Solar Exploration

With the expeditions of Richer, Halley, and Maupertuis, the exploration of the world had ceased to be the domain of adventurers, merchants, and missionaries. All through the eighteenth century, nourished by Carl von Linné's classifications and Buffon's grand descriptions, naturalists continued to dream of the marvels offered by distant seas and savage islands. In order to describe the strange curiosities of the prolific Nature, scientists embarked on journeys with famous captains, such as Cook and Bougainville. A project of astronomical observations to be effected in the Pacific, sponsored by the Royal Society of London, was the stimulus for Cook's first voyage in 1768. Adventure had long ceased to be foreign to the scientific world when Napoleon dispatched his troop of scientists into Egypt.

When the authority of science seemed to have no further boundaries, yet another vision of nature had appeared. Resuming J. J. Rousseau's quest for a simple and genuine society, Jacques-Henri Bernardin de Saint-Pierre (1737–1814), an engineer of bridges and roads, inaugurated the era of exoticism. In 1784 the five volumes of his *Studies of Nature* reinvented the world as a place governed by a Providence attentive to harmony and to sentiments, one that revealed enchanting tableaux charged with the picturesque. A naturalist such as Cuvier could not remain indifferent to this picture, and his success flourished in a scene that had been won over by romanticism. The Mosaic account had found an old-style bard in François-Auguste-René de Chateaubriand (1768–1848). "God might have created, and doubtless did create, the world with all the marks of antiquity and completeness which it now exhibits," he wrote in 1802 in *The Genius of Christianity*. In his poetic eyes, any signs of ancientness deciphered by naturalists were but the mark of the divine fabric: "Without this original antiquity, there would have been neither beauty nor magnificence in the work of the Almighty; and—what could not possibly be the case—nature, in a state of innocence, would have been less charming than she is in her present degenerate condition." Chateaubriand continued, "Man, the lord of the earth, was ushered into life with the maturity of thirty years, that the majesty of his being might accord with the antique grandeur of his new empire. And in like manner his partner, doubtless, shone in all the blooming graces of female beauty when she was formed from Adam, that she might be in unison with the flowers and the birds, with innocence and love, and with all the youthful part of the universe."

There was no need, though, of innocence, or of flowers, or of the Indian Ocean in order to discover exoticism. Right in Europe, the mountains and their grandiose sites continued to be terra incognita. By the end of the eighteenth century, people had begun climbing them or bowing before the majesty of their glaciers. For geologists and physicists alike, the mountains became a prodigious natural laboratory, in which the forces of nature revealed themselves with extreme violence. Driven by Saussure, meteorology was constituted as a branch of experimental physics: measuring the rates of moisture in the air or the speeds of the winds became simpler with hygrometers and anemometers, two of the numerous instruments he perfected. The lowering of temperatures at mountain altitudes was intriguing, presenting in a new manner the old problem of the effects of the sun on the climate. Because heat was better known, the question was posed in terms of quantifying the sun's radiance as it emitted heat around itself, and how the atmosphere acted as a filter for the heat that came to the earth. Exposing thermometers to the solar rays, as Saussure had done, did not give the right answers. What was necessary was to measure not elevations of temperatures but the actual heat radiated by the sun.

It was to this task that Claude Pouillet (1790–1868) applied himself. A former student of the École Normale Supérieure, he was a professor at the Faculty of Sciences in Paris and administrator of the Conservatoire National des Arts et Métiers (National Conservatory of Arts and Trades). He was close to King Louis-Philippe, to whose children he taught physics. Pouillet conducted research with his *pyrheliometer*, an apparatus that concentrated the sun's light to heat a quantity of water under well-controlled conditions. The flux of solar heat was then simply obtained from the observed heating and the time elapsed during the measurement. By comparing the values obtained for different heights of the sun in the sky, Pouillet was even able to determine how much light was absorbed by the atmosphere as a function of its cloud cover. There were many problems he was able to resolve in this manner: the "total quantity of heat emitted at each instant by the entire surface of the sun"; the "cause of the cooling at the higher regions of the air"; and the temperature out in space were but a few of them.

The results published in 1838 of several years of measurements indicated to Pouillet that the transmission of the sun's heat to the surface of the earth depends on two factors. The first varies from day to day and is characterized as the "clearness" of the atmosphere. The second is the *solar constant*, which is invariable and represents the intrinsic calorific power of the sun. Pouillet measured a constant flux of 1.76 calories per square centimeter per minute at the earth's surface (a value only 10% less than our modern measurement). Because such a bald result had no appeal to the imagination, Pouillet offered a useful comparison: he calculated that in one year the sun's radiation at the earth's surface would melt a layer of ice thirty-one meters thick. So it is indeed the sun that maintains life, as Fourier had found, because the flux of the earth's deep primordial heat is completely negligible. But the earth is a minuscule point lost in the immensity of space: it intercepts only an infinitesimal portion of the heat emitted by the sun, less than one-half-billionth. In just one day, our star would melt a seventeen-kilometer-thick layer of ice surrounding itself. The sun is indeed a blast furnace, and this conclusion was all the more convincing because it was independent of any conjecture regarding the constitution of the stars.

After this information appeared, it also became interesting to know whether the sun possessed "a source for its production of heat that would compensate in some manner, whether through chemical, electrical, or other reactions, for the losses due to its constant radiation of heat, because if these losses continue ceaselessly without compensation, then what would result would be a progressive diminishing, century after century, of the temperature, and this would be perceived on the terrestrial globe." Unfortunately the absence of knowledge about the pertinent parameter values, such as the thermal conductivity or specific heat of the sun, precluded any sort of reasonable discussion about the rates of cooling. These values demonstrated nothing more than the "limit of the uncertainties to which science is condemned," remarked Pouillet regretfully, perhaps too pessimistically. The question, still, would take on a considerable importance with the first law of thermodynamics.

The Stars and the Yoke of Thermodynamics

Before the notion of energy was established, the existence of the sun and the stars did not raise too many problems. Zeno of Elea (ca. 485 BC to ca. 430 BC), for example, who was famous for his paradoxes on the impossibility of movement, had noted that the sun would burn for as long as it had combustible material, but no limits were actually fixed for the quantity of heat a star could release. In the years immediately following Pouillet's works, the principle of conservation of energy brought about an abrupt change of perspective: the source of these inextinguishable flows of energy needed to be identified. The "first" law of thermodynamics, which was formulated after the "second," had a long and confused gestation; some dozen persons could have claimed to have formulated bits and pieces of it. One of its nebulous points of departure was the harmony that one naturally imagined as presiding over the material world: this harmony implied that the various known forces flowed from the same original cause, a postulate that still smacked more of metaphysics or aesthetics than of experimental physics. Until the end of the eighteenth century, the mutual transformations of various sorts of forces remained circumscribed within narrow frameworks. From this point of view, they did not evoke any particular interest. For example, the production of an electrical force by rubbing a cat's fur against a glass stem was only one of the numerous facets, and not the most remarkable one, of electricity.

The situation changed at the beginning of the nineteenth century when the repertory of known and cataloged interactions among various forces expanded rapidly into a vast gamut of original effects. It was observed, for example, that an electrical current produced heat when it passed through a conductor. A current also caused the needle of a compass to move, and it permitted the decomposition of water into hydrogen and oxygen. New terms such as *thermoelectric, electromagnetic,* and *electrochemical* enriched the physicist's jargon. Heat seemed to play a pivotal role: it appeared each time a force was put into play, such as during the displacement of electrical charges in conductors. This principle was already known for mechanical forces, because the friction between two bodies is always accompanied by some heat. And even vital forces paid tribute: cold was obviously a characteristic of death, as heat was of life. "Respiration is a combustion, actually a very slow one, but otherwise perfectly similar to that of coal," Lavoisier and Laplace had explained in 1783, after describing the consumption of oxygen and the release of water and carbon dioxide by living beings.

Conversely, the connecting rods of steam engines ceaselessly evinced, in the form of mechanical work, the heat that was released by the combustion of coal. In transforming work into heat and heat into work, the circuit was closed, and thus one had to conclude that work and heat are two different manifestations of the same agent. This agent was called *energy*, imported for the first time in 1807 as a term for the particular case of *vis viva*, by the English doctor, physicist, and Egyptologist Thomas Young (1773–1829). It was not enough to assert the possibility of such transformations, though. It was necessary to evaluate the factors of reciprocal conversion before coming to predicate that, just like matter, energy is neither created nor destroyed: it can only be transformed.

On this basis, four different authors have been recognized for the first law. Julius Robert Mayer (1814–1878), a doctor practicing in his native town of Heilbronn, in Würtemberg, declared that "Nature, science, and religion form an eternal union," but he did not have the satisfaction of seeing his merits recognized until the end of his life, which he passed at the fringes of the university system. While he was serving on a ship between 1840 and 1841, he noticed the lively color of a wounded sailor's blood in Java, and it caused him to suppose, through a bold association of ideas, some connection between the hot climate and a reduced metabolism and then between the chemical force of nutrition and the "combustion of the blood," on the one hand, and the heat and the mechanical work produced, on the other.

Less well known, the Danish engineer August Colding (1815–1888) became the inspector of roads and bridges for the city of Copenhagen. In 1843 he was inspired by spiritual considerations: for him, the immortality of the soul implied that the forces of nature were imperishable.

Of these four, only the Englishman James Prescott Joule (1818–1889) started out as a physicist. His measurements by various ingenious means of the mechanical equivalent of heat were begun in 1843 after he discovered the calorific effects of electrical current. His name has therefore been given to the modern unit of energy.

Hermann von Helmholtz (1821–1894) had wanted to study physics, but his father pushed him toward medicine. He was the regimental surgeon in Potsdam after his studies; however, he was actually best suited for the research environment. His point of departure in 1847 was a thoroughly solid analysis of the impossibility of perpetual motion. In order to communicate his notions to a large public, he compared work to money, explaining that to achieve perpetual movement would be tantamount to creating money out of nothing. Helmholtz was one of the last great dabblers in science. He was a genius, and he continued on after medicine to pursue a prestigious scientific career in the fields of his vast range of interest, including physiology, physics, and epistemology.

Curiously, it was Mayer, the doctor, who first perceived the astronomical implications of the first law. A striking comparison recapitulates the gravity of the enigma that arose: in one hour, one square centimeter of the sun's surface emits the same amount of heat as is released by the combustion of a ton of coal. Even if it were entirely constituted of the best terrestrial fuels, the sun would need less than five thousand years to consume itself entirely. But the history of humankind obviously testified that the intensity of the sun's radiation had not appreciably varied over such a period. One obvious conclusion was inevitable, that the sun did not draw its energy from chemical reactions, even if the source of this profusion of energy still remained unclear. After raising the question, the first law also led the way to its resolution. Based on the contemporary theory of the protosolar nebula, it appeared to Mayer that only the transformation of kinetic energy into heat could replace combustion as a source of energy for the stars. At the end of their long trajectory through space, small bodies or particles that had accelerated under the effects of gravitational attraction transformed their kinetic energy into heat in the act of attaching themselves to a star. This mechanism was remarkably efficient: a piece of coal falling from the earth to the sun would have a kinetic energy six thousand times higher than its energy of combustion.

Remembering that Kepler had said "there are more comets in the heavens than fish in the ocean," Mayer imagined in 1846 that a continuous flow of asteroids had maintained the solar heat since time immemorial. He even found it possible to assert that these asteroids were already in close orbit around the star; otherwise the mass of the solar system and thus the periods of the planetary revolutions would vary in accordance with the changing supply of material from the outside. Unfortunately, these ideas received only the most lukewarm welcome. They were far too new, and clarity of exposition was not one of the cardinal virtues of their author. Nevertheless, two observations received a simple explanation from Mayer: sunspots were the marks of the flux of asteroids, and the zodiacal light, that nebulous light of vast dimensions crowning the sun, was due to the asteroids concentrating at the center of the solar system. Mayer printed a more complete version of his dissertation *On the Production of the Sun's Light and Heat*. It ended up being produced at his own expense, because the Academy of Sciences—Arago and the mathematician Augustin Cauchy were commissioners for the occasion—did not want to publish it in Paris. Mayer had been suffering, in addition, from the political troubles of 1848 and from the total lack of interest aroused by his ideas, and then by disputes over priorities that put him into opposition against Joule regarding the subject of the first law. He had a serious depression punctuated by repeated confinements and attempted suicide in 1850.

Mayer's proposal that a continuing supply of material over the years maintained the sun's heat still presented its difficulties. An ingenious mechanism was proposed by Helmholtz in 1854 to resolve them: after the initial period of condensation, it was the contraction due to the progressive cooling of the sun's enormous mass that maintained its radiation. According to Helmholtz, "if the diameter of the sun were diminished only the ten-thousandth part of its present length, by this act a sufficient quantity of heat would be generated to cover the total emission for 2100 years." In this form, and as presented by Helmholtz, the solar theory was less shocking. One deduced from it that the sun could work from 40 million to 100 million years at the cost of only one degree of cooling per period of 7 to 7,000 years. Still, these ranges were rather large, owing to the uncertainties affecting the temperatures and the solar thermal parameters. It was also noted that the astonishingly sudden appearances and disappearances of stars, the novas, were now explainable: the collision of two great celestial fireballs could give rise to a star that would expire as soon as the heat produced by the shock had been radiated. By means of the first law, astrophysics thus owed to the steam engine a portion of the solid bases upon which it was built.

At the same time, the earth was introduced into the discussion by a young British physicist. William Thomson (1824–1907), who was the first to understand Joule, had just rigorously formulated the two laws of thermodynamics and had generalized the use of the word *energy* within the context of physics. In resuming one of Buffon's conclusions without realizing it, he declared in 1852 that "within a finite period of time past the earth must have been, and within a finite period of time to come the earth must again be, unfit for the habitation of man as at present conditions, unless operations have been, or are to be performed, which are impossible under the laws to which the known operations going on at present in the material world are subject." Although the geologists did not realize it, this seemingly innocent remark was like dropping a bombshell. And the act was hardly unwarranted, because Thomson had always considered our planet as an object worthy of attention. "For myself," he said later, "I am anxious to be regarded by geologists not as a mere passer-by, but as one constantly interested in their grand subject, and anxious in any way, however slight, to assist them in their search for truth." In order to establish this truth, William Thomson, better known nowadays by the name Kelvin, was going to make thermodynamics into an unlikely machine at the service of geology, and his doing so would be to the dissatisfaction of more than one geologist.

From William Thomson to Lord Kelvin

"Ego, Gulielmus Thomson, B. A., physicus professor in hac Academia designatus, promitto sancteque polliceor me in munere mihi demandato studiose fideliterque versaturum."[1] These vows that were never made by Fourier were solemnly pronounced by Thomson at the age of twenty-two. The ceremony took place on 13 October 1846 at the ancient University of Glasgow. It was one of the last stages leading to his nomination as professor of natural philosophy: the new graduate had just sworn his allegiance to the Scottish church and successfully defended a dissertation titled "De caloris distributione per terrae corpus" ("On the Distribution of Heat throughout the Body of the Earth").

William's father was a mathematics professor at the same university and played an important role in this nomination. For the elder Thomson it was the happy ending of an unflagging commitment and a result of the endeavors that he had pursued since leaving Belfast, where his children were born, to settle in Glasgow. Ambitious, hard-working, and fond of innovation, he had studied theology and medicine before turning to a successful career in teaching. One of the books he wrote, on arithmetic, was published in seventy-two editions. William, who was number five out of six children, lost his mother when he was six, shortly after his younger sibling was born. William was quick-witted and handsome and became the favorite child of his father, though the father was wise enough not to allow this preference to spark any sort of jealousies. Papa Thomson, who

1. "I, William Thomson, bachelor of arts, designated as professor of physics in this Academy, promise and affirm religiously that I shall apply myself studiously and faithfully to the function being conferred upon me."

was more ambitious for his children than for himself, served as their private tutor, taking care to ensure that nothing would be lacking in their solid education. The earliest years were passed at home. It was common practice in Glasgow to enter the university at the age of fourteen, but William and his older brother James, who also became a physicist, both began their university studies at the age of ten. And the event was not premature: William lost no time in landing his share of prizes in Latin, logic, and astronomy.

At the University of Glasgow at the beginning of the century, scientific instruction was based on the work of the great Parisian scientists. William was already immersed in the works of Lagrange, Legendre, and Laplace when he fell under the charm of John Pringle Nichol, a young astronomy professor who added Fourier and Augustin Fresnel to the list of worthy authors. For Nichol, the aim of education was "to draw man out into freedom and to establish a solid, practical harmony between him and the universe." In his later days, Thomson would say of him that "in his lectures the creative imagination of the poet impressed youthful minds in a way that no amount of learning, no amount of mathematical skill alone, no amount of knowledge in science, could possibly have produced." Nichol awakened in Thomson a remarkable intuition, enriched by a distaste for hollow speculation. He also protected him from invasive paternal pressure. William had found a domain in which he was able to develop complementary qualities that are rarely found together within the same person. He "was a great mathematician," said J. J. Thomson (not related to him) later, "and he used his physics to guide his mathematics and his mathematics to give precision to his physics." Another famous physicist, Ernest Rutherford, commented that William Thomson "combined to an extraordinary degree the quality of great theoretical insight with the power to realize his ideas in a practical form."

Thomson began displaying these qualities precociously. Under the direction of Nichol, he won the university prize at the age of sixteen for his "Essay on the Figure of the Earth," a dissertation that he would continue annotating episodically until his death. Shortly before winning the prize, Thomson had asked his master whether reading Fourier's *Analytical Theory of Heat* might be of some profit to him. Nichol, who admitted that he had not completely understood that work himself, encouraged him to do so. What an illumination it was! Thomson read it within fifteen days, even if he did not understand it entirely. According to H. I. Sharlin, one of his recent biographers, he found in it "what the telescope is to astronomers—a means of perceiving the physical world." And he quickly made good use of this telescope. Inspired by Fourier's theory, the adolescent published his first works in the following semesters. Beginning with plain proofs for theorems that had not been demonstrated by Fourier, he continued on to his own original applications for the analytical theory of heat. Of particular interest was his demonstration that the methods used by Fourier for the problem of heat conduction could be applied equally well to electrical problems. Without even realizing it, he had thus revealed that there is a deep analogy between temperature and electrical potential, even though by appearance the two phenomena have little in common. He was making a fine start.

In October 1841 Thomson went to Cambridge to prepare for the Tripos, the competitive examination concluding a cycle of three years of study that attracted the best among British students. His publications had provided him an unusual renown, so that he was the favorite in the competition from the outset. The candidates were trained—somewhat like racehorses before the course, including bets on the winners—by coaches whose reputations had been assured by the successes of their previous trainees. Thomson was admitted by the best known among those coaches, William Hopkins (1793–1866). Hopkins was a former gentleman farmer who had come late in life to physics, after being widowed and suffering his share of agricultural disappointments. As coincidence would have it, he and Thomson shared a solid interest in the earth. For about a decade, Hopkins had been applying his attention to providing physical bases for geology that would be as well-defined as those of astronomy. He was not always fortunate in his conclusions: once he demonstrated that rocks could not be transported by glaciers. Hopkins found better success in researching the earth's internal constitution. When Thomson arrived, Hopkins had just finished a long study of the precession of the equinoxes. The magnitude of the equatorial bulging owing to centrifugal force is determined by the nature of the globe's interior, and it would be much greater for a liquid body than for a solid body. In 1841 Hopkins presented the relation between precession, equatorial bulge, and the rigidity of the earth, in long equations and with adequate parameters; his findings indicated to him that "the minimum thickness of the crust of the globe which can be deemed consistent with the observed amount of precession cannot be less than one-fourth or one-fifth of the earth's radius." His new pupil, as an expert, deemed this use of the rotational movements for sounding the interiors of the earth to be a great idea.

Thomson finally discovered a modicum of social life in Cambridge, far from his family. Though he was not living extravagantly, he still had to request some money now and again of his father, putting a strain the family budget. He was not insensitive to the company of young ladies or to the honor of winning a regatta, and some evenings cornet music distracted him somewhat from his arid problems. But he never sacrificed his work. For the Tripos, even more than for the entrance examination to the French Grandes Écoles, speed of execution was more important than originality in physics; Thomson thus had the occasion to develop his mathematical talents. He also published a dozen or so original theoretical works during this time; that is to say, he did not forget to exercise his imagination. The outcome was disappointing, though: the favorite finished second. The reason was that, whereas the other candidates simply responded to the questions, he had submitted dissertations, as Hopkins explained after the test.

Thomson left for Paris in 1845 to finish preparing for his candidature for a university position. On a previous stay, he had been there with his family, but this time he established solid relations in the scientific circles. At the Collège de France, where he discovered experimental physics, Victor Regnault taught him "a faultless technique, a love of precision in all things, and the highest virtue of the experimenter—patience." And he was initiated into the results of the "immortal" Sadi Carnot, which he found "so practical, so far beyond the theoretical" that he placed them at the core of his own work in thermodynamics. His sojourn of less than five months completed his early readings and caused him to recognize a debt: France "truly is the *Alma mater* of my scientific youth, and the inspiration for my admiration for the beauty of science, which has led and guided me all throughout my career," Thomson admitted a half-century later, in French. In 1845 his career was on the verge of taking flight. During the months following his return to Great Britain, the chair of natural philosophy at the University of Glasgow was declared vacant. The Thomsons, both father and son, were ready and eager. William obtained the position and held it for the rest of his life.

An evaluation of Thomson's scientific life could be summarized by two large numbers: 70 patents and 659 articles and communications on various subjects, all from his own pen, published between 1841 and 1908—not counting the treatises and other books in which his papers were reprinted. He had no patience for wasting time, for idleness, or for negligence, the three sins Buffon scorned before him. Through his industrial activities Thomson acquired his renown before the public at large. He began to develop these in 1855, when the Atlantic Telegraph Company called upon him. With the aid of Fourier's methods, he founded signal theory and established the laws of the transmission of electrical signals in coaxial conductor cables. Furthermore, he met the need for detecting such signals by inventing the mirror-galvanometer for measuring low-intensity electric currents. This was an extremely sensitive instrument that also found many uses in laboratories. Not all of Thomson's suggestions were followed systematically, however. The first cable was laid in 1858 between Ireland and Newfoundland, and it operated only in a sporadic manner for three weeks. The famous message of ninety-nine words from Queen Victoria took sixteen and a half hours to transmit! After important new capital had been collected, Thomson was given entire responsibility for the technical aspects of a new project. A second cable was laid in 1865, and its success marked a brilliant triumph and consolidated Thomson's fortune. More impressed by the capitalistic success than by his theoretical physics, the English government made him Sir William Thomson.

Thomson's inventive fervor quickly extended to other applications of electricity and magnetism. It was necessary to produce, regulate, measure, record, and integrate all the electrical currents whose applications were multiplying. Innumerable problems arose in relation to measurements and normalization. Thomson had a passion for measuring and economizing. He invented apparatuses; he patented them; he had them manufactured in series at Glasgow; he pleaded continually for optimal exploitation of resources, whatever their nature. In addition to his numerous journeys related to university activities, he also traveled for his lucrative industrial occupations. He remained in contact with his laboratory via mail and telegraph, taking time in the trains between Glasgow and London to reflect on the physics problems he was studying. Wherever Thomson went, his most faithful confidant was the big green notebook he was constantly flashing out of his pocket. Because he did not care much for reading scientific literature, he relied on his conversations with friends to learn of the latest discoveries. Because of this habit, on occasion he ended up duplicating others' findings.

Thomson gave priority to mechanical analogies in his theoretical work. He tried his hand at representing atoms as vortices of a fluid whose permanence was accounted for by the laws of hydrodynamics. In the course of time, such conceptions were rejected by new schools that were oriented more toward abstraction. In spite of his diminishing influence on physics toward the end of his life, Thomson became the emblematic figure of English science. Staggering under the weight of his distinctions and medals, he found himself transformed, as it were, into a living statue. Some twenty universities had recognized him with the title of doctor honoris causa. And—from Rotterdam to Moscow, from the Batavian Experimental Physics Society to the Imperial Society of the Friends of the Natural Sciences of Anthropology and Ethnology—it seemed that few

Figure 7.2 Kelvin and his green notebook. From R. Strutt, "Some Reminiscences," 218.

scientific societies had failed to make him a corresponding or honorary member. He was appointed grand officer of the French Legion of Honor in 1889 and created baron three years afterward by Queen Victoria; Sir William Thomson became Lord Kelvin of Largs, honoring a branch of the Clyde River that meandered near his university, as well as the town where he had built a manor by the seashore. Kelvin, as we shall call him henceforth, never slowed in his activities and continued to maintain his simple manners.

In 1896 Kelvin celebrated his fiftieth year of professorship at Glasgow, the only university position he had ever had. Between 15 and 17 June, twenty-five hundred persons gathered for his jubilee. To celebrate his genius, there were technical expositions, conferences, discourses, and congratulations (by way of telegraph, of course). At the end of the nineteenth century, which was so fond of metaphor, homage was offered him repeatedly. "The entire scientific world takes its part in the honor that Lord Kelvin has rendered unto the human race. The light shining today upon the University of Glasgow is like that of the sun. It does not belong only to one country; it extends over all of humanity," proclaimed the Italian ambassador. Only Louis Pasteur (1822–1895), his elder by two years, had been made the object of a similar cult: "Marvel of Science, miracle of genius . . . Who can now say what human life owes you, or what it will owe you in the times to come?" the minister of public instruction had declared of him in 1892; "All of France glorifies you. Humanity blesses you." Along with Pasteur, Kelvin formed a diptych of a new species: the biologist and the physicist, each in his own manner, incarnated the triumph of science, from which nothing would any longer be able to escape.

Uniformitarianism Refuted

In a word, Kelvin already commanded an uncontested authority in physics in 1862, when he campaigned against the predominant geological conceptions. He was animated by a practical mind and was totally unfamiliar with metaphysical anxieties: the world was not to be reinvented; it was to be discovered, and it manifested order because it obeyed laws, the laws of physics. For Kelvin, there was no conflict here with his religious ideas, which were simple, free from dogmatism, and based on his unshakable faith in the existence of a creative design. Without such design neither life nor the universe would have any meaning or would behave consistently in accordance with universal laws. Certainly, the initial moment of the Creation escaped science, but the power of the scientist arose from knowing how to place strict limits upon his research. It was thus possible to decipher the history of the earth and the sun, and in this project, no concepts were pertinent other than those that could be verified by experience. Outside the limits of this framework, there was no science at all. All through Kelvin's career, the need to submit natural phenomena to quantitative methods provable by physics returned as a leitmotiv. Without quantification, science would be nothing but a modernized metamorphosis of Scholasticism.

For cosmology, the newborn thermodynamics was revealing itself as a tyrannical master. The strict rules imposed upon the sun by the first law of thermodynamics were not enough; the second was soon to prove just as demanding. Until then, the laws of physics had been symmetrical in relation to time: one could, for example, just change the sign of the time parameter in an equation for the movement of a body, and the fall of the object would become an ascent. By introducing the notion of irreversibility, the second law of thermodynamics had broken with this symmetry and accounted for the impossibility of any return to the past. With entropy, the intrinsic irreversibility of spontaneous phenomena even had its gauge. The German physicist who discovered the second law, Rudolph Clausius (1822–1888), formulated the rules imposed upon the universe since its creation in two short articles: "1. Energy is constant; 2. Entropy tends toward a maximum." It was this inexorable increase of entropy that gave history a sense of direction, though geology had not really taken measure of it yet.

Although they prided themselves on thermodynamics, geologists took very little notice of its laws. As will be seen in chapter 8, the temperatures measured in the mines were boldly extrapolated to give nearly 200,000°C for the temperature at the center of the earth. In spite of Fourier's works, the earth appeared to be formed of an immense ocean of incandescent lavas, surmounted by a solid crust only some fifty kilometers deep. Yet the diminishing of the geothermal gradient at the depths had been signaled by Fourier himself: "We would be quite wrong if we supposed that this increase [in temperature] had the same values as for great distances: it diminishes, certainly, in proportion as one goes farther from the surface." Furthermore, to prolong the temperature profile that had been obtained for the solid crust within an immense, moving, liquid mass was just plain absurd. And so, the estimations of internal temperatures had basically no foundation whatsoever.

As for the terrestrial heat flux, its well-established importance required finding an explanation for it. In 1853 Lyell, the herald of uniformitarianism, still attributed its source to chemical reactions produced in the center of the earth. He believed that this flux was able to create thermoelectric currents that ingeniously proceeded to dissociate the deep material by electrolysis, allowing it to recombine and then release new heat. Completely innocent of the consequences of his theory, Lyell thus had postulated the existence of perpetual motion at the interior of the earth. The first law was ten years old at the time; such a flagrant violation of it justified Kelvin's ire, and fifteen years later, Lyell still despaired at not being able to find the "divine artificer" that had commanded such perpetual movements.

To Kelvin, it was evident that the earth had to be considered as a thermal machine. Because solar heat only acts in a superficial manner, there was no need to take it into account here. The only motor for the terrestrial dynamic is the deep heat that is conducted from the interior (hot) toward the surface (cold) of the planet. As Carnot had explained, mechanical work could be produced by such differences in temperature: the explosive force of a volcano was the most spectacular manifestation of this fact. But as volcanoes showed equally well, such mechanical energy is, after all, dissipated in the form of heat. Even if it does so slowly, the earth always evacuates its heat through the surface. The day would arrive when the diminishing temperature of the globe would be uniformly equal to that of space; without a difference in temperature that could be exploited to produce mechanical energy, the earth would be condemned to eternal repose.

Uniformitarianism thus appeared unacceptable to Kelvin. Independently of Lyell's fanciful explanations, it presupposed a perpetual movement with cycles that had recurred without change since the dawn of time. Geological phenomena, on the contrary, are essentially irreversible. As the provision of primitive heat was being exhausted, the intensity of geological action could only diminish. Thermodynamics left no doubt: in contrast to what Hutton had proclaimed, the earth had a beginning and it would have an end. From 1862 on, Kelvin thus denied the sort of eternity inspired in geologists by the immensity of the terranes that were slowly transported from the depths of the seas to the summits of the mountains. It was not a new idea for him: "For eighteen years it has pressed on my mind that essential principles of Thermo-dynamics have been overlooked by those geologists who uncompromisingly oppose all paroxysmal hypotheses," he began his first official attack against uniformitarianism.

The Concordance of Times

Influenced by Fourier's ideas, Kelvin thought the cooling of the earth and the stars had initially been more rapid than at present. When the two laws were established, what had only been sentiment in 1844 became certitude, mentioned several times in the conclusions of his publications on thermodynamics. However, very concrete arguments were required in order to gain the attention of geologists. A first incursion into their camp occurred in 1855 when, subsequent to his interest in measurement of heat flow, Kelvin attempted to fix "absolute dates" for the volcanic intrusions, in accordance with their residual thermal effects. Seven years later, he broadened his consideration to the earth itself and the sun in a series of publications that marked the beginning of one of the most celebrated scientific controversies.

Kelvin did not intend to leave any space for doubt. He resumed the question of solar dynamics immediately. Without realizing it, he had agreed with Mayer completely in 1852 in supposing that the sun was nourished by dust particles that caused the zodiacal light. Because he had to recognize that the mass of this dust is too small to maintain the sun's energy, Kelvin came around in 1862 to Helmholtz's model of progressive contraction, accepting 10 million years as the lower limit of the time during which the sun could radiate. Offering broad margins, he advanced: "It seems, therefore, on the whole most probable that the sun has not illuminated the earth for 100,000,000 years, and almost certain that it has not done so for 500,000,000 years. As for the future, we say with equal certainty that inhabitants of the earth cannot continue to enjoy the light and heat essential to their life for many million years longer, unless sources now unknown to us are prepared in the great storehouses of creation." The conclusion was irreproachable: it included not only a large margin of error, but above all, a prudent reserve on the subject of energy from "sources now unknown to us." But Kelvin would tend to forget that afterward. With this exception, it thus appeared that the activity of the sun could only have diminished over the course of time.

After covering his bases, Kelvin followed Fourier by calculating the time for the cooling of a globe formed at high temperatures. He needed to establish beforehand that the earth was essentially solid, and not liquid, or Fourier's law would not be directly applicable. In examining more closely Hopkins's theory that predicated the existence of a thick rigid crust, Kelvin realized that it was mathematically incorrect. The respect he held for his old master, and the desire not to weaken his own position, inspired him to make a careful understatement, atypical from his pen: "Although the mathematical part of the investigation might be objected to, I have not been able to perceive any force in the arguments by which this conclusion has been controverted." In order to demonstrate this conclusion properly, Kelvin became interested in a much simpler effect, that of the rigidity of the earth's crust on the amplitudes of the inner tides.

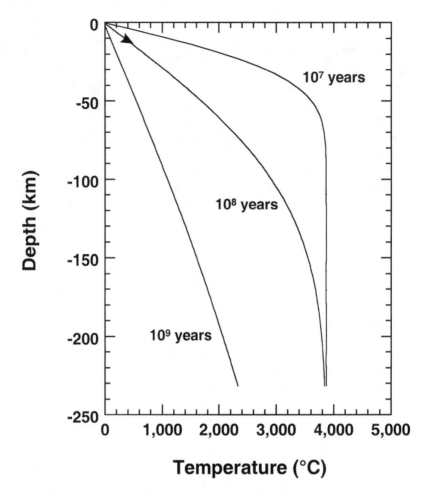

Figure 7.3 The age of the earth, determined according to a profile of temperatures at depth, calculated here for three different ages. With very little sensitivity, the date of the *consistentior status* is given by the time (about 100 million years) after which the temperature variation near the surface (*arrow*) attains the mean geothermal gradient of 37°C/kilometer. Note that even after 1 billion years, the temperature has hardly varied at depths greater than 900 kilometers. Initial temperature: 3,870°C; surface temperature: −18°C.

If the earth were essentially liquid, the masses of lava displaced by the attractions of the moon and the sun would be enormous, and the thin solid crust would be tossed about at the mercy of these deep inner tides. If such a thin crust were also able to deform itself freely, the waters of the seas would move along with it, so that there would be no sort of oceanic tides. Obviously, neither of these two effects had been observed.

On the contrary, the earth is essentially solid, and its "effective rigidity is at least as great as that of steel," Kelvin declared on the basis of one very simple calculation.

In order to determine the age of the earth, Kelvin postulated that our planet was initially a solid body brought to a uniformly elevated temperature. If so, he only needed to use Fourier's law and find the time after which the geothermal gradient calculated for the surface would have become equal to the average value observed, namely, 37°C per kilometer. As a result, Kelvin could "with much probability say that the consolidation cannot have taken place less than 20,000,000 years ago, or we should have more underground heat than we actually have, nor more than 400,000,000 years ago, or we should not have so much as the least observed underground increment of temperature." In accordance with an expression that later became dear to him, this age represented the *consistentior status* of Leibniz,[2] the moment at which a surface crust had been formed that was "fitted as an abode for life." Kelvin had not omitted from this capital cosmological problem a discussion of the validity of his hypotheses and the precision of the physical constants retained. The vast interval announced, of 20 million to 400 million years, attested not only to this caution, but above all to a remarkable consistency with the age of the sun that had been obtained by a completely independent method. In lending themselves so well to quantitative description, the formation and the evolution of the solar system boded well for Kelvin's program.

2. Leibniz wrote in his *Protogæa*: "Donec quiescentibus causis atque aequilibratis, consistentior emergeret status rerum" (When the perturbing causes have been exhausted and equilibrated, a more stable state will finally be produced). *Protogée*, 23. Among the various scenarios that he considered to describe the solidification of the earth, Kelvin retained the one in which crystallization was the cause that produced rocks, which, because they were more dense than the lavas, sank and accumulated at the depths. Owing to their agitation, maintained by the descending movements of these solids, the lavas kept a uniform temperature all throughout the crystallization, which progressed, therefore, through the surface of the globe. The sun would follow an analogous evolution, but by reason of its very great mass, it is only at the initial phase of complete fluidity now.

EIGHT

The Elasticity of Time

HALF A CENTURY OF CONTROVERSY   After Kelvin's foray into geology, animated controversies continued throughout the nineteenth century, mostly in England, in attempts to measure geological time using many different methods. Another fundamental problem was raised by the evolution theories of Darwin and Wallace, concerning the amount of time that was required to produce higher forms of life.

How to Convince?

The debate launched by Kelvin lasted more than a half century. A new feature of this controversy, compared with earlier ones, was that it could make use of the new forms of communication that had been invented during the first half of the nineteenth century. The new ideas were able to circulate without impediment, just as people could, owing to the long period of peace that followed the Napoleonic wars. It was no longer the era of scientific booklets that had to be printed at the author's expense, or of voluminous treatises drafted in the isolation of the scholar's study. The old practices had become inadequate for dealing with the rapid expansion and the new professionalism of science. Without waiting to resolve their questions completely, scientists published their interim findings in stages—a new method exemplified by Kelvin's 659 publications. There were no longer the delays of four to six years that had been required by the Royal Society of London and the Paris Academy of Sciences. Numerous modern scientific journals came into being by the end of the eighteenth century. People watched for them to appear; they admired the new intelligence; they secretly hoped to find such inspiration themselves—and they dreaded the last word of some rival. Through these journals, which their peers could not ignore, authors relentlessly seeking recognition found a way to retain at least some of their rights to fame.

In the same movement, scientific societies began to multiply. The maturing of geology into a discipline was marked by the creation of the Geological Society of England in 1807, twenty-three years before its French equivalent. In addition to the specialized societies, others targeted a much broader public, with the aim of promoting science in all forms, such as the British Association for the Advancement of Science, founded in 1831. Because popularization was one of their primary purposes, such societies sponsored lectures to which the highest of authorities lent their support by exhibiting the latest advances of science. Among the general public, geology enjoyed a remarkable vogue. "What is the cause for this powerful interest inspired by a science that previously was regarded as a simple study of sterile curiosity?" Nérée Boubée (1806–1863), a professor and popularizer, wondered. He gave his own prompt response: "It is because geology is finally being appreciated properly; because it has been recognized as being among the first in philosophical sciences." If "geology is the necessary introduction to history in itself," the reason is that it reveals "the history of the world since its creation until the appearance of humanity upon the earth." And above all, continued Boubée, the interrelations among archaeology, astronomy, and moral and theological studies, between natural and political or industrial and agricultural sciences, or even among economics, hygiene, and medicine, demonstrate that "geology is thus very obviously the first of the philosophical sciences today; it is for our century what *universal philosophy* was for the times of Plato and Aristotle, summarizing all of human knowledge; it is the necessary complement of all the special studies." Beyond the pompousness of this tribute to geology, it could indeed be seen, in England for example, that there were plenty of curious people ready to pay their entry tickets to improve their minds with the most recent discoveries and ready to admire the latest geological theories.

The new societies brought their members together in conferences wherever the railways allowed people passage. Year after year, scientists repeatedly connected with their contacts, met new colleagues, and updated their knowledge. And, just as at a sporting event, they saw the power struggles between schools and persons, knowing—blessed times!—that the implacable adversaries of a particular evening would often be the best of friends the following day. If the press became interested in the debates, the impact on the public was assured. Conferences had played a large role for Kelvin from the beginning of his career. It was at the meeting of the British Association that he met Joule and began a productive collaboration with him. As an active member of the Geological Society of Glasgow, of which he was president from 1872 to 1893, he subsequently had many occasions to preach his credo. Whenever a scientist of his stature initiated a debate, he benefited from a particularly efficient forum, since the traditional plenary lecture was given by the president of the inviting society. It would have been difficult to remain indifferent on some of these occasions, such as the day when Kelvin bluntly attacked the geologists from the floor, declaring, "A great reform in geological speculation seems now to become necessary!"

Heat as the Driving Force for Time

A reform was now required because there were simply too many scientists who had decided to ignore the requirements of physics, after the fashion of the great Lyell or of the Scotsman Sir Andrew Ramsay (1814–1891), director of the British Geological Service, who flattered himself by announcing one day in Kelvin's presence that he had no prejudice against a sun that would shine for a trillion years. He concluded the discussion exclaiming, "I am as incapable of estimating and understanding the reasons which you physicists have for limiting geological time as you are incapable of understanding the geological reasons for our unlimited estimates." Thirty years later, this outburst made by the holder of the highest geological post in the Empire still remained engraved in Kelvin's memory.

The question of time was therefore at the core of the reform that was required. In a sense, Kelvin's aim as a geologist was to reinvent linear time, to provide a contrast to the cycles that formed the backdrop of the eternal uniformitarian rebeginnings as well as to the violent, catastrophic convulsions. But rethinking time was not enough. Because geology is a history and there is no account without chronology, the discipline had to be founded on adequate timescales. Without submitting an account to dates, one is not authorized to reconstitute the past or to foretell the future. The earth could not be an exception; the inability to estimate its age precisely would leave "geology in much the same position as that in which English history would be if it were impossible to ascertain whether the battle of Hastings took place 800 years ago, or 800 thousand years ago, or 800 million years ago," muttered Kelvin, driving in the nail.

When Kelvin intervened, an important change in perspective had already been established for cosmology. From Newton's *Principles* to Laplace's *System of the World*, gravity had constituted the model of force that rendered the world intelligible. This force of attraction between masses not only accounted for the dances of the stars; it also constituted the historical motive for the entire universe. In removing from heat its disguise as a chemical element and in bringing it into the forefront of physics, Lavoisier, Watt, Fourier, Carnot, and the Industrial Revolution had thrust Newton's ideas into the shadows: before the first principle of thermodynamics was established, heat, though still enigmatic, had in its own turn begun to dominate cosmology. The sun's heat ordered activity on the earth's surface: the winds, the rains, ice, and even life owed everything to it. Where its influence diminished at modest depths, the primitive heat of the earth took up the relay: it was heat that caused volcanoes to burst forth and mountains to rise, a patient work that erosion, propelled from a distance by the sun, strained just as earnestly to put down.

Well before plate tectonics, heat provided a unified general schema for explaining geological activity. Its importance, asserted previously by Hutton, had to be recognized because of a former member of the Egyptian campaign, Louis Cordier (1777–1861), who taught geology at the National Museum of Natural History. Before becoming a state councillor, the president of the council of mines, and a peer of France—which made his critics remark that several more days ought to have been added to the week in order for him to fulfill his responsibilities—Cordier had established in 1827 that in all places where it had been measured, temperature rose regularly with increasing depths, typically by one degree for every thirty or forty meters. Terrestrial heat flux was not a local anomaly linked to subterranean combustions or pockets of lava, as Lyell had claimed. On the contrary, Cordier declared: "Our experiments fully confirm the existence of an internal heat, which is proper to the terrestrial globe, which has no influence whatsoever from the radiation of the sun, and which increases rapidly with depth." In accordance with the quite recent mathematical theory of heat, those experiments further indicated that "the interior of the earth has a very elevated temperature, which is particular to itself, and which has belonged to it since the origin." Not content to have justified Descartes, Leibniz, and Laplace, Cordier believed that he had settled another essential point, that of the solid or liquid nature of the interior of the earth, by deducing from these geothermal gradients the existence of a solid crust that was some hundred kilometers thick at the most and which rested upon an immense molten mass, at the expense of which it increased slowly. From this solidification owing to the loss of

COUPE DE L'ECORCE TERRESTRE

Figure 8.1 The stacking of terranes upon the "sea of fire," from the primordial granite crust to recent alluvia. The names of the terranes in English are (*from top to bottom*) Alluvial, Diluvial, Pliocene, Miocene, Eocene, Cretaceous, Jurassic, Trias, Permian, Carboniferous, Devonian, Silurian, Cambrian, Primordial source of granites and other igneous rocks, Sea of fire. From With, *L'écorce terrestre*, i.

the primordial heat, there ensued a slow contraction, and it was to this that the earthquakes could be attributed.

In order to avoid this apparent paradox of highly elevated temperatures at the depths, a mathematician who was second only to Fourier in renown advanced a different hypothesis in 1835. For Siméon Denis Poisson (1781–1840), an elegant solution consisted in denying the very notion of primitive heat. According to him, the heat would have been completely evacuated during the time of the earth's crystallization, beginning from its center, so that the flux of heat was nothing but the memory of an epoch during which the solar system had sojourned in a region of space that was much hotter. Such views were radically opposed by Léonce Elie de Beaumont (1798–1874), a highly prolix chief engineer of mining, who established the first complete geological map of France in 1841. For him, the loss of the primitive heat, which led to the cooling and contraction of the earth, was the cause "of the main geological phenomena, and particularly that of the division of the sedimentary terranes into successive formations and mountain systems." Pursuing the program of Werner, and still taking into account the catastrophes described by Cuvier, Elie de Beaumont thus attributed to this contraction the violent shocks that caused the mountain chains to rise along well-defined geometric lines. But while the pentagonal networks of mountains that he discovered all around the globe ended by discrediting his theory, uniformitarianism, as upheld by Lyell, gradually gained in popularity.

As this summary of ideas maintained during the first half of the century indicates, Kelvin was not advancing into virgin territory when he put on the geologist's cap. His adopted discipline was confronted with the formidable problem that physics itself had already resolved, namely, reconciling the first law of thermodynamics with the theoretical framework that governed its various branches. In an effort that has largely been forgotten since then, the foundations of physics had to be rebuilt in order to accommodate the conservation of energy. Within a Newtonian framework, the solar and terrestrial heat eventually appeared to be the last manifestations of the ancient gravitational attractions. After contributing to this grand task of the reconstruction of physics, Kelvin buckled down to geology. Fourier's law had allowed him from the outset to contradict Poisson. If the heating postulated out in space had been moderate, this would have been a very recent situation, and human history would have retained a memory of it. If it had been ancient, it would have to have been very great in order to produce the current heat flux, so great in fact that all forms of life would have been toasted at the surface of the earth. Paleontology, of course, excluded that hypothesis.

As for the thick, rigid crust indicated by Hopkins, the reality of its existence had not actually been established. On the one hand, such a crust supposed that the internal temperature of the earth increased with depth less rapidly than the melting temperature of rocks, which had not been demonstrated. On the other hand, a thin crust of fifty to one hundred kilometers was mechanically stable, and the vents that pierced it permitted lava to rise from the immense ocean of magma. Hopkins's scholarly mathematical theories were thus not really convincing when judged from a volcanic standpoint. But Kelvin believed that the solid or liquid nature of the depths was a matter of major importance. If the earth was largely liquid, it would be animated by powerful convection movements: the deep hot matter, less dense, would rise toward the crust, where it would cool upon contact; its density would then increase, causing it to fall towards the depths, there to take on new heat again. Such convective cycles represented a mode of propagation of heat that was considerably more efficient than the very slow conduction that could be exerted only in a solid. If a significant fraction of the earth was liquid, then the thermal history of the planet would not be subject to Fourier's theory. It is for this reason that the earth's rigidity had been an essential element of Kelvin's coherent—and, as he thought, persuasive—demonstration of the ages of the sun and the earth.

A Provocation: The Time of Evolution

Kelvin's demonstration actually represented a response to what he viewed as a threat. In the first edition of *On the Origin of Species by Means of Natural Selection, or the Preservation of Favoured Races in the Struggle for Life*, published in 1859, Charles Darwin (1809–1882) warned his public: "He who can read Sir Charles Lyell's grand work on the Principles of Geology, which the future historian will recognize as having produced a revolution in natural science, yet does not admit how incomprehensibly vast have been the past periods of time, may at once close this volume." Kelvin may have found this warning an appropriate reason not to read any further. The fact was that he was offended. Because he felt that his theses were threatened more by the new theory of evolution than by the old geological conceptions, he took advantage of an unexpected period of leisure, while recuperating from a broken leg, to send off some blazing backfires with his first series of "geological" publications of 1862. Perhaps without Darwin's provocation, he would not have taken the time to expound his views so completely and thus initiate the controversy.

Darwin had waited even longer before putting his ideas on evolution into form. They had germinated two decades earlier, during a five-year-long expedition in the Pacific. Darwin had been a mediocre student and had interrupted his medical studies, rejecting also his father's advice to become a pastor. "You care for nothing but shooting, dogs, and rat catching, and you will be a disgrace to yourself and all your family," his father had scolded him. He embarked on the *Beagle* in 1831 as an amateur naturalist, with no clear idea of what his future would bring. Darwin was completely disgusted by the Werner-style geology that he had learned during his studies, but nonetheless he was off, with the first volume of Lyell's *Principles of Geology* under his arm. This book was actually his private university: in his travels ranging from the Galápagos Islands to Australia, he used the book to associate theory with practical exercise and to learn the meticulous observation of nature and the strict interpretation of its characters. The notations on volcanoes, metamorphic rocks, atolls, and fossils that he sent back from the antipodes caused this autodidact to be recognized as a talented geologist. Upon his return to England, Darwin was admitted to the Royal Society of London at the age of twenty-nine and ended by becoming the colleague closest to Lyell, even if the latter was not yet entirely in agreement with his theses.

The idea of evolution had imposed itself upon Darwin little by little during the course of his expedition on the *Beagle*. Similarities among living and fossil species, or differences between the species populating the islands and nearby geographic zones, found a simple explanation once one admitted that the different species had descended from common ancestors. In September 1838 Darwin was inspired by the role of natural selection as a driving force for evolution after reading the *Essay on the Principle of Population* by Malthus, from whom he drew the expression "struggle for life." The inevitable shortage of resources that curbed the uncontrolled expansion of human populations was obviously found also in nature: the population of a species, whether animal or plant, could not increase faster than its food supply. In England, where teleology was still revered, such reflections were still too radical to be published. Darwin did not confide them to any but his closest colleagues, such as Lyell, and he did not undertake their development in the context of a longer work until the end of the 1850s.

Meanwhile, similar ideas had occurred to Alfred Russel Wallace (1823–1913), a Welsh naturalist, illustrating perhaps a convergence of the sort that would have been appreciated among biologists. Wallace, also self-taught, had journeyed long in South America and Malaysia, where he lost all his notes, and nearly his life, in a shipwreck. His route was somewhat similar to Darwin's. Ignoring all of Darwin's views regarding evolution, Wallace submitted his own view to him. Darwin was fair, but still he did not cherish the prospect of losing his twenty-year priority, and so he took council from friends such as Lyell, who decided to present the ideas of the two naturalists during the same meeting of the Linnean Society of London, on 1 July 1858. Darwin, who had abandoned his project of a great work, hastened to write *The Origin of Species*. The public had become impatient: the 1,250 copies of the first edition were all sold on 24 November 1859, the very day it came out.

As Darwin saw it, the evolution of species proceeded by the accumulation of imperceptible changes within isolated populations. But paleontology was capable only of showing discontinuous changes, which were, besides, one of the strongest arguments of the catastrophist school in postulating cycles of extinctions and creations. The theory of evolution was biological, however, not paleontological: it had been inspired by living beings, not by fossils. Evolution took on a sense that could be seen when passing through the whole of the stratigraphic column. Organization was pushed further and further by living beings, from the Cambrian to the Quaternary, and no longer appeared to be the expression of a divine will, but rather an expression of the period of time during which natural selection had been occurring. Thus, evolution and geology shared the same temporal bases, and Darwin, like Lamarck before him, presumed that the durations involved were immense.

In support of his speculations, Darwin ventured to estimate the time required for erosion to cut through the sedimentary formations between the North Downs and the South Downs in southern England, to form the great Valley of the Weald. By adopting a rate of erosion of 2.5 centimeters per century, he calculated that "in all probability a far longer period than 300 million years has elapsed since the latter part of the secondary period." Darwin had advanced a considerable age upon a very fragile basis, thereby committing an imprudence that Kelvin would never tire of calling attention to, even though Darwin actually removed the statement from subsequent editions of *The Origin of Species*. It is true, though, that Kelvin had rightly been suspected of not having read Darwin's treatise, any more than he had read Lyell's.

Kelvin's strong reaction did not result fundamentally from his rejection of evolution. Kelvin had even been among the first to consider that life on earth had germinated from seeds arriving out of space—quite a modernist attitude for an era in which spontaneous generation still had some supporters. The constant increase in degree of organization in living beings was obviously compatible with the directionality this gave to time. But Kelvin deplored that the argument of design had been nearly lost to sight in the recent zoological speculations, despite the "overpoweringly strong proofs . . . showing us, through nature, the influence of a free will and teaching us that all living beings depend on one ever-acting Creator and Ruler." As J. D. Burchfield has summarized, it was natural selection that Kelvin did not accept, and in a similar manner he refused uniformitarianism, "because he believed it to be directly contrary to the fundamental principles of nature. His objections were actually threefold: natural selection could not account for the origin of life; it required far more time for its operation than the laws of physics would allow; and its emphasis upon chance did not allow for the evidence of design in nature." For different reasons, the origin of life and the role of Providence were both points beyond the realm of science. Thus, Kelvin concentrated his efforts upon the question of time, in order to promote his natural philosophy. As one would expect, other biologists besides Darwin were involved in the controversy launched by Kelvin, because the key to the rate of the evolution of species was to be found in the sedimentary record.

The Message of the Sediments

Some individuals in the movement that produced the theory of evolution during the nineteenth century were tempted to go beyond the relative ages permitted by stratigraphy. Well before Darwin became interested in the Weald, naturalists such as Buffon, Soulavie, and Palassou had wondered how much time had been required to deposit and erode the sedimentary strata. But, in the same way that it caused Kelvin to proceed from vague, qualitative arguments to quantitative, physical models, *The Origin of Species* inspired the first attempt at a geological dating of "the *absolute* antiquity of the earliest inhabitants of the earth." The author of this work was John Phillips (1800–1874), a nephew of William Smith, the pioneer of stratigraphy, by whom he had been raised. Phillips was known as a field geologist and was the most ardent English propagandist for the use of fossils in stratigraphy. He had developed a meticulous map of Yorkshire and was a professor of geology at Oxford in 1860 when he published *Life on the Earth: Its Origin and Succession*, in which he attempted to depict the great changes that life had undergone within the ancient seas and lands. The book attacked Lyell's and Darwin's ideas, and in order to oppose them at their weak point, the long durations, Phillips made use of a dating method that was to play an important role in perpetuating the debate, one worth considering for a moment.

"The Geological Scale of Time," he explained, "is founded on the series of strata deposited in the ancient sea; if the forces tending to produce such deposits have always been productive of equal effects in equal times, then the thicknesses of the strata are exact measures of the times." It would be a simple operation, then, to extract a scale of time from the sedimentary message if one knew the total thickness of the stratigraphic column and the rate at which the sediments had been deposited. Phillips himself pointed to the difficulty that neither of these two parameters was easily determined. To begin with, the thickness of a given stratum can vary considerably from one outcrop to another. It can even be reduced to zero in an unconformity. Thus, in order to determine the height of the stratigraphic column, it was necessary to carefully combine the thicknesses surveyed for the different outcrops. Phillips noted that the fossiliferous strata obtained in this manner represented a height of 72,584 feet in England but only 52,800 feet in Wales. Even over such short distances, the thicknesses seemed to vary appreciably.

The rates of sedimentation presented even greater difficulties. They can vary quite a bit from one location to another, being fastest around estuaries and slowest in the presence of strong marine currents, so the rates cannot be measured directly. However, it is possible to determine the volume of the sediments discharged by rivers, starting from the flow of water and its burden of particulate matter. It would be easy to deduce the rate of sedimentation from these calculations if one knew over what surface the particles in suspension were deposited. But unfortunately, it is also difficult to evaluate those surfaces. The simplest hypothesis that Phillips could make, albeit an arbitrary one, was to suppose that the sediments were deposited over a surface equivalent to the continental zone from which they originated. In this case, the thickness of sediments would be equal to the thickness of the terranes that were eroded.

Because errors in calculation are smallest for the large river basins, Phillips decided to consider the Ganges River as his example. The order of magnitude for the parameters (which Phillips gave with an unfounded degree of precision) are instructive in themselves: The Ganges drains a basin of 777,000 square kilometers and annually evacuates 180,323,871 cubic meters of sediments. Erosion thus carries away an average of 0.23 millimeters of rock each year. Supposing a thickness of 22 kilometers of fossiliferous sediments in this basin and a rate of sedimentation equal to that of the erosion, Phillips concluded that nearly 95 million years had been required to deposit all of the sediments. By adjusting the rate of sedimentation to account for a more violent erosive action in the past, Phillips reduced this period to 64 million and even 38 million years. In any case, in his opinion, such periods were still "too vast and, we must add, too vague" to be conceivable. Phillips thus demonstrated upon a purely geological basis the absurdity of the 300 million years that Darwin had advanced for the erosion of the valley of the Weald. Moreover, he obtained durations that were compatible with the ages that Kelvin would vigorously maintain two years later. Kelvin was confident enough of his science that this agreement did not astonish him, but his correspondence with Phillips indicates that he was nonetheless contented about it.

The First Successes

At the beginning of the 1860s, Phillips was one of the few geologists who were attentive to Kelvin's arguments. He had not needed to be converted, but he was certainly not the most influential representative in this British golden age of geology. In order to propagate his ideas, Kelvin had to rely primarily on himself. Beginning in 1865, when the success of his telegraphic activities left him the leisure to return to geology, he began anew to attack uniformitarianism. He incorporated an original argument based on friction caused by the tides, which are due to the gravitational attractions exerted upon the earth by the sun and, especially, the moon. Because friction caused by the displacement of matter is dissipated in the form of heat, the mechanical energy of the system formed by the moon and the earth diminishes regularly over time. As a result the rate of the earth's revolution about its axis diminishes as well. Since the shape of the earth is close to what a liquid mass would have taken on in solidifying at the rate of the current revolution, Kelvin was inspired to unite these two observations and draw a chronological implication: although he was not capable of calculating precisely the angular deceleration of the earth since its consolidation, he could assert that "the present figure of the earth agrees closely with the supposition of its having been fluid not many million years ago."

This time the message was transmitted with greater force by Kelvin, and it would not be ignored. One of the first to receive it was the young director of the Scottish Geological Service, Archibald Geikie (1835–1924). He was a volcano specialist, capable of turning a geological description into a little piece of literature, and one of the strongest personalities in British science in the nineteenth century. He was responsible for one of the first histories of his discipline, and his *Textbook of Geology* rapidly superseded Lyell's *Principles* as a reference work. In 1868 Geikie decided to utilize the rates of denudation in the great river basins as a test of relevance for the durations authorized by Kelvin. For example, in the immense Mississippi River basin, Geikie found that erosion removes one centimeter of rock in 200 years. At this rate, North America would be flattened in 4.5 million years. In the small Po River basin, the rate of re-moving one centimeter of rock is only 24 years. At this rate, Europe could be leveled in less than 500,000 years. As Geikie had pointed out, these were only orders of magnitude, but still they demonstrated the rapidity of geological action. Kelvin's timescales could thus be deemed acceptable if one authorized a large number of episodes of erosion of the mountain chains. Geikie, who had worked in a bank for three years when he was an adolescent, concluded in the form of a parable: "Recent research in physics shows that the unlimited ages demanded by geologists cannot be granted. We have been drawing recklessly upon a bank in which it appears there are no further funds at our disposal. It is well, therefore, to find that our demands are really unnecessary, that even the facts of our own science do not require those exorbitant drafts upon the past."

Yet not everybody was willing to give up so quickly. The naturalist Thomas Huxley (1825–1895), who was nicknamed "Darwin's bulldog" because of his eloquent defense of *The Origin of Species*, became the spokesman of an enlightened opposition. The disagreement with Kelvin, spurred by the man who had invented the word *agnostic*, was all the more active in that it touched upon the very foundations of the scientific approach, and Huxley did not see any domain—including creation—in which reason ought to bow before faith. Huxley detected the fragility of Kelvin's argument in the discussion on time; it had been ill-concealed by the rigor of the mathematical apparatus involved. And he never tired of reminding Kelvin that "as the grandest mill in the world will not extract wheat-flour from peascods, so pages of formula will not get a definite result out of loose data." This first truth was but one of the arguments exchanged at a distance by the two protagonists during the geological conferences held in 1868 and 1869 in Glasgow and London, after which the all-too-vague interpretations of the sedimentary record had to yield, in spite of everything, to solid physical principles. A few extracts from Kelvin's and Huxley's separate works follow, combined here and presented in the form of a debate:

Kelvin: "British popular geology at the present time is in direct opposition to the principles of natural philosophy."

Huxley: "The critical examination of the grounds upon which the very grave charge of opposition to the principles of natural philosophy has been brought against us rather shows that we have exercised a wise discrimination in declining to meddle with our foundations at the bidding of the first passer-by who fancies our house is not so well built as it might be."

Kelvin (irked by the descriptor "first passer-by"): "It is quite certain that a great mistake has been made, —that British popular geology at the present time is in direct opposition to the principles of natural philosophy."

Huxley: "I do not suppose that, at the present day, any geologist would be found to maintain absolute uniformitarianism, to deny that the rapidity of the rotation of the earth may be diminishing, that the sun may be waxing dim, or that the earth may be cooling. Most of us, I suspect, are Gallios, 'who care for none of these things,' being of the opinion that, true or fictitious, they have made no practical difference to the earth, during the period of which a record is preserved in stratified deposits."

Kelvin: "The existing state of things on the earth, life on the earth, all geological history showing the continuity of life, must be limited within some such period of past time as one hundred million years."

Huxley: "*Some such period of past time as one hundred million years.* Now does this means that it may have been two, or three, or four hundred million years? Because this really makes all the difference."

Kelvin (triumphant): "Admission of such a limit as even worthy of attention [from modern British popular geology] is a sweeping reform."

In 1869 the millions of millions of years were clearly no longer in use. By allowing that one, two, three, or four hundred million years could be relevant, after all, the naturalist had indeed conceded the first match to the physicist.

Kelvin's Stranglehold

The path was clear for Kelvin to push his advantage further still. He was a man of pacific temperament and proverbial courtesy, and so he did it without any acrimony. And because he had no institutional power, he did it without any sort of reference to his scientific authority, but instead through the intrinsic force of his arguments. Nevertheless, the physical problem had not been resolved definitively as he saw it. Some initial efforts were begun by Hopkins and Joule in Manchester in 1851 to measure the melting temperatures of rocks and their thermal conductivities. Kelvin followed these efforts closely and assisted the experimenters with his counsel. With better data, which in spite of everything remained very fragmentary, he went back to his model of terrestrial cooling a number of times. A regular revision downward of the age attributed to the earth resulted from these calculations. From the bracket of 96 to 400 million years that he advanced in 1862, Kelvin's estimates went from 100 million years in 1868, to 50 million years in 1876, and to between 20 and 50 million years in 1881, before stopping at a mere 24 million years in 1893.

During this time, Phillips had competitors who benefited from more solid bases for their stratigraphic estimates. Geological exploration was going at a good pace; it had even inspired America. The grandeur of the virgin sites and the wealth of the mines aroused enthusiasm, particularly in the West. Nowhere except in the Colorado River's Grand Canyon and the surrounding lands could one see nearly four kilometers of the stacking of sediments—without the slightest unconformity. The Mississippi alone drained a continental area larger than all of western Europe. In providing evidence for the stratigraphic column and opportunities to evaluate erosion and sedimentation, America took its place proudly.

Date	Author	Thickness of sediments (km)	Rate of sedimentation (cm/1,000 years)	Duration (millions of years)
1860	Phillips	22	22.9	96
1869	Huxley	30	30.0	100
1871	Haughton*	54	3.5	1,526
1878	Haughton*	54		>200
1883	Winchell			3
1890	Lapparent	46	51.1	90
1892	Wallace	54	1.9	28
1892	Geikie	30	0.4–4.4	73–680
1893	McGee	80	0.5	1584
1893	Upham	80	8.0	100
1893	Reade	10	100.0	95
1895	Sollas	50	29.4	17
1897	Sederholm			35–40
1899	Geikie			100
1900	Sollas	81	31.1	26
1908	Joly	81	10.1	80
1909	Sollas	102	12.7	80

Source: Holmes, *Age of the Earth* (1913), 86.

*Determination consistent with results of calculations of the earth's cooling.

A short table (shown above) was published in 1913 by the English geologist Arthur Holmes. It allows us to compare the conclusions of the studies that integrated the observations made on the two sides of the Atlantic. Even though the durations advanced became shorter and shorter toward the end of the century, Kelvin's latest requirements proved to be too severe: with only a few exceptions, the ages obtained tended to be about 100 million years, not the duration that was finally recommended by Kelvin, one-quarter that amount of time. Using stratigraphy, which was its essence, geology found the way nonetheless to consistency with the short times originally suggested by Kelvin. What a long road to have traveled in just three decades!

The thicknesses of the sediments did not lend themselves to much discussion; it was the rates of sedimentation that were the principal causes for disagreement among the different estimates. As Phillips had already discussed, these rates, determined from rates of erosion, depended upon numerous factors that could have varied in the past. Because they could increase, for example, with the average altitude of the continents, or the atmospheric temperature, they ought to be viewed as quite malleable. A great oceanographic discovery resulting from the first efforts to lay telegraph cables, from 1873 to 1876, contributed to such a change in understanding: in exploring the flora and fauna of the abysses all around the world, the English expedition of the *Challenger* revealed through its dredging operations that the ocean floors were largely devoid of sediment. Sediments accumulated in the coastal zones, within areas that were more restricted than had previously been realized, and thus the rates of sedimentation were much higher than had been believed, justifying even shorter durations.

But even so, sedimentation was not completely coherent with Kelvin's views. Approximately the same types of deposits were found for all periods. As Huxley observed, the intensity of geological action did not appear to have diminished in any regular manner, at least not since the Cambrian Period, in spite of important changes in fauna, flora, climate, and relief, as are indicated by the types of strata observed in a given location. The agents of erosion and sedimentation had not changed in nature; the sea level had not varied more drastically than in the earlier ages; and volcanoes had always expelled the same types of lava. It was tempting to think of the adage "the more it changes, the more it remains the same." In a word, although uniformitarianism had been rejected in its extreme form, one could still postulate a uniformity in principle, if not in degree, and imagine that it had only been in the most obscure part of the Precambrian that the earth's outbursts had diminished in intensity. But in adding the duration of the Precambrian to the some hundred million years that have elapsed since, it appeared much more difficult to reach the 24 million years conceded by Kelvin since the *consistentior status*.

In England these controversies were raging, but in Continental Europe they found few echoes. In France, only Albert de Lapparent (1839–1908), a mining engineer who had graduated from the École Polytechnique, concerned himself with the question of time. In 1890 Lapparent was a professor of geology at the Catholic Institute in Paris when he reconsidered the usual terms of the discussions of rates of erosion and sedimentation, to propose an age of the earth between 67 and 90 million years. The lack of interest in establishing the age of the planet in France, Fourier's country, might seem surprising. The antiquity of the earth had already long been accepted there, but the data necessary "to apply rigorously the laws of heat propagation to our planet" were lacking, as was pointed out in 1847 by the physicists Antoine (1788–1878) and Edmond Becquerel (1820–1891) in their *Eléments de Physique terrestre et de météorologie*. They went on to say that, in addition to the thermal properties of the deep substances, "it is also necessary to know what chemical combinations are at work in the central parts, under the excessive pressure to which they are submitted, as well as the phenomena that took place at the surface of separation between the liquid mass and the solid crust." As noted in 1859 by Auguste Daubrée (1814–1896), who was chief engineer of mines and dean of the university at Strasbourg, "the forces of imagination are all the more to be doubted in the field of geology, as induction founded upon observation most often lacks any check." Daubrée avoided speculation and, like James Hall before him, he dedicated himself to experimental petrology. His influential colleague, Louis de Launay (1860–1938), who was a professor at the École des Mines in Paris, finally asserted in *La science géologique* in 1905: "When science is disarmed to such a point, it is better to recognize its powerlessness, because hypothesis is interesting only if it already has a solid basis of observation, or if one or another of its resulting consequences can be verified." As far as Launay was concerned, geological methods of dating thus had their epitaph written.

From the heralds of liberalism and noninterventionism, to Marx, Engels, and the other early socialists, the "struggle for life" had rapidly accommodated itself to all ideologies. While the limited timescales and the possible extinction of the universe left the great thinkers of the nineteenth century largely indifferent, Kelvin's influence had been felt through the efforts of the talented popularizers of the time. From the first edition of his famous *Popular Astronomy* in 1880, Camille Flammarion (1842–1925) adopted the new theories with his habitual enthusiasm. Flammarion was not against speculation. As a self-taught astronomer, he had written his five-hundred-page *Universal Cosmogony* at the age of sixteen. He judged that about 20 million years had sufficed for the geological record and imagined that, millions of years hence, the earth would "become cold in the sleep of death" because progressive infiltration of water into the depths of the planet would have deprived it "of the atmospheric vapor of water which protects it against the glacial cold of space." And even if it were to escape this unprecedented fate, the sun's extinction in 20 million years or so would lead to an analogous result. But it was not necessary to conclude that "in these successive endings the universe will one day become an immense and dark tomb," because, as Flammarion declared, "an intellectual law of progress governs the whole creation. The forces which rule the universe cannot remain inactive. The stars will rise from their ashes. The collision of ancient wrecks causes new flames to burst forth, and the transformation of motion into heat creates nebulae and worlds. Universal death shall never reign." In the form of a cyclical pattern, this conclusion was consistent with Kelvin's opinion that the final moment of history, like the Creation, would escape any sort of description.

Harmonies, but Also Doubts

In the same way that Kelvin's ideas moved some to erode the continents mentally or to stack strata one upon another in their thoughts, they encouraged others to study the new domains bordering on astronomy, geology, and climatology. The variations of the earth's orbital eccentricity induced by the gravitational attractions between planets had been calculated by Urbain Le Verrier, the astronomer famous for having discovered Neptune. Beginning in 1864, James Croll (1821–1890), a former carpenter who had become an attendant at the Glasgow museum after suffering from some bad business deals, began to attribute the great climatic changes to the orbital eccentricities. Croll developed theses that had, in fact, already been outlined by the English astronomer John Herschel, the son of William, and particularly by the Paris mathematician Joseph Adhémar (1797–1862), in *Révolutions de la mer* (*Revolutions of the Sea*). Even though Croll could not count on his *Philosophy of Theism*, published in 1857, to bring about a favorable scientific impression of him, he retained the attention of leading figures such as Lyell and Kelvin. According to Croll, the ice ages had resulted from complex climatic interactions, when the earth had a more eccentric orbit. The chronology of the ice ages could be established, and by counting the glaciations that had occurred, the earth's age could be estimated. When Lyell risked a duration of 240 million years, which he considered modest, upon this basis in 1867, Croll countered that the entire sedimentary record must represent less than 100 million years. In order to lengthen the duration of the sun's life to conform to the same scale, he increased its energy by supposing that it was formed from matter that was already hot. The rigorous chronological framework given by these astronomical calculations ensured an interest in the method and led the former manual laborer to seat himself beside Darwin and Kelvin upon the benches at the Royal Society of London in 1876.

Even more convincing, perhaps, was the result based on the lunar theory of George Howard Darwin (1845–1912), the fifth of Charles's ten children. A physicist who had always remained close to Kelvin, Darwin

Figure 8.2 The drama of the future thermal death of the universe: "Surprised by the cold, the last human family has been touched by the finger of Death. Soon their bones will be buried under the shroud of eternal ice fields." Image from Flammarion, *Astronomie populaire*, 101. Quotation from Flammarion, *Popular Astronomy*, 78.

the son had finished second in the Tripos before commencing his six years of law and registering at the bar. At that time he decided to return to Cambridge, to submit the most complex problems of geology and cosmology to the power of applied mathematics. He studied the influence of the moon on the rotation of a viscid and imperfectly elastic earth, paying particular attention to the friction due to the enormous tides that would have been produced in its mass, which would have slowed the earth's rotation upon its axis.[1] That friction also would have diminished the period of the moon's revolution around the earth. This effect, which was discovered by Halley in 1695 based on dates of lunar and solar eclipses, led to the increase in the distance between the earth and the moon. Proceeding backward in time from the current conditions, Darwin calculated that this distance would have been zero about 54 million years ago. It was thus tempting to imagine that the moon had been extracted from a fluid earth, as a result of a gravitational instability. Although it was an unexpected implication from a new theory published in 1879, this result had a remarkable consistency with the various ages given by Kelvin: such an agreement could hardly be fortuitous. But even though Darwin had stated that no precise estimate could be made of the rate of friction from the tides, people lost sight of the fact that 54 million years represented only the lower limit.

As for the 24 million years that constituted the latest age suggested by Kelvin, it was the result of a long effort begun in 1879, with the founding of the U.S. Geological Survey. The first director of the Survey was Clarence King (1842–1901), a geologist who had dedicated his career to exploring the western United States and who had displayed great foresight in his immediate creation of a laboratory for the measurement of the physical properties of rocks, the first of its kind in the world. Thus, in 1893, when he had already long ago quit his earlier functions, King could enjoy the "privilege of making geological applications of the laboratory results" that had been obtained. Thirty years after Kelvin's first calculations, it was finally possible to attempt a fine analysis of the earth's cooling and to buttress it with the support of experimental data. Now that he knew better about the melting temperatures of rocks, King investigated what initial temperatures would have allowed the earth to maintain its solidity at depth throughout all the geological ages. His conclusion was clear: temperatures greater than 2,000°C had to be

---

1. A familiar phenomenon illustrating the significant dissipation of energy as viscous friction can be seen by spinning a raw egg and a hard-cooked egg at the same time. The raw egg rotates for much less time than does the hard-cooked egg.

excluded; accordingly, the age of the earth had to be lower than 24 million years. Kelvin was so well satisfied with this result that he defended it on his own account. But as he pleaded incessantly for shorter durations, he lost sight of the fact that the elasticity of time had a limit. Although the sedimentary ages published at the beginning of the 1890s had agreed well with Croll's and George Darwin's theories, Kelvin's insistence on shorter durations aroused nothing more than a frank skepticism among some. For example, Geikie, who had converted quickly, recanted his approval in 1892, stating that the rates of sedimentation were actually so poorly understood that they led to ages that varied far too much, from 73 million to 680 million years.

New considerations were brought from geological observations to the model of the earth's cooling, which, most curiously, had escaped any rigorously critical examination until that point. As an irony of history, this questioning, which appeared shortly after the publication of King's calculations, was the work of John Perry (1850–1920), an "affectionate pupil" of Kelvin. Although he was thankful to the physicist for having "destroyed the uniformitarian geologists," who were, by that time "as extinct as the dodo or the great auk," Perry noted that ages much longer than the ones Kelvin had proposed could be obtained if one posited that the thermal conductivity increased with depth, rather than remaining constant. In an exchange of points of view published in January 1895 by the magazine *Nature*, Kelvin proved himself to be unexpectedly adaptable. "I thought my range from 20 million to 400 million was probably wide enough, but it is quite possible that I should have put the superior limit a good deal higher, perhaps 4,000 instead of 400 [million years]." But what Kelvin conceded on the one hand, he took back again with the other: he remained inflexible on the maximum time the sun could shine—a few scores of millions of years. Because actual measurements indicated that thermal conductivity diminished, rather than increased, with depth, Kelvin finally judged that Perry's criticism was not relevant.

Thus, the century was coming to an end in an atmosphere of uncertainty, when a completely new method of dating gave renewed value to Kelvin's very first estimates. The method was proposed by John Joly (1857–1933), a professor of geology at Trinity College in Dublin. With his solid background in physics, chemistry, geology, and mineralogy, Joly extended his curiosity to much of the world about him. He was known primarily for his calorimetric measurements on gases. Later, he was one of the first promoters of radiotherapy. Being a pleasure sailor, he was interested in plankton and sedimentation in the shore waters; that interest led to curiosity about the origin of the salt in the seas. Without realizing it, Joly had taken up Halley's long-forgotten thesis that the salinity of the oceans could give a measure of geological time.

Joly used chemistry to support physics. All throughout the nineteenth century, chemists had left their laboratories to discover the natural environment. Rocks, minerals, lakes, rivers, the atmosphere—nothing escaped their search for new elements and mass-balance estimations. With his passion for precision, not uncommon in his time, one of the great pioneers of this chemical geology was the German Gustav Bischof (1792–1870), who calculated in 1850 that 332,539 million oyster shells could be formed from the calcium carbonate poured into the sea each year by the Rhine. A half century later, Joly posed the following question: Starting from the time when the earth's primitive atmosphere condensed to form the oceans, how much time had been needed for the salinity to attain its current value if the influx of sodium from the rivers had been constant all through the geological ages? For a simple math problem such as this, the numerical answer is found in applying the rule of three: We know that there are 1.415×10^{16} tons of sodium in the ocean waters and that the rivers bring in 27,180 cubic kilometers of water each year, with an average sodium content of 5.787 grams per cubic meter. From these data Joly calculated a "raw" age for the ocean of 99 million years. After correcting for a few secondary factors, he reduced the result to a range of between 80 and 89 million years.

This age dated the moment when the primitive ocean had condensed, and thus the time when sedimentation began. The atmospheric temperature at that time had decreased to less than 100°C, a temperature that could only have been reached later than the *consistentior status* of Leibniz and Kelvin, when the molten matter, heated white-hot, had consolidated at the earth's surface. As Kelvin himself had expounded, the cooling of the surface crust by radiation was subsequently quite rapid. The various ages obtained could be compared among themselves: Joly's figure exhibited a striking agreement with the figure of about 100 million years that was based on stratigraphy. The two approaches, of course, had the drawback of appealing to the principle of uniformity, often problematic in practice, but the advantages of Joly's approach were obvious. Only two parameters were required, the current quantity of salt in the seas and the annual influx from the rivers, both of which could be determined much more accurately than could the thicknesses of the strata or the rates of sedimentation. Joly nonetheless had to predicate the important hypothesis that no sodium is incorporated into the sediments deposited. He justified this in noting that "the sodium contained in the sea, added to what is left over in the detrital sediments, would suffice to restore to the entire mass [of sedimentary rocks] a soda percentage almost equal to that in the eruptive, igneous, and crystalline rocks; the deficiency, about 0.4 percent, exists partly in rock salt deposits." Such a well-equilibrated mass balance naturally impressed his contemporaries.

Diverging Evolutions

Biologists and paleontologists were just as eager to determine the age of the planet as the geologists were. The domestication of plants and animals was an important subject of reflection for Charles Darwin, because it had demonstrated that the diversification of breeds could be an expeditious process, even within the brief scale of a human life, on the condition that selection remained pitiless. But this observation was not to be transposed to the evolution of species, because the transformations that came into play were much more marked, and the forces of selection at work were less rigorous. And paleontological observations were no more informative. Even though there were certain genera that had been maintained through the stratigraphic column practically all the way back to the Cambrian, there were others that had made only fleeting appearances. Between these two limiting cases, one often observed long stages of stability interrupted, especially toward the end of an era, by periods of more rapid evolution. Rates of evolution thus appeared to be characterized by an extreme variability. Besides, paleontology had the great handicap of remaining almost mute on the subject of life during the Precambrian. Because evolution appeared to proceed more rapidly at more complex stages of organization, one was pressed to suppose that very long durations had been necessary to prepare, amid the densest of mysteries, the evolving life forms that eventually burst out in the early Cambrian.

In order to make up for the deficiencies of biology and paleontology, intuition—that double-edged sword in the hands of science—was invoked as the principal judge of the time of evolution. Reactions to new concepts range from an often sterile rejection or passive resistance, to adhesion, or even audacious exploitation as a stimulus to major discoveries or resounding failures. It is interesting that the watershed imposed by Kelvin passed distinctly between Charles Darwin and Wallace. For Darwin, the views of Kelvin had long represented one of his "sorest troubles," not only because they precluded the immense durations postulated to account for evolution starting from the most primitive organisms, but also because they constituted a formidable weapon against the foundation of his theory, the principle of natural selection. Darwin was powerless to contradict physics, and he was shaken by Croll's theories, which led him to accept, grudgingly, the least short of the timescales. But the excessively low demands of the physicists finally forced him to become skeptical and to maintain with reserve the much longer durations.

By contrast to the silent opposition of Darwin, who had gradually been abandoned by his friends the geologists, Wallace quickly leaned to Kelvin's side. This position allowed him to shed new light on natural selection as he returned to Croll's theory of glaciations in 1869, to account for increasing rates of evolution. The successive advancing and receding of the glacial fronts would have intensified the effects of natural selection and the struggle for life by inducing animal and plant migrations; in the process new species would have been produced: "High eccentricity would therefore lead to a rapid change of species, low eccentricity to a persistence of the same form," Wallace advanced, "and as we are now, and have been for 60,000 years, in a period of low eccentricity, *the rate of change of species during that time may be no measure of the rate that has generally obtained in past geological epochs.*" Because high rates of evolution had the same astronomical causes as did the glaciations, they permitted him to extend the dating of the stratigraphic column over a period of 24 million years, well before the latest recommendations of Kelvin. "These figures will seem very small to some geologists who have been accustomed to speak of 'millions' as small matters." But Wallace added, "I hope I have shown that, so far as we have any means at present of measuring geological times, they may be amply sufficient." With the 100 million years conceded by Kelvin, there was indeed something that could "fully satisfy" Darwin, as he now had three times as long a duration for the Precambrian as for the total of the four successive eras. And Wallace, who was one of the most ardent defenders of the highest rates of sedimentation, defended the 24-million-year duration all through the following decade in the discussions of geological ages.

The Origin of Mutations

Without knowing the nature of the biological support of heredity and the causes of inheritable variations in this support, one could not understand evolution, and the rate at which it occurred remained necessarily indeterminate. What was even more serious for the evolutionary schools was that the short durations imposed by physics presented one of the major arguments to be used against them. Because Darwin rejected the possibility of the rapid evolution maintained by Wallace, he found himself entangled in his own hesitancy in this regard. "In this most discouraging state of things," the Dutch botanist Hugo De Vries (1848–1935) related, "I concluded that the only way to get out of the prevailing confusion was to return to the method of direct experimental inquiry." De Vries accepted Kelvin's timescales without reserve and, around 1885, undertook to discover a rapid mechanism of evolution that would prove more satisfying than the continuing changes proposed by Darwin.

Toward this end, De Vries turned his attention toward *mutations*, the term used for the sudden changes of strains that occurred from time to time in cultivated plants. In 1886 he chose the *Oenothera lamarckiana*, Lamarck's evening primrose, as a subject suitable for experimental study under strictly natural conditions. Through fifteen years of patience and perseverance, De Vries observed this primrose abruptly forming new species. He kept this part of his research secret, detached "from the need of publishing early and often unripe results." And, from the crossbreeds he produced between different varieties of evening primroses or of other plants during this period, he rediscovered for himself, in 1900, the laws of the transmission of parental characteristics, which were originally formulated by the Czech monk Gregor Mendel.

As De Vries observed, natural mutations could produce new species capable of reproducing and prospering in other locations. This conclusion was welcomed joyfully during the first years of the twentieth century by all those who had been wearied by the quarrels between the various evolutionary schools. For De Vries, mutations had an important implication: "Some thousand of them may be estimated as sufficient to account for the entire organization of the higher forms. Granting between twenty and forty million years since the beginning of life, the intervals between two successive mutations may have been centuries and even thousand of years. As yet there has been no objection cited against this assumption, and hence we see that the lack of harmony between the demands of biologists and the results of the physicists disappears in the light of the theory of mutation." Because of his knowledge of mutations and their frequency, De Vries thought he had determined the mechanisms of evolution completely. (It was discovered in the 1920s that the evening primrose had a complex genetic structure: the "mutations" had actually been new combinations of genetic factors already existing latently in hybrids.)

Whether long or short, evolutionary durations and geological durations supported each other reciprocally. But for different reasons, the ideas of Darwin, Wallace and De Vries took advantage of the coherence of circular reasonings. Fossils had led to the characterizing of the sediments and their classification in chronological order, but they could not reveal their absolute ages, and the uncertainty could only be reduced partially by the measurements of the thicknesses deposited. It was an important transition to be able to pass from relative time, cadenced by the fossils buried in the geological strata, to the "living" time of biology. A spectacular example was presented by the horse. From *Eohippus*, a modest quadruped the size of a fox, which roamed the North American plains at the beginning of the Tertiary, to *Equus caballus*, the noblest conquest of *Homo sapiens sapiens*, the horse has been one of those rare vertebrate species whose evolution could be followed closely across some dozen well-described species. One of the few Darwinists at the turn of the century, the paleontologist William D. Matthew (1871–1930) was a conservator at the American Museum of Natural History in New York. He established in 1914 a relative chronology of the horse's evolution beginning from the anatomical differences between successive species. After comparing the durations of the different periods on this basis, Matthew was still unable, of course, to produce concrete durations or to decide between 25 million and 1.5 billion years for the overall fossiliferous time. Without an independent clock, evolutionary time remained uncertain. And before such a problem, biology appeared to be stark naked.

An Impossible Verdict

Starting with Huxley, the participants in the natural-timescale controversy appeared to take pleasure in giving their reciprocal assaults in the form of "pleas for the defense." In that context, I am taking the liberty here, in summarizing the situation as it was in the very early twentieth century, of imagining a British tribunal court, located equidistant from Glasgow, Oxford, and Cambridge, to which the partisans of the short and long times have come to present their points of view. To preside over the tribunal, Reverend Osmond Fisher (1817–1914) has come from Harlton, a village near Cambridge, where he enjoys entertaining his colleagues. This pastor of priestly face, fringed with white whiskers, is an alert octogenarian with a sense of perspective: he was not present at the origin of the planet, but he has been around to watch geology grow up. (Even so, to avoid any sort of risk, the Geological Society of London waited until he was ninety-five years old to confer its highest distinction upon him.) Fisher was ordained after competing at the Tripos, and he is as familiar with fossils and flint tools as he is with the most devilish of mathematical equations. A pioneer of theoretical geology, he accepted Kelvin's first timescales, but he also demonstrated that he is entirely capable of refuting an incorrect idea or of welcoming criticism. In a word, nobody is better qualified than he to referee the debates. The case has already been exposed completely. Without taking account of the inevitably dramatic courtroom styles involved, we shall pass directly to the pleas.

In the role of prosecutor, we have the Scotsman Peter Guthrie Tait (1831–1901). Like many other distinguished physicists, he is a former pupil of Hopkins. Tait has been a professor at Edinburgh since 1860 and is close to Kelvin, with whom he collaborated on the *Treatise on Natural Philosophy*, a work that has been compared a bit too hastily by contemporaries to the works of Laplace and Fourier. Geology is hardly his specialty, but he is concerned with all matters relating to thermodynamics: he wrote a treatise on the subject, considered authoritative, and he has also written a very biased history of it, in which Joule is the hero and Mayer the victim. Because he attacked the laxity of geologists in 1885 during a series of lectures dedicated to energy, his indictment has been ready for a long time. He takes up the conclusion directly:

Thus we can say at once to geologists, that granting this premise,— that physical laws have remained as they are now, and that we know of all the physical laws which have been operating during that time,— we cannot give more scope for their speculations than about ten or (say at most) fifteen million years. But I daresay many of you are acquainted with the speculations of Lyell and others, especially those of Darwin, who tell us that for even a comparatively brief portion of recent geological history three hundred millions of years will not suffice! We say so much the worse for geology as understood at present by its chief authorities, for, as you will presently see, physical considerations from various independent points of view render it utterly impossible to grant more than ten or fifteen million years.

Kelvin himself was never so firm in requiring such a short duration. Like an unleashed prosecutor in the name of the second law of thermodynamics, Tait confirms the public's impression that he is behaving excessively as an interpreter of the great physicists of his time. And the defense takes note, especially, that in arriving at the "utterly impossible," all traces of the "premise," nonetheless duly emphasized, have evaporated. Fisher cannot help recalling what he said about mathematical physicists in 1882, in a discussion about the earth's crust: "Instead of seeking for an admissible hypothesis the outcome of which, when submitted to calculation, might agree with the facts of geology, they assume one which is suited to the exigencies of some powerful methods of analysis, and having obtained their result, on the strength of it they bid bewildered geologists to disbelieve the evidence of their senses!" The repartee is lively. "Reverend Fisher allowed himself that because he himself dealt with the mathematical equations of physics," whispers someone, knowingly, for the benefit of his neighbor. In his *Physics of the Earth's Crust* in 1881, Fisher demonstrated the existence of a plastic substratum beneath the solid crust. If convective movements were able to develop there, as he thought, then Kelvin's calculations would be invalidated.

On the plaintiff's bench, William Sollas (1849–1936), a geology professor at Oxford, requests the floor. With interests including crystallography, paleontology, anthropology, and biology, Sollas is not a man whose mind is limited to a simple narrow specialty. He belongs to the first generation of geologists who were exposed early to Kelvin's ideas, and he has never been seduced by uniformitarianism. He is a faithful representative of a school that seeks to integrate results coming from various origins. He attempts to stick to the facts: "(1) Time which has elapsed since the separation of the earth and moon, fifty million years, minimum estimate by Professor G. H. Darwin. (2) Since the *consistentior status*, twenty to forty million (Lord Kelvin). (3) Since the condensation of the oceans, eighty to ninety million, maximum estimate by professor J. Joly." It is easy for Sollas to remark "that these estimates, although independent, are all of the same order of magnitude, and so far confirmatory of each other." Passing to the total thickness of stratified rocks, Sollas notes subsequently that it was, "as we have seen, 265,000 feet, and consequently if they accumulated at the rate of one foot per century, as evidence seems to suggest, more than twenty-six million years must have elapsed during their formation." Because Sollas tends to prefer these short durations, he does not omit the mention of one last point of view in his peroration: "How far does a period of twenty-six million years satisfy the demands of biology? Speaking only for myself, although I am aware that there are eminent biologists who do not want to share this opinion, I answer, Amply."

Nothing is easier than for one geologist to respond to the well-developed demonstration of another geologist. John Goodchild (1844–1906) comes from an older school than Sollas, though he was not crowned by the prestige of Oxford or Cambridge. He is the curator of the Geological Survey Collections in the Royal Scottish Museum, and one of the last to continue arguing for the extremely long durations; "Nearly a living fossil!" one prankster in the audience chamber sniggers. Leaving it to authorities like Reverend Fisher to judge the physical-chemical aspects, for which he deems himself incompetent, Goodchild resolutely opposes Sollas regarding the sedimentary ages. According to his observations, 3,000 years are required to deposit a one-foot layer of sedimentary clay, and 25,000 for the same thickness of limestone. From these rates, which are considerably slower than those conjectured by Sollas and many others, Goodchild announces, quite candidly: "Total since the commencement of the Cambrian Period: 704,235,000" years. Various noises are heard in the tribunal, but Goodchild does not heed them, resuming: "Seven hundred million years back from the present to the commencement of the Cambrian Period may seem to many persons far too extravagant an estimate to be seriously considered for the moment." Still, he ventures to add, "For my own part, I am quite prepared to find that, as our knowledge of Geology and Biology advances, we shall feel it necessary to extend rather than to abbreviate the estimate I have put forward. Seven hundred million years carry us back to the commencement of the Cambrian period, but not to the commencement of life upon the earth." And before sitting back down, Goodchild concludes tranquilly: "To me, to whom Geology is of interest chiefly as it throws light upon the past history of life upon the Earth, this early appearance of highly-organized forms of Invertebrata suggests remarks sufficient to double the already considerable length of this Address."

The prosecutor is left speechless before such audacity. An incredulous murmur from the public meets this calmly declared billion and a half years. For the unfortunate judges, biology seems to lend itself just as easily as does geology to highly disparate readings. A stranger from the audience takes advantage of the confusion to launch a counterattack, aiming at Joly's methodology: "And the fishes from the sea during the Silurian, which are just as alike to their living counterparts as two droplets of water—seawater no less!—do you suppose perhaps that they swam in distilled water?" (Laughter and protests.) After restoring order in the court, president Fisher allows himself to remark: "The amount of salts in the ancient rocks being so near that in rocks now presumably being deposited, shows that the ocean was about as salt then as it is now, and consequently not much additional sodium can have accumulated in it during the long ages since Silurian times." Could the sodium have been recycled on a large scale by incorporation into sediments, contrary to the long justifications given by Joly, producing ages that were too short?

With so much confusion, some new information is obviously needed. A third geologist is called to witness, the famous Archibald Geikie.

Unfortunately, Geikie himself would prefer to see things more clearly. After confessing that physicists have done a fine service by reducing the "extravagant" demands of the early geologists, he attempts to take advantage of the residual perplexity: "For my own part, I can hardly doubt that there was some fault in the physical argument, though I do not pretend to be able to say where it is to be found. Some assumption, it seems to me, has been made, or some consideration has been left out of sight, which will eventually be seen to vitiate the conclusions, and which when duly taken into account will allow time enough for any reasonable interpretation of the geological record." With a touch of biology in his deposition, he supplements this declaration, which he included in his *Textbook of Geology* in 1892: "There is a general agreement among the geologists that, as far as the phenomena of sedimentation and tectonic structure are concerned, 100 million years would probably suffice for the completion of the geological record. But if on palaeontological grounds the allowance of time should be found too small, there appears to be no reason, on at least the geological side, why it should not be enlarged as far as may be found needful for the satisfactory interpretation of the evolution of organised existence on the globe."

With these words, the doubtful judges retire (alas!) to deliberate. Whispers in the audience chamber: Will the verdict side with Geikie? The age of 10 or 20 million years, in truth, appears to be too short. The durations Goodchild demands, a hundred times longer, recall the distressing excesses of the uniformitarians. And a voice notes that the very first studies of the propagation of seismic waves have given experimental confirmation of the deep rigidity of the earth. A reasonable majority tends toward the 100 million years initially proposed by Kelvin, a limit that is far higher than the durations later given by various different methods. Will the judgment confirm these opinions?

In fact, the judges remain undecided. Reverend Fisher recalls that Joly's "period of between 80 and 90 million years will perhaps satisfy geologists as being sufficient. The leading physicists, on the other hand, are disposed to grant us a good deal less time." If consistency is to prevail, one would tend to concur with the audience chamber and concede 100 million years for the earth. Still, doubts remain in favor of a much greater age. The doubts are vague, naturally, and poorly supported, but Geikie himself has never invoked a billion years. The embarrassment could continue. But, behold, the venerable bailiff of the court of appeals comes in with his two cents! From the shelves that alternate with the wainscoting along the walls of the deliberation room, the bailiff quickly makes his choice among the most learned of the works. Appearing facetious (or hard of hearing) and drenched with Continental skepticism (or confusing to a jumble within that space under his wig the thicknesses of sediments, the numbers of kings, thermal conductivities, and the durations of a reign), the bailiff retires (with footsteps he would rather were stealthier) across the squeaky parquet. In a well-lighted corner of the room, he leaves Diderot's *Encyclopédie* open to the entry "Chronology," perhaps offering a way out to the perplexed judges:

We must conclude these discussions with a reflection we owe the famous chronologists in the interest of truth and honor: that the greater part of those who reproach their differences in results do not appear to have understood the moral impossibility of the precision they require. If they had carefully considered the prodigious multitude of facts that must be combined, the spectrum of genius of the people by whom these facts have been considered, the scarce exactitude of the dates, inevitable for times when events are transmitted only through tradition, the mania for antiquity with which nearly all the nations have been infected, the lies of historians, their involuntary errors, the resemblances among names, which often has led to the diminishing of the number of personages, or their differences, which has led even more often to their multiplication—the fables presented as truths, the truths metamorphosed into fables, the diversity of the languages, that of the measurements of time, and an infinity of other circumstances that all conform to produce obscurity—if they had, I say, carefully considered these things, they would have been surprised to learn, not that they had found so many differences in the *chronological* systems that they had invented, but that such a system could actually ever have been invented.

NINE

The Pandora's Box
of Physics

THE ORIGINS OF RADIOACTIVITY   Most geologists had been convinced by Kelvin that geological time was not as long as they had thought. But then a revolution was triggered in physics by the study of cathode rays and the discovery of X-rays. This occurred at a time when Martians and spiritualism had become fashionable, along with a renewed interest in strange and occult phenomena.

Prelude to a New Age

The year 1895 played a pivotal role in the resolution of the great enigma of the earth's antiquity. Our contemporaries who are satisfied with their fill of discoveries have forgotten that this year began one of the most prodigious decades ever from a scientific point of view. The coming twentieth century was to inaugurate the atomic age and the electronic age, but in 1895 the atom and the electron were merely abstractions, or nearly so. It was then that the real upheavals began to happen in physics, and the subsequent shockwaves also reached the natural sciences. The geologists enamored of eternity had for the most part beaten a hasty retreat, even though the shorter and shorter timescales imposed by Kelvin were not unanimously accepted. Although by this time the controversy initiated in 1862 was more a question of degree than of nature, it was by no means over. Kelvin was hardly worried. In March of 1895, he took delight to end his debate with Perry by making his own the conclusion just drawn by the geologist King that "the concordance of results between the ages of sun and earth certainly strengthens the physical case, and throws the burden of proof upon those who hold to the vaguely vast age derived from sedimentary geology."

Kelvin was right on one point at least: geology was incapable of determining its own timescales with precision. Since geologists could not furnish the rigorous time clocks desired, physicists had to find them in their own bag of tricks. It was an irony of history that physics was now hastening to undo what it had just established. While Kelvin was sticking to his own brief timescales, a revolution was getting ready to happen, and astonishing concepts were about to undermine the foundations of his theory.

The starting point was in February 1896, with Becquerel's discovery of the strange radiations emitted by uranium salts. This was the beginning of nuclear physics; and it was from here that the rigorous solutions were to come for the problems of time and of the dynamics of the earth and the stars. After the discovery of isotopes, entire sections of the earth's history would progressively be rewritten with the help of radioactivity. The debate about time in nature and the antiquity of our planet was to come to its final conclusion, to the disadvantage of Kelvin and his followers, and it would soon be seen that the unsuspected complexity of matter would allow scientists to depict nature's past with a previously unimagined abundance of details. However, the road leading to these solutions was quite tortuous.

Contrary to some common opinions, Becquerel's discovery was no accident. It followed immediately after that of X-rays, crowning a productive century for the study of electrical and magnetic phenomena. A clear definition of its origins would somewhat resemble the events during a long game of blind-man's-bluff. To limit it to just the main milestones, we shall begin at the end of the eighteenth century, with the elements of arbitrariness and simplification that are inherent to such accounts. In conformance with a rather well-established rule, the new science that emerged at the dawn of the twentieth century was built largely by newcomers. So we shall make a brief detour behind the scenes, to encounter Ernest Rutherford and the Curies, at the time when those young actors were preparing to occupy the front stage. This rapid overview will indicate, once again, how the paths of science are often impenetrable from a larger perspective: the great discoveries rarely appear where they are expected and can even result from false reasoning. With the physicist Arthur Schuster, we will conclude that "the ideas which induce us to conduct an experiment are often of small importance compared to the results which, if carried out in a philosophic spirit, may be reached through it." And the picture would not be complete unless some of the dead ends encountered along the road by certain of the protagonists were described. Without mentioning the "psychic force" or the birth of the Martians, we could not really put the waning nineteenth century into perspective.

From Frog Legs to Radio Waves

As a preamble to a succinct genealogy of radioactivity, it is perhaps useful to recall that even though electricity had been known since antiquity, it was still understood only in a rudimentary fashion at the end of the eighteenth century. In the salons, the enlightened company amused themselves by transmitting hand to hand the electrical discharges produced by the first electrostatic machines. The spectacular aspect of these collective experiments had something to do with the popularity of the first "electricians." Behind this superficial facade, it had taken nearly two hundred years of efforts to cast the foundations of electrostatics. Du Fay, who was Buffon's predecessor at the Royal Botanical Garden, had identified in 1733 two kinds of electricity, which he called *vitreous* and *resinous*, because they were collected by rubbing upon stems of glass and of amber. These two repelled or attracted according to whether they were of the same or of different nature. It was also known that bodies could be classed into two categories, the *conductors* and the *isolators*, by whether or not they permitted the movement of electrical charges. And even though the more subtle modes of the movement of electrical charges, such as induction, had been identified, the currents that could be produced were restricted in character to violent, irreversible discharges, which made them a sort of worldly sensation.

Besides these spectacular features, advances in understanding of the laws of electricity rapidly served important practical needs. The lightning rod, for instance, was invented by a self-taught former printer who had earlier proved, through his clever experiments, the "sameness" of lightning "with the electrical fluid." This invention was made by none other than Benjamin Franklin (1706–1790), who later became a statesman. A contemporary eulogy written in France attests to the dramatic interest it raised:

Il a ravi le feu des Cieux
Il fait fleurir les arts en des climats sauvages
L'Amérique le place à la tête des sages

La Grèce l'auroit mis au nombre de ses dieux.
[He stole the fires from the heavens;
He brought the arts to blossom in wild countries;
America placed him at the forefront of wise men;
Greece would have numbered him among the gods.]

Not a chance discovery, the lightning rod was the outcome of systematic experiments performed from the mid-1740s by Franklin in Philadelphia to determine how electricity was transferred through conductors and accumulated in devices like the so-called Leyden jars. In the long run, however, of even greater importance was the fact alluded to in the eulogy: that fundamental research had just set foot in North America.

It was a few decades later, during a properly revolutionary period, that Luigi Galvani (1737–1798), an anatomist and obstetrician from Bologna, made a startling discovery. Galvani was a bone specialist who had just expanded his interests to the physiology of nerves and muscles. During his experiments aimed at determining the influence of static electricity on his studies of frogs, he discovered in 1791 and 1792 that the thighs of the poor beasts could contract without any apparent electrical stimulation: to produce a contraction it sufficed to connect a nerve and a muscle to pieces of different types of metals. Galvani deduced from this observation that the electrical circuit thus formed allowed the discharge of a form of electricity that was present in living beings. He dedicated the rest of his career to this animal electricity, although he greatly overestimated its intensity. His efforts were vain, and his last years were darkened by the loss of his position as professor owing to his opposition to the Cisalpine Republic of Napoleon. At the time he died, the point of view of his adversary, Volta, had already superseded his own.

Among the many reactions aroused by galvanism, that of Alessandro Volta (1745–1827) stood out in particular. He was a physicist of Pavia, already famous for his discovery of methane and his efforts in relation to electrostatics. He was also the nephew of a Dominican, a canon, and an archdeacon, the son of a former Jesuit, the brother of two nuns and three ecclesiastics, and himself a fervent Catholic, yet none of these facts prevented Volta from taking advantage of life and, in the words of one of his friends, he "understood a lot about the electricity of women." When he learned of Galvani's experiments, Volta repeated them and concluded, with rare insight, that the frog's spasm was actually stimulated by electrical charges originating from the metals utilized by the experimenter to connect the nerve and the muscle. In the same manner that the tongue perceives the weak current produced by contact between a filling and a piece of aluminum foil, the animal had only been playing the role of a highly sensitive detector, reacting to the feeble exchange of electricity between the two metals of the circuit.

This conclusion produced a lively opposition between Galvani and Volta. Volta began in 1795 to study the currents produced by contact between different metals and to create an "artificial electrical organ," analogous to those in electric eels, which had so long fascinated the electricians. This organ was the *battery*, constituted of a simple stacking of disks of tissue and two different metals (for example, silver and zinc), all of which was bathed in a tube filled with saline water so as to enhance the exchanges of electrical charges. Modern electricity had been born. The battery was described to the Royal Society of London in 1800. The following year it aroused the enthusiasm of one of the influential members of the Paris Academy of Sciences: Napoleon, who was on the verge of becoming king of Italy, hastened to accord tidy rents and titles of nobility to Volta, his new compatriot, friend, and colleague of the Institut de France. It had become possible using the battery to produce continuous currents whose intensity and voltage could be controlled, rather than merely the previous irreversible discharges. The classic laws of electrical currents were quickly determined, and the chemical effects of these currents were soon discovered: electrolysis also dates from 1800. Current language correctly retains the memory of these two Italian fathers of electricity, as we speak of "galvanizing" a muscle or of the "voltage" of a battery.

The next stage was developed by the Dane Hans Christian Oersted (1777–1851), a professor at the University of Copenhagen. He was a self-taught physicist who had trained as an apothecary and was saturated with the philosophy of Kant. He founded a polytechnic school modeled on the French style and was also a poet and friend of Hans Christian Andersen. In conformance with the syncretism existing in the decades preceding the formulation of the first law of thermodynamics (to which he contributed indirectly, through his pupil Colding), Oersted was convinced that the forces of nature flowed from one single essence. Inspired by the long-known fact that lightning could disorient a compass, he postulated that "heat and light consist of the conflict of electricities." The lightning-compass association also caused him to suppose that electricity and magnetism were closely connected and that an electrical current ought to have an influence upon magnetic phenomena. And indeed, Oersted observed in 1820 that an electrical current could deflect a compass needle. By demonstrating that an electrical current creates a magnetic field, this famous experiment inaugurated the study of the mutual influences of electricity and magnetism, the relationship between which is indicated by the term *electromagnetism*.

After Oersted, scientists performed innumerable experiments probing the interactions between electrical and magnetic forces; among these, André-Marie Ampère (1775–1836) in France and Michael Faraday (1791–1867) in England distinguished themselves particularly. From the electrical motor to the dynamo and to telegraphy, practical applications marked the progress of the new physics. Kelvin, better than anyone else, symbolized this productive alliance between science and its applications. In the same way that mechanics and thermodynamics gradually found their elaborate mathematical equations, so did electromagnetism. As Fourier believed, mathematical analysis appeared to be a preexisting element of the natural order, since it "brings together phenomena the most diverse, and reveals the hidden analogies that unite them," and it is able, furthermore, to supplement experiments by foretelling effects that escaped observation.

Nonetheless, scientists had to resort to experimentation in order to decide between the great syntheses of electromagnetism that had been born in the second half of the century in Germany and England. One of the theories propounded by the Scotsman Clerk Maxwell (1831–1879) triumphed as a result of such experimental tests. It reduced the laws of electromagnetism to a set of four elegant equations that interconnected electrical and magnetic fields at each point in space to the position and movement of the electrical charges. Any perturbation of these fields propagates as a wave. These waves were thought to shake the *ether*, an imponderable fluid that was postulated at the beginning of the century to fill all space and permeate all matter so as to act as the medium for the transmission of light and electricity. Maxwell calculated that electromagnetic waves traveled at the same speed as light. A marvel of the theory was that the transverse vibrations of the electromagnetic field also represented light! With Maxwell, the ancient science of optics had become a branch of electromagnetism.

Visible light from red to violet represented only a narrow interval of wavelengths for which evolution had produced highly sensitive detectors: the eyes. At each side of the visible range, ultraviolet light (emitted by the sun or electric arcs) and infrared light (emitted by hot bodies) extend over wavelength ranges that are scarcely any broader. One can detect ultraviolet radiation with photographic plates and infrared radiation with thermometers. In the range of wavelengths a billion times longer than those of these lights, the radiations foretold by Maxwell's theory were extremely difficult to detect. Heinrich Hertz (1857–1894) was a young professor at the University of Karlsruhe in 1888 when he detected such long-wavelength lights created by the extremely rapid oscillations of electrical charges. Seven years after this majestic crowning of Maxwell's theory, Guglielmo Marconi (1874–1937) learned of the existence of these "Hertz" waves. The twenty-year-old Italian student took immediate interest in them for the purpose of transmitting signals: he devised the wireless telegraph. The first messages he sent in 1895 were received at nearly two kilometers distance. In 1902 Marconi was able to send messages across the Atlantic. In one long century, human communications had passed from the speed of a horse or a boat to the speed of light.

Kelvin's Failure

One decade before the end of the century, the panorama of physics had been considerably clarified. Optics, electricity, and magnetism at this point were considered parts of one single framework. From another point of view, thermodynamics had been constituted around the two laws of the conservation of energy and the tendency of entropy to increase within the universe. Its authority was no longer limited to steam engines; it even extended to distant disciplines such as geology, whose wild chimeras' necks Kelvin had wrung. Paradoxically, the immense successes of physics aroused a certain pessimism. Some people were beginning to doubt that any more great discoveries could be near. In 1871 Maxwell was preoccupied by the theory that "in a few years all the great physical constants will have been approximately estimated, and that the only occupation which will then be left to men of science will be to carry on these measurements to another place of decimals." It seemed for some that the end of physics was in sight, as others today boldly suggest. But Maxwell did not think so; he added: "I might bring forward instances gathered from every branch of science, shewing how the labor of careful measurements has been rewarded by the discovery of new fields of research, and by the development of new scientific ideas."

Besides, nobody denied that there were large questions that were beyond any sort of response. For example, the existence of atoms had become the object of animated controversies from the beginning of the nineteenth century, but no one could foresee what experimental methods would be appropriate to detect and investigate these elementary particles of matter. The formidable difficulty of such problems led researchers to believe that their resolution would have to await the competence of future generations. Among a certain few, perhaps too demanding, the progress that had been achieved only emphasized the incomplete state of science. Kelvin was one of these. He affirmed, before the gathering for the Scientific Jubilee in June 1896: "One word characterizes the most strenuous efforts for the advancement of sciences I have made perseveringly during fifty-five years, and that word is FAILURE. I know no more of electric and magnetic forces or of the relation between ether, electricity, and ponderable matter, or of chemical affinity than I knew and tried to teach to my students of natural philosophy fifty years ago in my first session as professor."

Despite the humility displayed by Kelvin, the omnipotence vaunted by science was not always well received. During the second half of the nineteenth century, scientists had claimed a general right to propose explanations that threatened the foundations of faith. As a distant echo of the romantic protest, there was a reactionary current rooted in the vogue of animal magnetism at the end of the eighteenth century as well as in the long mystic tradition of Christianity. This current attempted to find answers—especially in England, where the physicists were principal actors—to such questions as whether it would be possible to reconcile the scientific method with the ancient foundations of religious beliefs, and whether one could scientifically establish the immortality of the soul or the intercession of invisible beings in human lives. Although they were not conclusive, some of the results announced were spectacular. In a certain way, they prefigured the cataclysmic upheavals caused by radioactivity, quantum physics, and relativity.

Another Novelty: The Psychic Force

In 1847 at Hydesville, a modest village in the state of New York, unusual sounds and strange voices were being heard. Spirits of the dead had begun to manifest themselves to the living, according to mediums who quickly multiplied and who soon had their European counterparts. The tables had already turned. It was understood that the sounds and voices were another manifestation of souls who were searching for their reincarnation and for whom the past, the present, and even the future had no secrets whatsoever. In France these errant souls ordered the doctor and pedagogue Hippolyte Rivail (1803–1869) to change his name to Allan Kardec. Under this name, Rivail published the *Livre des Esprits* (*The Book of Spirits*), the first of a series of works that would consecrate him as the grand theologian of spiritualism and the pope of a chapel who, even today, keeps vigil over his tomb at the Paris cemetery of Père Lachaise.

Under the domination of spirits, haunted houses proliferated, objects dematerialized or were moved about spontaneously, and bodies levitated or mysterious blows were felt intermittently. In England, all of these manifestations, so precious to the actors of the séances, aroused the interest of many unprejudiced scientists. Wallace, for example, even though a materialist, a socialist, and an advocate for women's rights and for the collectivization of lands—a progressive in every way—was one of the first to be intrigued. After affirming with Darwin that humans were not exempt from the rules of natural selection, he remarked that intellectual capacities, manual dexterity, the organs of speech, and the loss of hirsuteness had appeared too early in the process of humanization to have been the fruit of such selection. Because function had preceded usage, Wallace saw the appearance of these essential traits as the mark of higher intelligences that manifested themselves to living beings after having guided human evolution. Beginning in 1865, he became an active proselyte for the spiritualist movement, into which he attempted to draw Darwin and Huxley. For some eminent physicists, the miracles that occurred during séances suggested that captivating fields of study were ready for scientific investigation. By uniting their spiritualized fluid to the vitalized fluid of mediums, the spirits endowed the latter with magnetic or telepathic powers that could lead to great discoveries. Was this a coming revolution, without precedent since Copernicus, touching upon the very idea of the nature of the universe? William Crookes was the physicist and chemist who applied himself most completely in this quest. He dedicated his time between 1870 and 1874 to innumerable séances directed by famous mediums.

Crookes (1832–1919) remained outside of universities for his entire life. He was knighted in 1897 and presided over the Royal Society of London between 1913 and 1916. He was one of the most original scientists of the nineteenth century, as well as an unparalleled experimenter with penetrating and productive intuition. We shall cross his path more than once. His father was a prosperous tailor who had five children by a first marriage and sixteen by his second. Crookes, the eldest of the second marriage, had ten children of his own. After an irregular schooling, he was engaged as an assistant to a chemistry professor who trained him on the job. The precocious talents of this practically self-taught man attracted the well-meaning attention of the great Faraday, who himself had started as a bookbinder's apprentice. By the age of twenty-four, Crookes had already established his own laboratory at home and was able to support his growing family from the fruits of his efforts. He achieved scientific renown early: at twenty-nine, his discovery of a new chemical element, thallium, provided for his admittance to the Royal Society of London. Shortly before, he had ventured into scientific publishing when he founded the weekly *Chemical News* in 1859, which continued until 1932. In addition to his fundamental research, Crookes was also interested in countless practical questions, ranging from fertilizers to gold mines, the dying of tissues, beet sugar, livestock prophylaxis, wastewater treatment, and electrical lighting, all of which themes had the common twofold goal of relieving the difficulties of humanity and gaining a certain financial comfort for himself and his family.

Crookes was already a highly visible scientist in 1867, when the tragic death of one of his brothers plunged him into a painful distress. His desire to communicate with the dead youth aroused his interest in spiritualism. With his customary enthusiasm and ardor, Crookes attempted to apply the experimental method to the study of spiritualistic manifestations. He left aside the religious connotations of the phenomenon in order to concentrate on the physical causes. Results came rapidly. Crookes conducted experiments in 1870 in which he observed levitations during séances, where the best mediums were able to produce an "alteration in the weight of bodies." These experiments appeared "conclusively to establish the existence of a new force, in some unknown manner connected with the human organization, which for convenience may be called the psychic force." By contrast with the forces regulated by the implacable determinism thus far categorized by physics, this had the innate possibility of choosing, deciding, and manifesting according to its own consciousness: the mediums' power was thus simple to explain. In a scientific environment that usually welcomed new phenomena, this great discovery was nonetheless received less than warmly. Crookes, who knew his classics, had anticipated such a reaction. In the epigraph of his second publication dedicated to the new force, he called Galvani to the rescue, quoting Galvani's own lament: "I am attacked by two very opposite sects—the scientists and the know-nothings. Both laugh at me—calling me 'the frog dancing-master.' Yet I know that I have discovered one of the greatest forces in nature."

In fact, Crookes's authority in animal magnetism, spiritual theology, and magic psychology labored with difficulty for recognition beyond spiritualistic circles. Considered on one side as an archetypal genius whose indisputable observations collided with the rancid forces of scientific conservatism, Crookes was judged by his opponents to be an honest scientist who was nonetheless naive and abused by mediums who were both clever and without scruples. In spite of four years of efforts, his numerous publications never left the ghetto of spiritualistic literature. Crookes could not show his colleagues the great things he had perceived. Maxwell, who was among the rebels, had immediately rejected all the strange psychic manifestations as obvious nonsense. Kelvin vehemently denounced spiritualism as a "wretched superstition" to which science ought to remain a stranger. And Huxley and the Darwins, both father and son, saw the mediums as a band of swindlers. Fatigued by all the sarcasm and quarreling, Crookes decided to get a fresh start. He quit participating in séances in 1874, but he would not repudiate his opinions later on, any more than Wallace would.

Even if his position was extreme, Crookes had opened a road that was later paved in concrete by the creation in London of the Society for Psychical Research. This group was composed of an enlightened public who were curious about telepathy, clairvoyance, premonition, and other obscure faculties. To note only the few physicists mentioned in the following pages, Oliver Lodge, Joseph J. Thomson, Lord Rayleigh, and Robert Strutt were members of this society, along with Crookes, Wallace, of course, the future prime minister, Lord Balfour, and some illustrious foreigners, such as the French philosopher Henri Bergson. An American society was created in 1884 with the same name, and it counted among its founders a fine lot of Harvard University professors. But no decisive results were presented before the scientific societies. These astonishing psychic phenomena had, meanwhile, been given a physical explanation: it was thought they occurred through the medium of ether, the consubstantial fluid of electromagnetic actions. Oliver Lodge (1851–1940), who was an expert on the question, noted that "we know that ether and matter interact"; and from this understanding he postulated the existence of an analogous interaction between the spirit and ether, which would be "of a closer and more fundamental kind than any indirect action between mind and matter." Unfortunately, Einstein's theory of relativity was soon to relegate ether to the level of a superfluous hypothesis in physics.

At the end of the century there was renewed interest in all forms of psychic research. When the theories of Charcot and Freud began to gain ground, the writings of Arthur Conan Doyle, the creator of Sherlock Holmes, and those of Camille Flammarion contributed to maintaining the practice of spiritualism. In Paris, fashionable society members still gathered at séances. Flammarion regularly saw the Nobel Prize winner for medicine, Charles Richet (1850–1935) in such settings. Fifty years after Crookes's experiments, Richet dedicated to him an eloquent treatise on *metapsychism*, in which spiritualism, expurgated of any sort of religious aspect, was raised to the level of a new scientific discipline. For example, the treatise compared the materializations of "ectoplasmic formations" with the "condensation of nebulae." Richet did not miss the chance to emphasize in his preface the mobile nature of the frontier separating magic from reality: in 1875, who could have seriously foreseen the possibility of transmitting voices over a distance, or of photographing a living person's bones, or of cultivating the germs of contagious diseases in a jar? But Richet had scarcely any followers. The requirements of science appeared to be radically incompatible with those of spiritualism. The great majority of scientists rejected these phenomena as charlatanism, because they could only be glimpsed on the condition that one checked one's good principles—and the proven methods of experimental science—at the spiritualistic cloakroom.

Electricity and the Void: Cathode Rays

As Maxwell had foreseen, the real revolutions came from problems that appeared less profound and whose solutions seemed to be within reach. What we shall describe here has largely been forgotten in our day. Nevertheless, it was basic to the discoveries of X-rays, radioactivity, and the first subatomic particle, the electron. The importance of these experimental works is attested to by the fact that four of the first six Nobel Prizes in physics were awarded to their various authors: Röntgen (1901), Becquerel and the Curies (1903), Lenard (1904), and Joseph J. Thomson (1906).

The initial problem had to do with the conduction of electricity within a gas at very low pressure. One reason for interest in this was the direct connection with the properties of a vacuum and of ether. In order to illustrate the diversity of the ideas that were debated at the time, it suffices to mention that some thought a complete vacuum ought to be a perfect conductor of electricity, because the absence of matter would permit the ether to flow without any resistance.

Since such hypotheses, and ether itself, are now nothing but distant memories, we shall pass directly to discharge tubes, which were utilized to study electrical conductivity in gases. Distant descendants of these apparatuses include cathode ray tubes for televisions and mass spectrometers; the latter were to play a crucial role in the measurement of geological

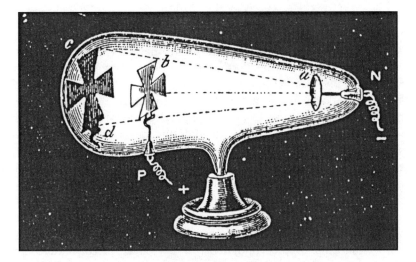

Crookes tube

Figure 9.1 Crookes tube. Matter as radiation in a discharge tube. *P* = anode; *N* = cathode. From Crookes, "Sur la matière radiante," 212.

time. From another point of view, the study of the phenomena that were produced in these tubes considerably facilitated the identification of the radiation that was emitted by radioactive elements.

A discharge tube consists simply of a glass tube and two electrodes at some distance from each other. When voltage is applied to the electrodes, the current that passes through the gas is signaled by a luminous emission. After Faraday's first experiments, conducted in 1833, the study of this phenomenon continued, especially in Germany. Nearly fifty years of efforts had demonstrated the existence of a new type of radiation that could be deflected by magnetic fields, and whose nature depended strongly upon the quality of the vacuum. We have all observed the familiar blinding electric welder's arc, which occurs in air. When a good vacuum is produced in the tube, a bluish glimmer extends from one electrode to the other, and a greenish light begins to appear at the end of the tube opposite from the cathode. The expression *cathode rays*, naturally, was applied to this emission.

Crookes had invented a pump that would produce a particularly high vacuum (about one-millionth of an atmosphere) in order to observe cathode rays better, and he had just confirmed that they were emitted perpendicularly to the cathode before reaching the opposite extremity of the tube, where an intense phosphorescence appeared at the point of impact. Crookes observed, in addition, that a metal leaf positioned to intercept the trajectory of the cathode rays produced a shadow of the same form. For him, these were indications that the rays were particles of matter, "radiant matter," which traveled in a straight line, rather than being the manifestations of a wave, as Hertz, for one, believed. With a remarkable intuition, Crookes suggested in 1879 that a new type of radiation, corpuscular, could exist independently of electromagnetic waves, and he foresaw that "in studying this fourth state of matter we seem at length to have within our grasp and obedient to our control the little indivisible particles which with good warrant are supposed to constitute the physical basis of the universe." He noted that with this radiant matter, "we have actually touched the border land where Matter and Force seem to merge into one another," and we dare "think that the greatest scientific problems of the future will find their solution in this Border Land, and even beyond; here it seems to me," he concluded, "lie ultimate Realities, subtle, far-reaching, wonderful."

A long decade elapsed without any crucial experiment able to settle the differences between the prophetic views of Crookes and the interpretations of the German school, according to which it was a new type of electromagnetic radiation that was being observed. At the beginning of the 1890s the situation had begun to unblock. In Germany, Philip Lenard (1862–1947), a former pupil of Hertz at Bonn, observed that the cathode rays could traverse thin sheets of metal that were impermeable to gas, as well as opaque to light, and that their absorption, curiously, did not depend upon the density of the obstacles placed in their paths; furthermore, they produced images on photographic plates. Jean Perrin (1870–1942), who became better known for his advanced experiments on Brownian movement, collected cathode rays in a Faraday cylinder and demonstrated in Paris in 1895 that they were negatively charged with electricity. (We shall note, by the way, that these experiments led Lenard and Perrin, independently, to suggest the correct structure of the atom, which is essentially made up of a void).

At Cambridge, the physicist Joseph J. Thomson (1856–1940) considered cathode rays to be a phenomenon of secondary interest. Lenard's experiments subsequently convinced him of their corpuscular nature. By deflecting the rays using electrical fields, he demonstrated that particles lighter than atoms were involved. In order to identify them, he finally borrowed from his compatriot Arthur Schuster (1851–1934) a technique for deflecting electrical charges by electrical and magnetic fields. Thomson applied these two types of fields around a discharge tube to deviate the corpuscles and determined, in 1897, the relationship between their

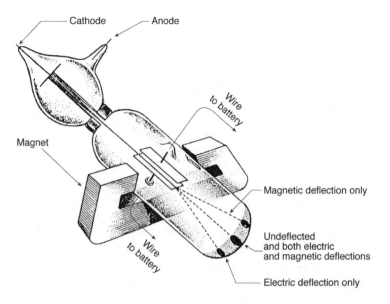

Figure 9.2 Deflection of electrically charged particles by an electric field (produced by electrically charged plates) and a magnetic field (created by a magnet). After G. P. Thomson, *J. J. Thomson and the Cavendish Laboratory*, 50.

charge and their mass. In 1899 other experiments allowed him to measure their mass and their charge separately. A remarkable fact appeared: these negatively charged corpuscles were always the same, regardless of the metal of the cathode. And another extraordinary fact followed: their mass was two thousand times less than that of the lightest ion, hydrogen (namely, a proton), even though they had practically the same electric charge. A dogma had collapsed: the atom was not the smallest unit of matter! These corpuscles quickly took on the name proposed in 1891 by the Irishman George Stoney (1826–1911) for the fundamental unit of electric charge, the *electron*.

With the increase of the vacuum that could be produced experimentally, the visibility of the arc indicating the complex interactions between electrons and the rarefied gas began to diminish. Other than the walls of the discharge tube, there was no longer any major obstacle to the electrons' movements.[1] A connection had been established between

1. Interpretation of the phenomenon: the gases present in trace quantities are ionized in the discharge tube; the positive ions produced are accelerated toward the cathode, against which they collide; the electrons that are then ejected from the cathode are displaced at great velocity until they are stopped by the opposite wall of the tube, provoking the emission of X-rays from the point of impact.

electricity and matter; the "classical" problem of conducting electricity through a gas had unknowingly given rise to subatomic physics. In the hands of experimenters of the first order, such as Crookes, the humble vacuum pump had particularly well illustrated the adage that the history of discoveries is largely the history of scientific instruments. As for the electron, the enviable location of its discovery was Cavendish Laboratory in Cambridge. In 1884 J. J. Thomson had been chosen, amid general surprise, to direct the laboratory at the age of 28, succeeding the famous Maxwell and Lord Rayleigh. After becoming a physicist for lack of sufficient money to undertake engineering studies, this young theoretician had accepted a position that had twice been declined by the great Kelvin, who did not wish to leave Glasgow. J. J. Thomson quickly justified the bet that had been placed on him. Although he was a remarkable inventor of apparatuses, he was actually a clumsy experimenter; he was known for concentrating on new phenomena, leaving others the trouble of determining the "second decimal." He had taken charge of one of the most famous schools of experimental physics, ready to seize whatever opportunities physics would present.

The Knight of X-rays

The rapid progress of the electron has taken us to 1899. A four-year backtrack is now necessary for us to meet Wilhelm Röntgen (1845–1923). At the age of fifty, Röntgen was the director of the Physics Institute of the University of Würzburg. He was a solitary and secretive experimenter, meticulous to the point of fabricating his own apparatus, and he applied himself to detecting the most delicate of effects. Seven years earlier, his experiments in electromagnetism had already won him a certain renown. Röntgen had also been interested in cathode rays since 1893. In autumn of 1895, his efforts were crowned, far beyond every expectation.

Because Röntgen burned all his notes, the circumstances of his discovery have been the object of numerous speculations. According to his account to an American journalist, the story began one day when Röntgen was working in his laboratory with a discharge tube lent him by Lenard, which had been enveloped with opaque cardboard. Röntgen remarked that a fluorescent screen placed at some distance from the tube had registered a feeble luminescence. Perhaps he was aided in his observation by the fact that he was color-blind and that his handicap was compensated by a remarkable sensitivity to contrasts. The fluorescence became more intense when the screen was brought closer to the tube, raising questions as to the nature of this radiation that passed through the tube's glass, as well as the cardboard. A second, even stronger surprise resulted when Röntgen by chance placed his hand between the tube and the screen: he could see the image of his own bones on the screen! Considering how long cathode rays had been studied, it was strange that such observations had never been made before. With the specter of spiritualistic séances, so dear to Crookes, hovering about in the room, Röntgen hastened to fix the image of his own bones upon a photographic plate, to prove that he had not been hallucinating.

Whatever it was, Röntgen did not take long to understand the importance of his discovery. He used the name *X-rays* to christen these mysterious radiations that ignored walls and penetrated bodies. He camped in his laboratory, dedicating to them long weeks of experimentation in absolute secret. The X-radiation occurs at the location where a phosphorescence appears when the cathode rays are intercepted by an obstacle, such as the wall of the discharge tube. By contrast with visible light, X-radiation is not refracted when passing from one medium to another. And by contrast with cathode rays, it is not deflected by magnetic fields. It is more penetrating than any known radiation, and still it is partially absorbed by all substances in varying degrees. From this fact the medical applications for this radiation became immediately apparent.

At Christmas, Röntgen drafted his results, out of fear of being outstripped. He asked one of his friends to publish them in the proceedings for the meeting of the Society of Physical Medicine at Würzburg, which were already at the press. He also sent them to a learned group of eminent physicists in Germany and abroad, upon whom the first radiographs produced an effect just as profound as they had upon him. An Austrian physicist who received the radiographs immediately forwarded them to his father, the editor of a Viennese newspaper, and they were reproduced on the front page on 5 January 1896, even before their scientific publication. The fascination evoked by the image of a "living" skeleton and the distribution of the related photographs caused an extraordinary impact on the greater public, abroad as well as in Germany. Because of this publicity, Röntgen was summoned to appear on 13 January before the Kaiser, who was curious to observe in person the effects of these practically miraculous rays. Because Wilhelm II was satisfied with the demonstration, he made Röntgen a knight of the Second Class of the Order of the Crown of Prussia, on the spot. This was a high distinction within a class composed of four orders, and the first of a series of distinctions that culminated in Röntgen's receiving the very first Nobel Prize for physics in 1901.

Naturally, the small scientific world did not remain isolated from the surrounding enthusiasm. In France alone, an avalanche of more than three hundred communications, dedicated to the most varied effects of X-rays, descended upon the Academy of Sciences during the first half of 1896. One of the first, on 27 January, illustrated the new role of the press in scientific communications, with its advantages and disadvantages: "I shall admit straightaway that I know nothing regarding Professor Röntgen's discovery, other than some rather vague notions drawn from the daily newspapers, and that I am still ignorant of his actual experiments," began one author, the physicist Jean Perrin, before describing manipulations resembling those made by his German colleague. Throughout the world physicists and chemists, as well as doctors and biologists, had their own observations. The tubes began to diversify in accordance with their various applications at a speed that would have astounded even the most ardent evolutionists. The first attempt at a treatment for cancer was made in February. Four months after Röntgen's discovery, the renowned American inventor Edison boasted of his "X-ray apparatus of all kinds for professionals and amateurs," which he had already put on the market. And Röntgen was already bargaining tube prices with Siemens, his supplier. Industrials such as the brothers Auguste (1862–1954) and Louis (1864–1948) Lumière, the newly famous cinematographers, also contributed to the discussion: they presented two notes to the Academy of Sciences in February 1896 with their concerns about producing film that would be more sensitive to X-rays than their plates used for photography, of which a certain Becquerel was making use. In spite of an abundance of scientific activity, the true nature of X-rays still remained elusive. Fifteen more years of effort would demonstrate that these rays are actually light of a very high frequency.[2]

It was not only scientists who were inspired by the spectacular properties of X-rays: humorists also exploited the theme, in caricatures and amusing stories. For example:

She: "I wish some photographs taken."
Photographer: "Yes, Madam, with or without?"
She: "With or without what?"
Photographer: "The bones."

2. Following the example of J. J. Thomson, a number of insightful physicists had remarked that X-rays seemed to be related to light waves of very short wavelength. Not until 1912 did the diffraction of X-rays through crystals prove that these rays are a hard radiation located well beyond ultraviolet, on the opposite of the visible spectrum from the soft Hertzian waves.

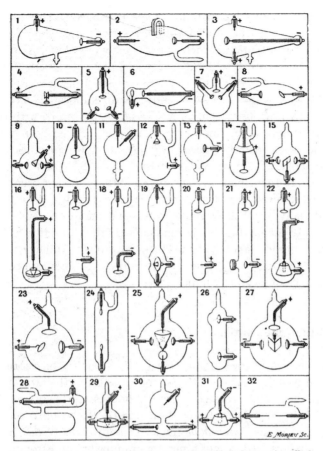

Nᵒˢ 1 à 32. — Divers modèles d'ampoules pour radiographie et fluoroscopie. — Nᵒˢ 1
et 2. Ampoules de Crookes. — Nᵒ 3. Ampoule Séguy. — Nᵒ 4. Ampoule Wood. —
Nᵒ 5. Ampoule Séguy. — Nᵒ 6. Ampoule Chabaud et Hurmuzescu. — Nᵒ 7. Ampoule
Séguy. — Nᵒ 8. Ampoule Tompson. — Nᵒ 9. Ampoule Séguy. — Nᵒ 10. Ampoule
d'Arsonval. — Nᵒ 11. Ampoule Séguy. — Nᵒ 12. Ampoule Puluj. — Nᵒ 13. Ampoule
Séguy. — Nᵒ 14. Ampoule d'Arsonval. — Nᵒ 15. Ampoule Le Roux. — Nᵒˢ 16,
17 et 18. Ampoules Séguy. — Nᵒ 19. Ampoule de Rufz. — Nᵒ 20. Ampoule
Crookes. — Nᵒˢ 21, 22, et 23. Ampoules Séguy. — Nᵒ 24. Ampoule Röntgen. —
Nᵒ 25. Ampoule Brunet-Séguy. — Nᵒˢ 26, 27. Ampoules Le Roux. — Nᵒ 28. Ampoule
Colardeau. — Nᵒ 29. Ampoule Séguy. — Nᵒ 30. Ampoule Colardeau. — Nᵒ 31.
Ampoule Séguy. — Nᵒ 32. Ampoule Röntgen.

Figure 9.3 Imagination at the service of Röntgen's tubes, during 1896 alone. From G. Séguy,
"Étude expérimentale des ampoules utilisées en radiographie et fluoroscopie," *La Nature* 1225
(1896): 385.

But some people of more serious mind wondered whether these rays that made light sport of walls would pose serious threats to privacy.

An obvious question, which is not merely anecdotal, is, Why had X-rays not been discovered earlier? Crookes had noted in 1879 that the photographic plates he kept in the laboratory were mysteriously fogged. Even strange shadows with well-defined outlines appeared on the plates without any obvious reason. Other physicists, beginning with Lenard, had long observed the fluorescence that was induced near discharge tubes. Because Lenard had taken a succession of positions that required him to change locations between 1894 and 1898—from Bonn, to Breslau, to Aachen, to Heidelberg, to Kiel—his conditions had not been the best for allowing him to outstrip Röntgen. Hence, the discovery of X-rays escaped many scientists such as Crookes and the inconsolable Lenard, who finally became Hitler's scientific surety in the opposition to "Jewish physics." The fact was that scientists generally were too intensely focused on the cathode rays to consider the shadows and fluorescence, which seemed to be of secondary importance.

The Birth of the Martians

We must not allow the sensation created by X-rays to conceal the other great events of the year 1895. One of these was the birth of the Martians, to whom the public accorded undisputed honors, especially after the English novelist H. G. Wells featured them in his *War of the Worlds*. After Fontenelle and his *Conversations on the Plurality of the Worlds*, Martians had been mentioned various times, for example by the astronomer Herschel, who affirmed in 1784 that the inhabitants of Mars "probably enjoy a situation in many respects similar to ours." These were not merely speculations by any means, because the Milanese astronomer, Giovanni Schiaparelli (1835–1910) had essentially confirmed the existence of Martians. During the course of his important research involving comets, Schiaparelli began to study Mars in 1877 and kept on until 1890. Augmenting the quality of the innumerable observations and data for the red planet, he identified "seas" and "continents" and confirmed the existence of long rectilinear marks, called "canals," which had the surprising characteristic of multiplying or disappearing between one observation and the next. Because Schiaparelli was a punctilious astronomer and aware of the difficulties of interpretation, he refrained from proposing fixed conclusions. He wrote in 1893 that "in spite of the almost geometric appearance of their whole system," these canals "are the product of the evolution of a planet, much as the English Channel or the Channel of Mozambique are on the earth." But this assertion notwithstanding, he reported two pages further along that "their singular aspect, and the fact that they are drawn with absolute geometric precision, as if they were the product of rule and compass, have induced some people to see in them the work of intelligent beings, inhabitants of the planet. *I should be very careful not to combat this supposition, which involves no impossibility.*"

The unusual nature of Mars had effectively been recognized, as the indefatigable Camille Flammarion revealed in 1892 in *The Planet Mars and Its Habitability Conditions*. What remained now was to make contact with the inhabitants! This project was practically a fait accompli for the American astronomer Percival Lowell (1855–1916), whose love for Mars and whose remarkable talent as popularizer made him the veritable father of the Martians. He was born into a patrician family of New England and studied mathematics before turning to business and developing a passion for the Orient. He had lived in Japan and Korea, where he worked as a diplomat, and he wrote accounts of his travels. As Lowell was approaching his forties, he turned his attention to astronomy, with a success that by now is well known. He predicted the existence of Pluto, the farthest-away planet of the solar system known at that time. He recognized the advantage of installing telescopes on mountains far away from the atmospheric difficulties encountered near cities. In 1893 Lowell's intrigue for the Martian canals inspired him to build an observatory from scratch at more than 2,000 meters altitude, near Flagstaff, Arizona. His initial observations were published in 1895, in *Mars*, the first of a series of books he wrote, with lively pen, about the planet's geography and the customs of its inhabitants.

The existence of life on Mars was widely accepted until the middle of the twentieth century. According to a long-standing belief, the contrasts between the "seas" and the "continents" marked the limits between desert and vegetation. Lowell pushed the reasoning even further. A close examination of the Martian geography led him to postulate that the canals represented a vast irrigation system that had been excavated by intelligent beings so as to distribute the water, which was abundant only at the poles. Lowell even suggested that the Martians could be three times larger and twenty-seven times stronger than their terrestrial neighbors—no longer the odd little green men! And so Martians were born in 1895, the inhabitants of a very old planet, and doubtlessly they were beings "in advance of, not behind, us, in the journey of life." Giving a scientific basis to what previously had amounted to nothing but gratuitous speculations, some dating as far back as the "Chaldean shepherds" of antiquity, Lowell noted that "man is but a detail in the evolution of the universe," and that "though he will probably never find his double anywhere, he is destined to discover any number of cousins scattered through space."

Over the years, through the support of new observations, these hypotheses came to seen as scientific certainties, presented by Lowell in his successive editions entitled *The Evolution of Worlds, Mars as the Abode of Life*, and *Mars and Its Canals*. The debate broadened in scope. In *Mars and Its Mystery*, the zoologist Edward S. Morse (1838–1925) supported Lowell and the comparisons he made with the earth to elucidate the Martian mysteries. However, in *Is Mars Habitable?* the famous evolutionist Wallace concluded that "Mars, therefore, is not only uninhabited by intelligent beings such as Mr. Lowell postulates, but is absolutely UNINHABITABLE." The majority of astronomers remained skeptical of life on Mars. It was not until 1960 that the American space expeditions of the *Mariner* established the absence of evolved life upon Mars and showed that the canals were actually fortuitous alignments of dark zones. Meanwhile, in 1914, at the Lowell Observatory in Flagstaff, the relative rates of the separations of galaxies were measured for the first time. The now classic Big Bang theory and the age of the universe thus owe something, at least indirectly, to the Martians.

The Surprise of Noble Gases

The last great discovery of the year 1895 was that of the noble gases, even if it might have seemed somewhat less spectacular. The first of these gases, in fact, had been isolated one hundred years earlier by Henry Cavendish (1731–1810). A great and extremely wealthy aristocrat of an almost pathological timidity, Cavendish was the first person to synthesize water and was the discoverer of hydrogen. In 1785 this talented physicist observed, after the violent oxidation of air, a gaseous residue that represented 1/120th of the original volume of the nitrogen, but no one ever demonstrated any interest in this tiny residue. By a strange coincidence, Cavendish's observation returned to the foreground just shortly before radioactivity studies expressed a need for it.

This discovery of the noble gases had an unusually strong impact, considering it was a finding made in the realm of chemistry. For more than one hundred years, scientists researched the strangest of minerals in the hope of extracting some new chemical elements from them. And indeed, in January 1895, it was announced that a new element had just been isolated, quite simply, from the air. This new gas, which we all breathe without realizing it, was argon. It owes its Greek name to its remarkable chemical inertness. The English physicist Lord Rayleigh and the Scottish chemist William Ramsay found that the air contains nearly 1 percent argon, which is thirty times more than the familiar carbon dioxide. By measuring atomic masses, Rayleigh discovered that atmospheric nitrogen is slightly denser than the nitrogen liberated from chemical reactions. Ramsay later read an old dissertation by Cavendish and wondered whether another, denser gas might be present in the atmospheric nitrogen.

The third baron Rayleigh, whose ancestors had made their fortune in the milling trade, was John William Strutt (1842–1919). He distinguished himself in many branches of physics throughout the course of a long career. In 1871 he resolved an enigma as old as humanity itself, the reason for the sky's blue color, which is due to the diffusion of light by the gaseous air molecules. His theoretical work on the sound propagation is still considered authoritative today. With the exception of a short period in Cambridge, when he succeeded Maxwell, he performed his experiments in a laboratory installed in his mansion at Terling Place in Essex.

Somewhat younger than Rayleigh, William Ramsay (1852–1916) was a professor at the University College of London. Ramsay had proved quite helpful in separating argon from atmospheric nitrogen. He was the descendant of doctors on his mother's side and of seven generations of dye-makers on his father's side. He was good company, a constant optimist, a great walker and swimmer, and an accomplished musician; much later his flair for languages influenced his idea of conceiving a hieroglyphic Esperanto. His mother wanted him to become a pastor, but he was inclined from the beginning toward chemistry, organic at first, before focusing his interests on the properties of gases.

Three months after having isolated argon, Ramsay struck again. A London mineralogist had just attracted his attention to a curious phenomenon described in 1890 by William F. Hillebrand (1853–1925). This chemist from Washington, D.C. had observed that uranium-bearing minerals, on which he was a specialist, released nitrogen when they were dissolved in appropriate hot solvents. He wondered whether this unexpected nitrogen could actually be argon. Ramsay then hastened to purify the gases released by these curious minerals. He did indeed find a bit of nitrogen and also of argon, but at the same time he isolated another inert gas, which proved to be helium, a second member of the family of noble gases. Helium had previously been unknown on the earth, even though it had been identified through the observation of the sun's light spectrum made during an eclipse in 1868 by the astronomers Jules Janssen (1824–1907) and Norman Lockyer (1836–1920). Hillebrand himself had noted the presence of some unknown lines within the spectra of the sun's gases, but he did not take the observation any further. How was helium a connection between uranium-bearing minerals and the sun? This bizarre question quickly proved to have considerable importance. Ramsay worked along with a young chemist named Frederick Soddy (1877–1956) to resolve the question. Like Ramsay, Soddy had been raised in a Calvinist environment, and also like Ramsay, he was destined for science since childhood. In 1895 Soddy was eighteen and was preparing for his studies at Oxford University, so even though he had already authored a note dedicated to the effects of ammonia on carbon dioxide, it is still early to say more about him.

The Year 1895: Vignettes from the Fireside before the Battle

With wireless radio, cathode rays, X-rays, Martians, and noble gases, the year 1895 had been stamped with the seal of scientific effervescence. New discoveries were made in every sector of physics, while Crookes's followers dreamed of even more staggering feats. As the century drew to a close, industrial activity was in full expansion; it transformed people's lives and required many hands to keep it going. Even though young engineers were trained by the thousands each year, fundamental research engaged only a relatively small number of scientists. The number of university chairs for physics throughout the entire world was in the hundreds. The holders of these chairs knew each other well enough that it was not necessary for them to include their addresses beneath their names at the beginning of a publication, and in the best journals one could find paragraphs mentioning a friend who had recently gotten well, the promotion of a comrade, or the death of a teacher. Within this microcosm, old Europe still prevailed, with England and Germany in the lead and France in the decline, but the United States was emerging as a great scientific power. Following Harvard and Yale, which were already old universities, new institutions were created all through the second half of the century, and these, from MIT to Stanford University, were not long in becoming famous as well.

The young Pole Maria Sklodowska (1867–1934) had long dreamed of France. In 1895 she married Frenchman Pierre Curie, a newly promoted professor at the École Municipale de Physique et de Chimie Industrielles in Paris. Maria was the youngest of five children; by the age of eleven, she had already lost her mother, who was a school principal. Her father was a professor who was made to suffer because of his nationalist and liberal opinions; his post was located in the portion of Poland that had been annexed by Russia. In order to help one of her sisters to go to study in France, Maria had worked for more than three years as a private tutor out in the countryside. By 1891, when it was her turn to go to Paris to pursue her scientific studies, she had saved only a meager sum. She was true to herself: proud, independent, obstinate, and brilliant. Through hard work she was able to fill the gaps occasioned by her irregular schooling. Her devotion to science played a crucial role in the love that her future husband felt for her.

At twenty years of age, Pierre Curie (1859–1906) had decided to "make life a dream and to make a dream become reality." He held to his purpose. His father was an inveterate republican, a anticlerical confident in the future of science, and a nature lover, who in the end drew but a modest income from his medical activity. It is an interesting fact that Pierre never went to school. He was instructed at home, and it was there, with his father, that he learned the techniques of observation and experimental science. After his physics studies at the Sorbonne, he became a respected theoretician and experimenter. His thesis, which he defended in March of 1895, laid the foundations for modern magnetism, and it would prove to have fundamental implications much later for geology. Kelvin, who had originally been seduced by Curie's precocious experiments on piezoelectricity, helped him obtain a position at the École de Physique et de Chimie. If Curie had been more interested in his career, he could have had a university chair much earlier, but like his wife, he disdained honors and money, after the fashion of Faraday, who had declined the enviable honor of becoming baronet, because "a title that could not teach him anything was of no use to him."

The Curies were idealistic, but they were lucid and careful in their practical applications, as well as relentless workers who were deeply devoted to science and inimical to hasty publications or quarrels about priorities. Their legendary summer bike rides have no need of any further mention. Not even scientific fashion had any hold on them: during the hubbub about X-rays, Pierre concentrated on the growth of crystals, and Marie was preparing for the competitive examination for physics teachers before commencing her first research the following year on the magnetization of steel.

In 1895 Ernest Rutherford (1871–1937) was at the other end of the world. He was working in the family vegetable garden in New Zealand when his mother came to inform him that he had won a research scholarship in England. "That's the last potato I'll dig!" was his response. He was about to celebrate his twenty-fourth birthday. The fourth of twelve children of bold pioneers of the New Zealand countryside, Rutherford had demonstrated early his great practical cleverness as well as his marked interest in science. He had a harsh upbringing, and he was ambitious, with an energetic and direct character. From scholarship to scholarship, he asserted his talents in mathematics and physics to the point where they finally led him to Christchurch. He made his scientific debut at a university college with seven professors and 150 students, on a theme that was also familiar to Marie Curie, the magnetization of steel.

This brilliantly managed work bought Rutherford his ticket to England. The long journey was financed by a loan. He had to leave his fiancée and would not see her again until five years later. At the end of summer in 1895, Rutherford arrived at Cambridge, where he was welcomed by J. J. Thomson, age thirty-nine. As Rutherford kindly depicted him in one of his numerous letters to his beloved, he found Thomson "very pleasant in conversation and not fossilized at all." At Christchurch his work led him to invent, before Marconi, an adjustable Hertzian wave detector, which brought him a flattering reputation among his new English colleagues. Thomson pressed Rutherford to improve his invention immediately. The waves in question, called *radio waves*, could travel through any sort of weather. Even if their range is only on the order of a kilometer, they must have some practical applications, thought Rutherford. Lighthouses could emit them; boats could receive them; treacherous reefs could be avoided; and Rutherford himself could draw substantial benefits from his apparatus. But then the X-rays appeared. J. J. Thomson discovered that they enhanced the electrical conductivity of gases, and he assigned his new student to study the phenomenon. Rutherford abandoned his radio wave detector, forgot about the potential shipwrecks, and set himself to elucidating the ionization of gases with Thomson. It did not take him long to justify the reputation he quickly developed at Cambridge: they said, "We've got a rabbit here from the antipodes and he's burrowing mighty deep!"

Now all we need do is to return to Paris to meet Henri Becquerel (1852–1908)—hardly an unknown figure at the time. After graduation from the École Polytechnique, he occupied the chairs of physics at that institution and at the Conservatoire National des Arts et Métiers, as well as at the National Museum of Natural History. He succeeded his father and his grandfather in the Museum chair, and he was even born in the Museum house, where his parents and grandparents had lived. In 1859 chance had it that Pierre Curie was born just across the street, which was named Cuvier. Like his two ancestors and the father of his late first wife, Becquerel was a member of the Academy of Sciences. He had been elected to the Academy in 1889 at the precocious age of thirty-six, the same age at which Pierre Curie had just defended his thesis. Becquerel had engendered appreciation for his experiments on the passage of light through crystals.

On 20 January 1896, Becquerel was attentive when the mathematician Henri Poincaré (1854–1912) hastened to demonstrate to his colleagues at the Academy of Sciences the X-ray photographs he had received from Röntgen. The X-rays are emitted at the location where the discharge tube becomes phosphorescent under the effects of cathode rays. "Thus, it is the glass that emits the Röntgen rays," Poincaré specified, "and it emits them as it becomes fluorescent. So then, can we not ask whether all bodies with sufficiently intense fluorescence will emit Röntgen's X-rays, in addition to light rays, *regardless of the cause of their fluorescence?*" This was the question that Becquerel decided to study during the winter of 1896.

According to popular wisdom, a well-presented problem is half solved, and Becquerel did not contradict the adage. He proceeded simply to determine whether exposing phosphorescent crystals to intense lights would also induce radiations that had, until that time, gone undetected. Since phosphorescence was an old family affair, Becquerel found the extensive Museum collection available for his choice of appropriate crystals. He needed only to irradiate these and watch for the appearance of a fogging on the photographic plate. Unfortunately the first experiments were negative; even when submitted to the violent sparking of electric arcs, the zinc blende and fluorite crystals left no traces on the plates. But Becquerel was not a man to beat a retreat after the first upset: he waited until the return of the fine uranium salts that had been lent to one of his physicist friends, and then he resumed his experiments.

TEN

The Sun, the Earth, Radioactivity— and Kelvin's Death

THE STRANGE TASTE OF A NEW PHYSICS The nineteenth century ended momentously for the scientific world, with the discoveries of "uranium rays," radioactivity, and radioactive transmutations, by Becquerel, the Curies, and Rutherford-Soddy, respectively. The first radiometric datings were presented by Rutherford, Boltwood, and Strutt, but their crude methods failed to convince the geologists, who had become very wary of physicists after the dismissal of Kelvin's ideas.

Becquerel and Uranium Rays

In the year 1896, February 26 and 27 were two days on which the sun appeared only intermittently over Paris. In the near-rustic environment of his laboratory at the National Museum of Natural History, facing Curie's house, Henri Becquerel had stored some photographic plates in a drawer, and on these he had deposited a thin and transparent crust of uranium potassium sulfate. On March 1, the sky was still overcast, and the physicist renounced his plans to expose the uranium salts to the sunlight. Without further delay he developed his plates, expecting to see nothing more than the faintest of images. Surprise! The black outlines of the crystals were clearly visible on the plates. Without any sort of any previous exposure to light, the uranium salts had produced a radiation that passed through the opaque carton protecting the plates, producing an image on the gelatine silver bromide.

On March 2, Becquerel described to the Paris Academy of Sciences his discovery of "invisible radiations emitted by phosphorescent bodies." His observations of the evening before were supplemented by other experiments that demonstrated the penetrating power of these radiations that could pass not only through the carton, but also through plates of glass or copper, or even a sheet of aluminum. The obvious conclusion was that these invisible radiations resembled the X-rays of Röntgen's recent staggering discovery. Becquerel's finding, however, received no publicity at all. A dry three-page note on the effects of uranium potassium sulfate crystals on a gelatine silver bromide plate was scarcely something to excite the crowd's imagination.

Becquerel's difficulty, of course, was that he was in terra incognita. X-rays had just come out of limbo, while the electron still took its delight there, and the atom remained a controversial entity about which there was only a single point of agreement, its indivisibility (a-tom = that which cannot be cut). After centuries of alchemical and chemical activity, the problem was still unresolved. Amid the public's general indifference, Becquerel buckled down to his work, which clearly promised something out of the ordinary.

In the beginning, Becquerel's experiments produced their share of errors. The most striking of these was communicated to the Paris Academy of Sciences on 24 February 1896, when Becquerel related that after exposure to the sun, the uranium salts emitted a radiation resembling X-rays. But this statement was completely false. Becquerel had no proof that the observed radiation had been excited by sunlight. He made a definitive test in the following days when the sky was overcast, and he confirmed that neither the radiation nor its intensity was affected by the amount of previous exposure to sunlight. The unprovoked radiation of uranium salts was established as a new discovery, and in a week the author corrected his error.

Within one short trimester after his false start, Becquerel determined the principal physical effects of the radiation. He especially recognized its penetrating power and its capacity to render air electrically conductive; this property, which is also characteristic of X-rays, made precise measurements possible for the quantities of radiation emitted. In the end, it was clear that radiation is an intrinsic property of uranium itself and that it is independent of the chemical combinations of the element. Uranium thus represented "the first example of a metal exhibiting a phenomenon of the order of an invisible phosphorescence." In light of this interpretation, Becquerel found himself in a context that would not have disoriented his father, who was the most eminent specialist on phosphorescence of the nineteenth century. He later returned to other subjects and did not publish anything about uranium rays between April 1897 and March 1899.

Radioactivity's Early Childhood

Lost in the flood of communications dedicated to X-rays—some thirty odd papers arrived each semester at the Paris Academy of Sciences a-lone—there were a few publications that registered an echo of Becquerel's work. In the prestigious *Annalen der Physik*, from which Einstein later set off his first shockwaves, H. Muraoka, a physicist of Kyoto, reported that glowworms emitted rays resembling uranium radiation. Despite the spec-ulation that batteries of these worms might replace Crookes's tubes in the burgeoning field of radiology, Muraoka's discovery was quickly found fallacious: the photographic plates had not been fogged by radiation, but by the chemical action of the vapors that the little creatures released.

Among the rare experiments that were solidly founded, there was one regarding the electrical conductivity induced in gases by means of radia-tion; this was the work of that seventy-three-year-old physicist who was always on the lookout: Kelvin. Another on the same theme was signed by Ernest Rutherford. The Germans Johann Elster (1854–1920) and Hans Geitel (1855–1923) also entered on the scene; they were two great spe-cialists of electrical phenomena in gases, including the atmosphere. They were nicknamed the Castor and Pollux of physics because of their long collaboration, which lasted some forty years. They taught mathematics and physics at the *gymnasium* in the small city of Wolfenbüttel, east of Hanover. They had their laboratory set up in their own home, between the telescope and a menagerie of small exotic animals, and this was where Elster and Geitel confirmed the astonishing persistence of ura-nium radiation. Even more importantly, they discovered subsequently that this radiation was unaffected by any sort of external action, whether chemical or physical, and it displayed an equal disdain for temperature variations or bombardments by cathode rays.

In comparison with the frenzy surrounding X-rays, the field of ura-nium radiations remained scantily explored. The literature dedicated to this radiation was meager indeed when Becquerel took up his pen again in March 1899 to review three years of study. But this was the inevitable calm preceding the storm. Five years later, a close competition was taking place on both sides of the Atlantic, and Rutherford and Soddy had written treatises on the subject. Between the two of them they had only sixty-two years of age, and they had developed into uncontested authorities on these new phenomena. What happened then? The storm had been let loose by the Curies.

If one of the marks of a great scientist is knowing how to identify a fruitful path amid the confusion of new observations, then here, just like Volta one hundred years earlier, Pierre and Marie Curie displayed intuition of the first order. Although uranium radiations were still poorly distinguished from X-rays—even in the tables of contents of the proceedings from the Academy of Sciences—their manifestation suggested a very simple question to the Curies in the autumn of 1897: Is uranium the only element that emits these radiations? That is, could one determine "whether bodies other than uranium compounds can make the air electrically conductive?" Marie Curie began taking systematic measurements right after the birth of her first daughter, the future physicist Irène Joliot-Curie, and found that only one other element was active. This was thorium, the heaviest element in the periodic table after uranium. Gerhard Schmidt (1865–1949) had just published this same result independently in Germany, but Marie Curie took it much further, with her attempts to determine with her electrometer the connection between the intensity of the radiations and the uranium content of the compounds that were studied. Using such quantitative measurements, she discovered two uranium-bearing minerals, pitchblende and torbernite, that were more active than uranium itself. She concluded from this unexpected observation that these minerals contained an unknown element that was even more active than uranium.

The search for this element has become one of the legends of science. Marie treated increasingly larger quantities of uranium ore residues with aggressive chemicals to isolate the active portions. Pierre abandoned his work on crystal growth—temporarily, he thought—to help his wife. Napoleon's support for physics had unfortunately faded, and the Curies did not even have a laboratory worthy of the name. And so it was in "an asphalt-paved shed, with a roof that did not even keep out the rain, like a greenhouse in the summer, and poorly heated by a cast-iron stove in the winter," that "we passed the best and the happiest years of our existence," as Marie wrote nostalgically after Pierre's accidental death in 1906. Their destitution was indeed the context for some authentic pleasures. "As we had no furniture for storing the radiant products that we obtained, we placed them on the tables or on planks, and I remember the enchantment we felt on just coming into our workplace at night, when we perceived, all around us, the faintly luminous outlines of the products of our work," she added. Their self-denial, the massive irradiations they endured, and the trying conditions of their labor are all too well-known to recount yet another time.

Three new radioactive elements were then identified in rapid succession. Polonium was described in July 1898, in a note titled "On a New Radio-Active Substance," and it was here that the qualifier *radioactive* appeared for the first time (without definition and only in the title) to designate elements that spontaneously emit radiation.[1] Five months later, radium was identified, causing even more excitement. Actinium came in rapidly to join the other two. It was announced in October 1899 by André Debierne (1874–1949), who was a student of Pierre Curie. To obtain a few tenths of a gram of radium chloride, the Curies had to treat several tons of pitchblende. These elements responsible for the ore's high radioactivity were thus about one hundred thousand times more active than uranium. After these results were announced, an immense interest in radioactivity became widespread. Beyond X-rays, radium came to captivate both the press and the public at large, as "the most wonderful and mysterious metal in the world." A "radio-mania" developed to the point where a product's radioactivity became its publicity argument. And as the medical needs for radium climbed, speculation overwhelmed the pitchblende deposits. This development was much to the displeasure of Crookes, who sighed: "Were I younger I would pay all off and buy the mine myself and work up the radium and become a rival to Rothschild in wealth!" But geologists showed no interest yet in radioactivity. It is true that radium and polonium had been found in mineral substances, but pitchblende was a rarity, not something of which entire mountains were made.

1. The impossibility of perpetual motion had already been admitted for a good century when the Paris Academy of Sciences published Becquerel's and the Curies' notes. From the point of view of conservation of energy, uranium and thorium presented a redoubtable enigma in relation to the source of the energy that was inexorably dissipated by their radiations. The phosphorescence invoked by Becquerel did not pose the principal difficulty: the energy released by the radiation had previously been provided to the element under the form of a light that remained to be identified. Similarly, Marie Curie suggested that space is "constantly being traversed by rays analogous to Röntgen's rays, but much more penetrating, and they cannot be absorbed, except by certain elements having higher atomic weights, such as uranium and thorium." Sklodowska Curie, "Rayons," 1103. The activity of these elements induced by radiations coming from all about in space is, as a consequence, called *radio-activity*. Although she had anticipated cosmic rays, Marie Curie was wrong about her explanation. In spite of everything, though, something of her hypothesis remains even today.

The Heat of Radiation

Less than one year after the discovery of radium, Thomas C. Chamberlin (1843–1928), a geology professor in Chicago, demonstrated his remarkable intuition. Responding to what was to be Kelvin's last intervention on the subject of geology, he questioned Kelvin's conclusions about the age of the sun. He noted that "what the internal constitution of the atoms may be is yet an open question. It is not improbable that they are complex organizations and the seats of enormous energies. Certainly, no careful chemist would affirm either that the atoms really are elementary or that there may not be locked up in them energies of the first order of magnitude." And Chamberlin concluded that one could not "deny that the extraordinary conditions which reside in the center of the sun may set free a portion of this energy." Obviously Chamberlin could not guess that these prodigious energies were soon to appear in their full light, and upon our good old earth.

The discovery was announced in Paris on 16 March 1903 by Pierre Curie and his collaborator Albert Laborde. It concerned a phenomenon that by now was well known, radioactivity. Using the large samples of radium that had been so carefully purified, Curie and Laborde were finally able to measure the heat produced when the radiations were absorbed by matter. This heat is so significant that radium maintains itself at a temperature higher than the rest of the matter surrounding it. The profusion of energy liberated over the course of the years is mainly due to alpha rays (see box below), which no chemical reaction had been able to explain. Curie and Laborde deduced from this fact that a "profound transformation" such as the "modification of the radium atom itself" was involved. No other calorimetric measurement had ever had such success before the general public. And observing the marvel aroused by these inexhaustible reserves of energy, the Cassandras were ever watchful: could some instrument be invented one day that would pulverize the earth at the touch of a button and bring on the end of the world? The fear was not new; it had already appeared in large headlines in the newspapers, such as the *St. Louis Republic*, a year before the World's Fair held in the city—which had radium as the chief attraction.

RADIATIONS AND RADIOACTIVITY

It had become obvious quickly that ionizing radiations have features that vary from one radioactive element to another. Many physicists in England and elsewhere sought to identify them at the same time as Rutherford, Becquerel and the Curies in France, and Elster and Geitel in Germany. Shortly before departing for Montreal in 1898, Rutherford distinguished two broad categories of these radiations, which "he termed for convenience" *alpha rays* and *beta rays*.

The electron from cathode rays had already become a familiar figure. Becquerel recognized that beta rays are electrons that are expelled from an atom at velocities 20 percent to 90 percent of the speed of light. Alpha rays are positively charged and several thousand times heavier than electrons; they resembled other particles that had been observed in discharge tubes. Following J. J. Thomson's experiments, it was suspected that they were hydrogen or helium ions. Rutherford and Soddy proposed the latter hypothesis in their theory of radioactive disintegrations in 1902. Even though it was justified in the following year by Ramsay and Soddy, the hypothesis was not rigorously demonstrated until 1908 by Rutherford. The highly penetrative gamma rays were discovered in Paris in 1900 by Paul Villard (1860–1934), who recognized their relationship with X-rays. Gamma rays, in fact, even further extended the domain of electromagnetic radiations on the side of extremely short wavelengths, well beyond X-rays.

The heat created by radioactivity is due to absorption of the kinetic energy of the particles expelled. Alpha rays, which are feebly penetrative and high in mass, are the essential cause for this phenomenon. Before Curie and Laborde's measurements, the thermal effects estimated, corresponding to the number of alpha rays detected outside of a radioactive compound, were far too low, because the majority of the alpha rays emitted are absorbed by the emitting compound itself.

Radioactivity was seven years old, and its consequences on the thermal budget of the earth and the stars had not been foreseen, because its effects had appeared far too feeble. After Curie and Laborde's experiments, physicists and astronomers were ready to conclude that radioactive elements like radium brought back into question the validity of Kelvin's and his followers' calculations. These ideas received experimental foundation with Elster and Geitel's systematic measurements of radioactivity in natural environments, made between 1900 and 1905. They found traces of radioactivity just about everywhere, in the soils and waters, even in the mists of Niagara Falls. These traces were due mainly to radium and its emanation radon (which had not yet been identified at that point), whose higher contents in caves and grottos had been noted previously by Elster and Geitel.

These results led Robert Strutt (1875–1947), a young specialist in alpha radiation at Cavendish Laboratory in Cambridge, to begin measuring the radium content in rocks and minerals. Strutt had no inclination toward theory, by contrast with his father, Lord Rayleigh (John Strutt); it was experimental physics that interested him. He had great manual dexterity and enjoyed working alone, and he often set up his experiments using makeshift means. He was moderately conservative, an excellent orator, and curious about everything. His taste for minerals led him to develop expertise in gems. And in the footsteps of his father, he became president of and benefactor to the Society for Psychical Research. In 1905 the Royal Society of London had just admitted him when his first experiments on natural radioactivity indicated that the radium contents within rocks can vary considerably. For example, basalt and granite differ in radium content by a factor of ten. More important still was the observation that even the lowest levels of radioactivity measured were ten times greater than the average value for the earth that would account for the heat flux at the earth's surface. In order not to obtain too high a flux, Strutt had to suppose that radium was absent at depth. If radium was not limited to a superficial crust of about 75 kilometers in thickness, one would have to conclude that the earth is not in a cooling stage, but in a heating stage. It was clear that the heat being given off by the earth was not a result of its initial condensation. The situation had now turned around completely: physics could no longer set any limits for the earth's age.

Even before attention turned toward the earth, the sun was the object of rapid speculations. These were not made by Chamberlin, whose prophetic views in 1899 on the sources of solar heat had been forgotten, but by Rutherford and Soddy. Two months after Curie and Laborde's announcement, they produced a brief note to the effect that "the maintenance of solar energy, for example, no longer presents any fundamental difficulty if the internal energy of the component elements is considered to be available, i.e., if processes of subatomic changes are going on." Before George Darwin disclosed that radioactivity could augment solar energy by a factor ranging from ten to twenty, the astronomer W. E. Wilson stated that it would suffice to imagine a concentration of radium of 3.6 grams per cubic meter to explain this emission of energy. Of course, the presence of radium was doubtful because there was no evidence of the characteristic lines of this element in the sun's light spectrum. Helium, however, is abundant in stars, and it was in stars that it was identified by spectral analysis for the very first time. Since its discovery on the earth in 1895, it had never been isolated anywhere except in uranium ores, a fact that suggested some connection between helium and radioactivity.

July 1903 represented another important date in this regard, because Ramsay and Soddy demonstrated in London that the disintegration of radium actually produces helium. The presence of helium in the sun thus suggested the existence of radioactive processes, which remained, however, to be elucidated. Rutherford noted in 1904, "It is not improbable that, at the enormous temperature of the sun, the breaking up of the elements into simpler forms may be taking place at a more rapid rate than on the earth. If the energy resident in the atoms of the elements is thus available, the time during which the sun may continue to emit heat at the present rate may be from 50 to 500 times longer than computed by Lord Kelvin from dynamical data." From the 24 million to 100 million years that had been proposed by Kelvin's school, an audacious hypothesis now led to a bracket of 0.6 to 5 billion years. After Kelvin had put order into the muddle of old geology, the new physics seemed to be permitting anything, or nearly so. Only the last issue remained certain, as Rutherford emphasized in 1905: "Science offers no escape from the conclusion that the sun must ultimately grow cold and this earth must become a dead planet moving through the intense cold of empty space."

The New Alchemy

Radioactivity had some more surprises in store. At the same time as Curie and Laborde's inexhaustible sources of heat were appearing, absolute chronometers were being discovered, prepared by Rutherford and Soddy. Rutherford had been a professor at McGill University in Montreal since 1898. He had been exempted from his courses so that he could dedicate himself to his research, which he performed in a laboratory lavishly endowed by the generous but spartan tobacco producer MacDonald, who professed that smoking was a "filthy habit." Because Rutherford was quite a smoker himself, he needed to be able to aerate his laboratory hastily and conceal all traces of tobacco whenever his patron's arrival was announced. Rutherford found the solution to his search for a chemist's collaboration in Soddy, who was also a young, energetic, and ambitious researcher. Soddy was twenty-three and had recently graduated from Oxford when he was taken on as a demonstrator at McGill in 1900. He was to elucidate the origin of helium with Ramsay in London in 1903. He remained in Montreal for only three years, but this was long enough to assist Rutherford in formulating the theory that was to cast down one of the most solid dogmas of chemistry.

In the early days of radioactivity, one of the greatest difficulties was that the radiations emitted were the only tangible traces of the phenomenon. To all appearances, no new element was seen to appear through the course of a radioactive transformation. Only the curious *emanations*, which themselves emitted radiation, had been observed. But their infinitesimal quantity escaped all analysis and left doubts about their materiality. And so, Pierre Curie saw in them "centers of condensations of energy located between the gas molecules."

As research continued, the true nature of radioactive transformations was revealed in 1900 in an unexpected manner. Crookes thought that the apparent activity of uranium and thorium was due not to these metals themselves, but to their impurities. Through the process of repeated chemical purifications, he observed that natural uranium could be separated into two portions, and that the activity came from the residue of the extraction, rather than from the principal portion that was exclusively uranium. He named this impurity uranium X. The situation became more complicated when, after making independent observations of the same phenomenon, Becquerel noticed that in 1901, after a year, the purified uranium had regained its radioactivity, while the uranium X had lost what it had had.

These changing elements truly seemed to infringe upon the dogma of atomic immutability. The resounding explanation was revealed by Rutherford and Soddy in 1902: radioactivity represents the transmutation of one element into another, accompanied by an enormous release of energy. These transmutations are produced in cascade: uranium gives rise to uranium X, uranium X to another element, and so forth, until a stable element constitutes the end product of the radioactive chain. Radium is one of the elements of this chain, and—a crucial fact—the rate of each of these transmutations is characterized by the *half-life* of the element that disintegrates (see box below). Bold as they were, Rutherford and Soddy took their time in putting their theory to the test. Alchemists did not have a good reputation.

RADIOACTIVE TRANSMUTATIONS

By contrast with Becquerel, Rutherford and Soddy followed without interruption the changes of the radioactive activity of the different fractions. They observed that the activity of thorium X diminishes by half in the same amount of time needed for the activity of thorium to reduce to half of its value before purification (that is, in 3.6 days), and so they concluded that thorium X begins disintegrating immediately after being isolated. Its quantity diminishes by half

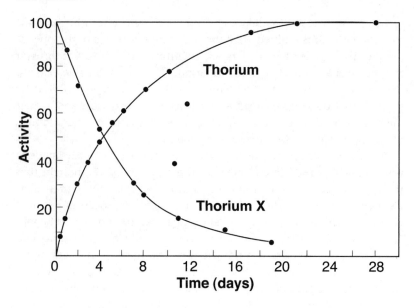

Figure 10.1 The half-lives of radioactivity: in 3.6 days, the activity of thorium X (isotope 224 of radium) is reduced by half. From Rutherford and Soddy, "Cause and Nature of Radioactivity," 381.

every 4 days, until its activity eventually vanishes. After purification, thorium begins to disintegrate very slowly, producing thorium X. As soon as thorium X begins to accumulate, it starts its own disintegrations, at the rate of one out of two atoms every 4 days. At a certain point, the quantity of thorium X disappears at the same rate as it is being produced by thorium. Its quantity and its activity stop increasing, so that they remain constant in time, having reached *radioactive equilibrium*.

Thorium X is therefore unstable. With its own half-life, it disintegrates in turn to give another element (thorium A), which produces the subsequent elements (thorium B, C, etc.) of a long radioactive chain. The same holds true for uranium. At fixed intervals of time (the half-life), the quantity of a radioactive element diminishes by one-half. After a period equal to three half-lives, the quantity of an element is one-eighth of the initial quantity; after ten half-lives, only about one-thousandth remains, and so forth. However, the element with the longest half-life functions as a bottleneck: the number of disintegrations of the elements situated downstream from that one is limited only by the rate at which those elements are formed. It follows that if the first element is by far the most stable, then the number of disintegrations all along the chain is constant, to the point of producing equilibrium. This is the case for the chains of uranium and thorium, whose half-lives are on the order of a billion years.

Radioactive Clocks

Before Rutherford and Soddy, Pierre Curie had observed that the gaseous emanation produced by radium lost its activity in an exponential manner, diminishing by half every 3 days, 23 hours, and 42 minutes (the half-life of radium). At a 1902 meeting of the French Physics Society, he made the capital observation that these laws gave for the very first time an absolute "standard of time," *independent of astronomical observations*: any variation of the quantity of a radioactive element is, in effect, a measure of the time elapsed. With his theory of radioactive disintegrations, Rutherford could proceed further and attempt the first geological datings. Such attempts were audacious for an era in which the mechanism of disintegrations were understood only in a very summary fashion, in which the concept of the isotope had not yet been formulated, and in which radioactive chains were only partially understood.

Rutherford's dating method was based on the quantity of helium produced by radioactive disintegrations. Rutherford and Soddy suspected that the alpha rays emitted took on the electrons they lacked in order to form helium atoms. This hypothesis was consistent with the fact that helium had not been isolated on the earth except within uranium ores. As we have seen, Soddy joined Ramsay in London in 1903 to demonstrate its validity for the case of radium disintegrations. Other elements of the radioactive chain emit alpha rays as well, so that the total quantity of helium produced is proportional to the number of disintegrations occurring from one end of the chain to the other, and therefore proportional to time elapsed. For the very first radiometric dating that he attempted, Rutherford took as his guinea pig fergusonite ($YNbO_4$), a mineral rich in uranium and whose uranium and helium contents had been measured. Only one more parameter remained to be determined before calculating an age, and that was the annual rate of production of helium from uranium. With the rate he estimated in 1905, Rutherford calculated an age of 140 million years. His mineral was older than Kelvin's earth! A simple rule of three seemed to have ruined the clever calculations of the greatest living physicist.

Of course, a single determination did not permit arriving at a conclusion. Between 1905 and 1910, Robert Strutt took up the relay in studying the quantity of helium included in various minerals containing uranium or thorium. Unfortunately, the ages obtained suffered from the uncertainties affecting the rates of production of helium by these two elements and the losses of the gas from the minerals over the course of time. Strutt discovered that nodules of phosphate and fossilized bones had high radium contents, and so believed for a time that these would give more convincing results, because their relative ages could be given by stratigraphy. Unfortunately, he found a poor correlation between the radiometric and geological ages. Some highly refractory minerals such as sphene and zircon in Precambrian rocks had ages as great as 700 million years. To obtain more precise results, Strutt directly measured the rate of helium production by minerals rich in uranium or thorium; he found rates close to those he had estimated, and once more obtained great ages, specifically, 250 million and 280 million years. At this point he ceased dating efforts.

Another method, one less subject to error, was used by Bertram Boltwood (1870–1927), an American chemist with whom Rutherford collaborated after Soddy's departure for Oxford in 1903. Boltwood started as the head of a private laboratory and then became a professor at Yale University. He was interested in considering the earth as a natural laboratory, in the sense that the immensity of time had provided for the accumulation in measurable quantities of those much-sought elements of the radioactive chains. For him, datings were only a by-product of his chemistry work; he had undertaken them only at the entreaties of his friend Rutherford.

Boltwood's first observation was that uranium-bearing minerals are almost always rich in lead. Further, the ratio between the quantity of lead and that of uranium increases in proportion with the geological age of the minerals, and this ratio's variation has very little dependence on uranium content. These correlations suggested that lead ought to be the stable end product of the chain that began with uranium. The age of a given uranium-bearing mineral could then be deduced from its lead content. It "sufficed" to know the annual rate of formation of lead produced by uranium (assuming, of course, that the initial lead content was zero and that no lead had been lost over the course of time).[2] Boltwood waited to understand more precisely the rate of lead production from uranium before publishing his analyses in 1907. The ages he obtained ranged from 410 million to 2.2 billion years, and—an important fact—they

---

2. Uranium disintegrates much too slowly to allow a precise measurement of the rate of formation of lead. Boltwood got around that difficulty by passing on to the best-known element of the uranium chain, namely, radium, whose half-life is short enough to be measured. He deduced that the annual rate of disintegration of radium was 2.7×10^{-4}. At radioactive equilibrium, uranium and radium undergo the same number of disintegrations, and that number is equal to the number of lead atoms that are formed (the presumed end product of the chain). Analysis of minerals indicates that there are 3.8×10^{-7} radium atoms per uranium atom. Because the number of atoms of lead formed each year per atom of uranium is $(3.8 \times 10^{-7}) \times (2.7 \times 10^{-4}) = 10^{-10}$, it suffices to divide the ratio Pb/U measured for a given mineral by 10^{-10} to obtain its age (in millions of years).

were classed in an order that was consistent with geological data. His method confirmed that the minerals could have ages greater than 100 million years, but in his caution, Boltwood did not make the slightest comment about his datings. To the times that had elapsed since the formation of the minerals, it would be necessary, of course, to add the years that had passed since the very first moments of the earth. Like helium, lead thus indicated an antiquity for the earth that far surpassed the durations granted by Kelvin. Regardless of the uncertainties, the unit of measure for radiometric dating remained on the order of a billion years. Nonetheless, there were no guarantees that these results would be enough to convince geologists of the validity of the new methods.

First Datings, New Debates

To its credit, radioactivity not only provided an instrument for dating geological phenomena, but it also explained the earth's antiquity: our planet is much older than Kelvin's calculations indicated, not because those were based on incorrect data, but because they postulated an absence of internal heat sources within the earth. Indeed, the radioactive elements with long periods, such as uranium and thorium, were what constituted the thermal engine of the earth. Without these sources, the mountains would not have risen, the volcanoes would have been reduced to a fossilized state, and nothing more than a few faded scars would now be left from the earthquakes. Without radioactivity, the earth would indeed have died, as Kelvin calculated, after some hundreds of millions of years at the most, even before life had had the time to make its appearance.

To its discredit, radioactivity had upset the results of the old physics by manipulating strange principles, methodologies in their infant stages, and uncertain data. Half-lives were subject to continual modifications, and the first mineral considered by Rutherford, in fact, had seen its age increase from 40 million to 400 million years between 1904 and 1906. Boltwood, who was somewhat less rash, had wisely waited to achieve greater precision regarding the rate of the formation of lead before publishing his datings. He also had good reason to leave it to others to comment upon the 2.2 billion years of his oldest mineral: modern analyses concede only 400 million years for its age. The disagreement came principally from the fact that the existence of isotopes makes the chains of disintegrations much more complicated than Boltwood and Rutherford had ever dreaded.

Within a few decades, geologists had submitted to the authority of thermodynamics and had adjusted their timescales in order to conform to its rules. Was all their patient work to be called into question now because of the ages obtained by two people from a few radioactive minerals? Better think twice.

In a debate that was rather long in getting started, the position of the American geophysicist George Becker (1847–1919) summarized the reductio ad absurdum produced by the adversaries of the radiometric methods. This demonstration is all the more interesting because it was the work of a pioneer of new physical-chemical methods in geology. Using Boltwood's method, Becker obtained ages in 1908 ranging from 1.6 billion to 11.5 billion years for the constituent minerals of rocks whose stratigraphy was well known—obviously an unacceptable result. From another point of view, the radium contents postulated by Strutt accounted for the earth's global heat flux, but postulating radium's concentration in a thin crust resulted in a geothermal gradient that was too high near the surface: rocks would then have to melt at depths that seemed much too shallow. Becker concluded that radioactivity must be confined to an even thinner layer, or that it was locally concentrated and its contribution did not exceed 10 percent of the terrestrial heat flux. Becker maintained that radioactivity played an obscure role, although most probably minor, both for ages and for heat flux, and that it ought not to be considered unless it provided results comparable to the ages given by purely geological methods. And Becker confirmed these ages by fine-tuning the method of Kelvin, his "great master in geophysics," according to which the best data could not change by more than 5 million years the age of 60 million years that he had calculated for the earth.

On the other side of the Atlantic, Joly's position is interesting for another reason. Four years after having dated the earth according to the salt content of the oceans, he was the first to point out the possible impact of radioactivity upon geological activity. He participated in the animated discussions of 1904 about the origin of radium and was the only one other than Strutt to apply himself to the measurement of radioactivity in terrestrial materials. On this occasion, he discovered in 1907 that highly radioactive minerals produce around themselves colored halos owing to the defects created through continuous bombardment by alpha rays. Comparisons made in collaboration with Rutherford between natural and artificially irradiated minerals permitted him in 1913 to utilize the intensity of these halos as a criterion for dating the samples under consideration. In 1909 he wrote a book entitled *Radioactivity and Geology*, in which he discussed the new methods of datings using radioactivity and provided a general review of the implications of radioactivity on geodynamics.

If Joly preferred the ages deduced from rates of sedimentation and the saline contents of the oceans, this was because the fragile datings provided by radiometric methods seemed to call into question the very foundations of geology. If the oceans were 1.4 billion years old, rather than 100 million years, the calculated sodium influx would be fourteen times higher in our day than was the average across geological times. And no alteration of climate, or of surface level, or of the altitude of the continental masses could explain such an anomaly. Furthermore, it was not possible to imagine that the Precambrian had lasted two or three times longer than all the successive eras together, when the 25 kilometers of its deposits represented only one-quarter of the total stratigraphic column. Even though Strutt had been able to attribute 222 million and 715 million years to the same mineral by using different analyses, Joly did not contest the principles of these datings, nor did he doubt that half-lives remained constant without regard to temperature or pressure. In order to resolve the conflict between Strutt's and Boltwood's results, he suggested that the rates of disintegration had been higher in the past, and this hypothesis seemed to be supported by his observations of the colored halos.

Obviously, authorities such as Becker and Joly influenced their geologist colleagues, who were novices in the field of radioactivity. The skeptics, such as Geikie, said nothing. And Chamberlin, the prophet of new energy sources, did not even mention radioactivity in the discussion of terrestrial heat sources that he published in the autumn of 1904; once again, in 1914, he passed over it in silence in his treatise on geology.

As for the physicists and the chemists, their interest in the exchange was not lively. Datings were merely a pastime for Rutherford, who had settled in Manchester, where he founded nuclear physics. And Boltwood spent his time deciphering the radioactive chains, while Strutt busied himself with his electrical discharges. None of them shared Kelvin's profound interest in the earth's history or his ardent desire to convince the geologists; the study of the earth did not open any obvious paths to great discoveries in their disciplines.

While watching these radioactive clocks of changing humor, the majority of geologists adopted an attitude of wait-and-see, along with indifference, based on the reasoning that the physicists, led by Kelvin, had already deceived them once and that there was nothing to keep them from doing it a second time. This attitude prevailed in Continental Europe, where Kelvin's influence was less pronounced than in England or in the United States. A representative case is that of Louis de Launay who, again, said nothing about radioactivity in his *Science géologique* in 1905. He did not even mention Kelvin's calculations, except incidentally, when he noted that "if it were not for the names of the scientists who established them, these figures would not have merited any attention, inasmuch as they have been gathered from improbable hypotheses." Geology had been constituted as a solid discipline that concerned itself with quite different problems. Although geological timescales remained of fundamental importance, the legendary question of the age of the earth was relegated to the background. The mountain chains and sedimentary basins were already vast enough in scope to occupy the attention and talents of geologists.

Æpinus's Revival

What right had the physicists to tell the chemists that their atoms could disintegrate? However much a few grumbling chemists might have protested, the theory of radioactive transformations was rather quickly accepted, beginning with Becquerel and the Curies. In a few months, it put a welcome order into a subject that had rapidly become quite confused. And above all, it was impossible to deny the results of the experiments that anyone could repeat, as desired. From this point of view, there was an obvious contrast with the vicissitudes of the astonishing psychic force. After assisting Rutherford and Soddy in publishing their theory, Crookes hastened to popularize their experiments, proclaiming that "a few decigrams of radium have undermined the atomic theory of chemistry, revolutionized the foundations of physics, revived the ideas of the alchemists, and given some chemists a bad case of 'swelled head.'" The popular success was assured with these alchemical reminiscences and with the introduction of the word *transmutation*, which established itself on its own, even though Rutherford and Soddy had taken pains to avoid it. Meanwhile, radioactivity was preparing other surprises, and alpha rays, in the hands of Rutherford himself, would soon be bombarding atoms for the purpose of delving into their highly mysterious structure.

In regard to the theory of radioactive disintegrations, Kelvin continued playing the role of the implacable opponent. He had been one of the early participants in the debates about radioactivity, and he quickly aroused fear in those who dared to contradict him. According to one famous anecdote, he and Rutherford nearly had a confrontation at a conference on radium in London in 1904. Recounted Rutherford:

I came into the room, which was half dark, and presently spotted Lord Kelvin in the audience and realized that I was in for trouble at the last part of my speech dealing with the age of the earth, where my views conflicted with his. To my relief, Kelvin fell fast asleep, but as I came to the important point, I saw the old bird sit up, open an eye and cock a baleful glance at me! Then a sudden inspiration came, and I said Lord Kelvin had limited the age of the earth, *provided no new source was discovered*. That prophetic utterance refers to what we are now considering tonight, radium! Behold! the old boy beamed upon me.

Perhaps Rutherford embellished the story somewhat, but the fact is that Kelvin gave no evidence that he had converted to the radioactive theory in his written discussion that followed the conference. He asserted that it was quite simply impossible for radium to emit so much energy without being supplied by some exterior source, in this case, the ether that filled all of space. In order to explain the different forms of radiation, Kelvin defied various ideas that had already been accepted, by asserting, for example, that alpha rays are radium atoms or molecules of radium bromide. Using other equally provocative ideas, he explained the differences in the penetrative capacities of rays using an "electro-ethereal hypothesis," and proposed "an atomic resuscitation" of the doctrine of the German electrician Franz Æpinus (1724–1802)—whom Kelvin was no doubt one of the last to have studied—to conclude that all substances must be radioactive, whether liquid, solid, or gaseous.

The experiments proposed by Kelvin to test his speculations never needed to be performed; by the end of the year 1904, the patriarch of physics had abandoned his own theories. Doing so did not prevent him two years later, in a letter to the *Times*, from contesting anew the principle of transmutations and denying the role of radioactivity in terrestrial heat flux. To one of his former pupils who asked him in 1906 whether radioactivity could have undermined his calculations, he responded that there was not "any serious probability . . . that either the heat of the Sun, or the underground heat of the Earth is practically due, in any considerable proportion, to radioactive matter." He died peacefully the following year on 17 December at the age of eighty-three and was buried with solemnity on the twenty-third, in Westminster Abbey beside Newton and near the place where Rutherford's ashes were to find their repose three decades later.

And so, Kelvin never recognized his error. He did nevertheless admit it in private, when he confided to J. J. Thomson "that before the discovery of radium had made some of his assumptions untenable, he regarded his work on the age of the earth as the most important of all." But why was he so determined to deny the evidence in physics and to combat the very principle of transmutations? As Becker had maintained, the neglecting of radiogenic heat sources might not have had any significant consequences for geology. It had taken some years for the geologists to become doubtful, and it took twenty-five years of the efforts of a younger generation, led by the Englishman Arthur Holmes, to establish the imposing antiquity of the earth in a definitive manner.

On the level of principle, Kelvin had been right without a doubt: the earth and the stars submit to the laws of physics, but new concepts ought not to be postulated until the accepted laws have demonstrated their failure. This principle of parsimony is indeed one of the canons of science. Kelvin had nonetheless failed to adhere to it, because the physical parameters he used in his model were too poorly established to justify the smaller and smaller uncertainties that were announced. The durations that he finally conceded were undoubtedly limited by the age presumed for the sun, which appeared to be founded on more solid bases. Kelvin had better experience with practical experimentation than any other theoretician had, and he knew the limitations of his models. He ought perhaps to have grasped, more quickly than most geologists, their lack of pertinence in relation to radioactivity. The result is that he left an undeservedly unpleasant image of himself in natural history, even though it was he who had undertaken, well before anyone else, to tell its tale, using the language of physics.

The Resistible Career of N-rays

In an ambiguous period that saw the strangest of concepts being made into laws, the discoveries of X-rays and radiant matter were followed rapidly by the appearance of other enigmatic radiations. Almost every day some obscure follower of Röntgen submitted photographic plates with mysterious images to the brothers Lumière. In January 1896 the amateur physicist Gustave Le Bon (1841–1931) outdid Becquerel and his uranium rays by announcing his discovery of *black light* to the Paris Academy of Science. Le Bon was also a doctor and a sociologist who published a classic, *The Crowd: A Study of Popular Mind*, in 1895. He was close to Poincaré and related to Carnot. He discovered a type of radiation intermediate between light and electricity that testified to the dematerialization of matter. His discovery inspired men of vastly different backgrounds—such as Edouard Drumont, a theorist of anti-Semitism, and Jean Jaurès, an eloquent socialist—and later led him to dispute with Einstein over who discovered mutual transformations between matter and energy.

The experiments of Professor René Blondlot (1849–1930), an honorable physicist of Nancy, reverberated even more. He was a correspondent of the Academy of Sciences who twice received awards from the Academy for his experiments in electromagnetism. In February 1903, Blondlot observed a new characteristic of X-rays, that they could be polarized; but then he corrected himself and announced the discovery of a new type of radiation that passed through opaque bodies. In honor of his hometown, Blondlot named these N-rays. There were more than one hundred publications between 1903 and 1906 describing their astonishing properties. After distinguishing several types of N-rays, of different wavelengths, Blondlot observed that, like radium, they emitted radiant matter. Jean Becquerel, the son of Henri and a novice physicist at the time, observed later that the intensity of the rays decreased when the metallic emitters were anesthetized with chloroform. Interest in these rays finally culminated with the discovery by Augustin Charpentier, a medical professor of Nancy, that N-rays were also emitted by the upper nerve centers of the human body and that, furthermore, these rays enhanced visual or olfactory sensitivity. One particular medium, who thought spiritualism was about to receive a serious explanation on the basis of N-rays, got into an argument about the matter with Charpentier. The priority claim was settled at the expense of the medium by the Academy of Sciences, however, which intended to reward Blondlot for his great discovery.

Although Becquerel discussed N-rays in his courses, the rays were never glimpsed outside their place of origin. In his very first publication on X-rays, Röntgen had been eager "to avoid deception." He wrote: "wherever it has been possible, therefore, I have used photography to confirm every important observation I have made on the fluorescent screen." Unfortunately photographic plates were not the sort of detectors to which the N-rays liked to reveal themselves. Indeed, Blondlot even insisted on the difficulty of seeing them: "In fact, the observer should accustom himself to look at the screen just as a painter, and in particular an 'impressionist' painter, would look at a landscape. To attain this requires some practice, and is not an easy task. Some people, in fact, never succeed." R. W. Wood (1865–1955), a well-known professor of experimental physics at Johns Hopkins University in Baltimore, cultivated the same rigor as did Röntgen. He invited himself to Nancy in 1904. His various attempts there did not permit him to detect the N-rays, though he noticed that Blondlot, oddly, could observe them even when the experimental setup had been deactivated without his knowledge! Abroad, where skepticism reigned, there was nothing more to be said.

In France, Henri Piéron (1881–1962), the future director of the Sorbonne's laboratory of physiological psychology, was another man for whom the N-rays escaped observation. On his initiative, the *Revue scientifique* decided to sound out French physicists. Jean Perrin raised a strong protest, claiming fraud; the great Poincaré defended Blondlot, his colleague from Lorraine; and a majority abstained, following the example of Pierre Curie, who was surprised that a distinguished physicist such as Blondlot could have not only made a mistake but also led astray so many experimenters from various disciplines. The doubts eventually prevailed, even though Blondlot, Charpentier, and Jean Becquerel did not recant. The question was closed in 1907 by Piéron, in an exemplary study of the "belief" in N-rays, which appeared in the *Année Psychologique*. For Piéron, the moral of the story was that "N-rays have shown us how in a great mind an idea engendered from reflections on previous discoveries" had misled their author "in a domain where the subconscious plays an immense role." While some physicists had wanted to annex psychic activity to their domain, others had instead gotten caught in the nets of psychology. And Piéron added that N-rays had furthermore demonstrated "how, without an effective suggestion, the notion of authority caused many to admit to something that could not be seen, even if it meant attributing to oneself an incapacity, a veritable infirmity." The geologists had stayed away from this debate. Nonetheless, Piéron with this observation seemed to have expanded it into quite different controversies.

Long Decade of Marvels

In June 1896 Kelvin referred to the great, unresolved problem of physics as a failure. Retrospectively, what seemed failure to him may have represented only the fact that he had been a witness to the upheavals that physics underwent as he was nearing the end of his days. The decade between 1895 and 1905 was inaugurated to the fanfare of Röntgen's X-rays and concluded with Einstein's relativity. It also experienced the sudden appearance of radioactivity and quanta. Few periods could rival this turning point in terms of innovations and discoveries, when physics was laying the foundations upon which it would continue reconstructing itself throughout the twentieth century. With X-rays and quanta on one side and automobiles and aviation on the other, it was clear that successful fruits of long effort had come in due course. X-rays had a ready public even before they were discovered because of the interest aroused by cathode rays. Discharge tubes were apparatuses of relatively simple design, and they were accessible to a large audience, including physicians, as soon as the first X-ray photographs made their way around the world. People were as ready for this innovation as they were for aviation in the months following the Wright brothers' first flight in December 1903, when the sky was conquered by new aircraft.

Radioactivity did not distinguish itself from this perspective. Henri Becquerel recognized this in declaring: "If my father had been alive in 1896, he would have been the one who discovered radioactivity." Becquerel was rooted in a family tradition that began in 1838, after a few experiments by his grandfather Antoine, when the Museum created a chair of physics and established a small laboratory to promote the application of the discipline to natural history in the broadest possible sense. A good century after Buffon, the decision was doubtless judged by geology to be fruitful. Radioactivity had suffered a difficult upbringing in the shadow of X-rays, but afterward it was in a position to challenge the atomic dogma chemists had labored earlier to establish, as well as the incorrectly brief times that Kelvin had wanted to impose. At a time when physics was being shaken upon its foundations and was conjuring up an abundance of disconcerting radiations, it ought not to be surprising that the geologists were indifferent to the new concepts. They had in fact followed the bishop Basil, who, to justify his own disinterest for science, had entreated a millennium and a half before, "We ask the Greek sages not to mock us before they agree among themselves."

If radioactivity had been discovered earlier, it could not have benefited from the results and methods that had been honed for the study of cathode rays. An alpha ray is expelled at a speed of 30,000 kilometers per hour, and it produces by itself alone nearly 100,000 ions in the air. This ionization of gases induced by alpha radiation offered an extraordinarily sensitive method of measuring, one that relegated photographic plates to the level of antiquated instruments. It allowed the detection of trace materials in dilutions on the scale of one part per 100 million, a concentration impossible to detect by any chemical analysis. Such detectability is another measure of the colossal energy of radioactivity, and it is also what allowed the search for radium in the waters of ocean bottoms and in the cores of minerals. Another related immensity, the vastness of the half-lives of uranium and thorium, did not raise metaphysical questions.

Nonetheless, the half-lives did raise the question whether they were commensurate with geological or astronomical time, and they renewed the question of how these elements were formed within the stars. A strange element, this uranium.

A Brief History of Uranium

During the course of the unforgettable year of 1789, a new element had been born to chemistry. It was isolated in rare ores coming from Bohemia and was called uranium in honor of the planet Uranus, newly discovered by Herschel. The planet itself had been named after the muse Urania, who was the beloved of Apollo and who presided over astronomy and geometry. Uranium took its place in a company of fewer than thirty chemical elements. Sodium, aluminum, and the great columns of the periodic table were not admitted until later, and meanwhile, uranium remained the heaviest of the elements. It was discovered by Martin Klaproth (1743–1817), the son of a poor tailor of Wernigerode, a town situated about seven leagues from Wolfenbüttel (later the home of Elster and Geitel). Klaproth started as an apothecary and later became a chemist so eminent that he was eventually one of the six foreign scientists admitted into the Institut de France. History reveals, though, that he had come to his conclusion too quickly, for Eugène Péligot (1811–1890), a specialist in alcohols and beet sugars, demonstrated in 1841 that what Klaproth had isolated was uranium oxide (UO_2) rather than the pure element, which Péligot himself isolated.

Uses were found for uranium, both in oxide and in metallic form. When used as a pigment for coloring glass or enamel, the oxide produced a fluorescent yellow. The metal was used beginning in 1865 for the toning of photographs, among other purposes. The process involved replacing the silver that makes up the black in a negative by another compound so as to obtain a more stable image or a pleasanter tone. When treated with uranium, a photograph is brown; when treated with ferrocyanide, it is blue or sepia.

Daguerreotypes appeared in 1838, four years before uranium. Photography had been utilized since that time in laboratories as well as by the general public, and uranium had made its entry into photographic laboratories. The discovery of radioactivity was a result of the juxtaposition of a photographic plate and a morsel of uranium salt. Why had it not been discovered earlier? At the end of the 1850s, Abel Niepce de Saint-Victor (1805–1870), a military officer and the inventor of photography on glass, had made note of the particular action of uranium salts on sensitive paper, but he did not distinguish this effect from the action of tartaric acid. Perhaps one day the plates were slightly fogged, and no one noticed. History does not say, remembering only the discovery by Becquerel—though it could have remembered another, too.

Professor Silvanus Thompson (1851–1916) was busily occupied on 26 and 27 February 1896, in his laboratory at Finsbury Technical College on the left bank of the Thames, north of St. Paul's Cathedral and London City Hall. Thompson, at forty-five years of age, was a respected physicist. He was also a practicing and courageous Quaker, an authentic scholar, a watercolorist, and a bibliophile in his spare time. He was warmly loved by a large circle of friends. At the end of the 1870s, his luck did not serve him well when he prematurely stopped the experiments he was conducting, ten years before Hertz—just short of detecting electromagnetic waves. As a fellow of the Royal Society of London, he was familiar with Crookes and other distinguished English and Continental physicists. He was an effective pedagogue and the author of classic manuals on electricity. Toward the end of his life he also wrote biographies of Kelvin and Faraday. He was hoping "to be an old man in a garden, with a pipe" one day, but his days were too short for him to be able to bring his innumerable activities to conclusion. According to one of his biographers, his eclecticism may have cost him an important place in the history of physics.

Like Henri Becquerel, or others whose names have been forgotten by the discipline, Thompson was intrigued at the beginning of 1896 by the apparent relation between phosphorescence and the emission of X-rays. In midwinter, the time when we look in on him, he had been developing photographic plates that "were left for several days upon the sill of a window facing south to receive so much sunlight (several hours as it happened) as penetrates in February into a back street in the heart of London." Aluminum sheets that had been covered with fluorescent crystals had been placed upon the plates of gelatinous bromide. After developing them, Thompson observed that some "photographic action" had taken place through the aluminum sheets, underneath the crystals of uranium nitrate, ammonium fluoride, and uranium. He had discovered *hyperphosphorescence*. He informed one of his colleagues, who told Thompson that two days earlier, a similar phenomenon had been presented to the Paris Academy of Sciences by a physicist of the Museum. Silvanus Thompson waited three months before drafting his own observations and turning his attention to other matters. He was left unremembered in connection with uranium salts.

ELEVEN

The Long Quest of Arthur Holmes

A RELUCTANT GEOLOGY Progress in radiometric dating methods, resulting mainly from the efforts of Arthur Holmes in England, achieved success in the 1920s, with the consensus that the earth's age is a few billion years. But then a new problem presented itself, because the earth had become older than the sun and the universe! Before long, however, new astrophysical theories by Eddington and Jeans would yield incredibly great ages, on the order of 10^{12} years for stellar evolution. Hubble's observations of the expansion of the universe would eventually set astronomical timescales back to the "terrestrial" billion years.

For Want of Anything Better, Some More Lead

"It is perhaps a little indelicate to ask of our Mother Earth her age, but Science acknowledges no shame and from time to time has boldly attempted to wrest from her a secret which is proverbially well guarded." Thus began the book published in 1913 by a young man of twenty-three, upon his return from a geological expedition to Mozambique. The title of this work was *The Age of the Earth*. By contrast with the stream of dissertations previously published under the same title, this one by Arthur Holmes (1890–1965) already radiated somewhat of a modern flavor. It prefigured an oeuvre that would extend over nearly a half century and that would make its author not only the veritable father of geological timescales but also one of the most productive explorers of terrestrial dynamics. Confronted with skepticism and even open hostility from a majority of geologists, Holmes at the end of his book deplored the fact that "the surprises which radioactivity had in store for us have not been always received as hospitably as they deserved." With the advent of radium, the old controversy about the immensity of time had been settled, to the detriment of the physicists, he asserted, but "the pendulum has swung too far, and many geologists feel it impossible to accept what they consider the excessive periods of time which seem to be inferred."

The fact is that the geologists would have been more easily convinced if the new methods had not been such easy targets for criticism. The first difficulty, of course, came from the fact that the elements involved were fidgety, as Joly rightly pointed out in his book *Radioactivity and Geology* in 1909. Being a gas, helium aspires to the great open spaces, but instead it finds itself caged in wherever some disintegration produces it within some mineral. Because of its tiny dimensions, it can slip in among the atoms of its host and then escape once an occasion presents itself, for example, with the slightest increase of temperature. As for radium, continued Joly, ever since the measurements of Elster and Geitel, it had been known that it cruises around everywhere. When dissolved in natural waters, it is easy for it to infiltrate into minerals and to leave a bit of helium there, as a sort of signature of its installation. In Joly's view, God alone knew the respective contributions of these two opposing effects on the quantity of helium extracted from a mineral. Strutt's method was difficult to put into practice, and Boltwood's, Joly noted further, lent itself to similar criticisms. Lead is found everywhere, especially in magma, and is incorporated into minerals when they crystallize. One cannot suppose therefore that all the lead in a given mineral has been produced by the disintegration of thorium or uranium. It was not at all astonishing to find ages that appeared to be extremely long. And when Joly postulated that half-lives increase with time, he found himself in good company with a camp that was doubtful about the invariability of disintegrations. On the basis of analogies with chemical thermodynamics, the Swedish chemist Svante Arrhenius (1859–1927) supposed that radioactivity was a reversible phenomenon. Like uranium, "radium will be formed from its decomposition products, if present in sufficient quantities" at the great temperatures and pressures at the interiors of the earth and sun. And so, added Arrhenius, we ought not to be surprised that "radium has not been discovered in the solar spectrum."

Holmes was not flustered by such arguments. He was the son of a cabinetmaker and a schoolmistress, and he had been intrigued since childhood by the date given for the Creation, in the margin on the first page of

Genesis, which was 4004 BC, rather than the round figure of 4000. As an adolescent he had read Kelvin, who had opened up for him immeasurably vaster horizons. Although he began with the study of physics, he found his path in geology when the implications of radioactivity had established themselves, inspiring him with fascinating perspectives. He was a student of Strutt and by then already established at Imperial College of London; because of his vocation, he was more concerned than either Strutt or Boltwood to seat his work within a geological perspective: one of his persisting aims was to buttress the stratigraphic column by means of radiometric ages. In 1911 Holmes published a study that had been well prepared against the objections of Joly. Even if Strutt had deserted the project of datings in 1910, the relay had quickly been taken up again.

In his first study, Holmes admitted that Strutt's method was not convincing, because the fate of helium to be found in minerals is unpredictable. However, he was certain he could demonstrate that Boltwood's method would yield correct results, provided it was utilized rigorously. Toward this end, he devised a test of elegant simplicity. The quantity of lead incorporated within a given mineral through the course of its formation could certainly be other than zero, but different minerals within the same rock would normally differ by their initial lead contents. If, in spite of everything, these minerals still gave the same ages, it would be a clear proof that their initial lead contents were negligible in relation to the quantities produced later by radioactivity. To put his test into practice, Holmes considered unaltered igneous rocks (nepheline syenites) found around Oslo in Norway, which were well known to be from the Devonian Age and which contained a dozen or so uranium-bearing minerals. Furthermore, he resumed Boltwood's analyses for minerals whose geological contexts were already known. The results were conclusive: the minerals of the same geological periods gave similar radiometric ages. Upon this basis, Holmes proposed his draft for a timescale (in millions of years), which is presented in the first column of the table below.

In this way, Holmes confirmed that the geological eras were to be dated by the hundreds of millions of years. For some periods, comparisons with later studies show that the timescales derived by Holmes in 1911 were essentially correct. As for the earth, there were some Precambrian minerals that demonstrated 1.025 billion to 1.640 billion years, indicating ages even ten times greater. In that same year, 1911, Joly came back to the attack with his sodium content of the oceans. Ages of less than 100 million years had been attributed to the earth independently by other geological authorities, such as Becker, the geophysicist from Washington whose views have already been summarized, and Sollas, the Oxonian

Period	Holmes (1911)	Barrell (1917)	Holmes (1927)	Modern Ages
Tertiary		65	60	65
Jurassic		150	150	200
Carboniferous	340	370	330	359
Devonian	370	420	420	416
Silurian	430	460	450	444
Cambrian		700	600	542

geologist who had submitted to the shortest of Kelvin's timescales. Sollas nonetheless concluded that the new radiometric method "is of great promise, but a long series of concordant observations will be required before we can feel absolute confidence in its results."

It would have been easier for Holmes to convince his colleagues if other methods had been able to confirm the ages deduced from the radioactive chain leading from uranium to lead. But there were not many other radioactive elements with long half-lives. The first of these, thorium, was not particularly accommodating. The end product of its radioactive chain remained an object of controversy, and the ages determined in accordance with the quantities of the helium it produced were clearly incorrect. Joly did not hesitate to declare that these uncertainties cast doubt on the very principles of radiometric methods.

So the search for other possibilities with more promise continued. It had been known for some years that the two alkalis potassium and rubidium are weakly radioactive elements, both with very long half-lives. The fact that potassium is a thousand times more abundant than uranium or thorium made it more difficult to deny the effect of radioactivity on the thermal budget of the earth. Otto Hahn (1879–1968), with whom we shall become better acquainted further along, began to take interest in this element in Berlin in 1913. He correctly supposed that potassium transmuted directly into calcium, and rubidium into strontium. He even succeeded in estimating the half-life of rubidium. Unfortunately, the techniques of the times would not allow the quantities of calcium or strontium formed by disintegration to be measured.

Thus, the problem of thorium had to be resolved. Uranium, for its part, kept raising new enigmas. And it is important to remember that, for the time being, only minerals were being dated; the age of the earth itself remained cloaked in mystery. In another domain, the situation was even more confusing: from a larger perspective, one had to admit that the thermal effects of radioactivity, which were better known, appeared insufficient to account for the dynamics of the stars throughout the vast periods indicated by radiometric methods. Even if the sun was formed completely of uranium and its by-products, it would radiate only half as much energy as it actually does. The bursts of optimism from Rutherford and company therefore subsided again quickly. As was summarized in 1915 by the English physicist Frederick Lindemann (1886–1957), who was also a former tennis champion and later to become a very close adviser to Churchill, "the origin of the sun's heat cannot be referred to any known cause." The physics of the stars had to be reconsidered completely. Even more serious, right after the worst of the uniformitarian excesses had been justified retrospectively, the universe was now threatening to become younger than the oldest of the terrestrial minerals.

Tell Me How Much You Weigh ...

A new physical-chemical interlude is needed here. Even though we may have lost sight of them in 1903, the radiochemists had not been inactive. Three radioactive chains had been distinguished, those of the three heaviest elements: uranium, thorium, and actinium. In proceeding through the labyrinths of these chains, radiochemists had trapped some thirty radioelements whose half-lives varied between a fraction of a second and thousands of years or more. Radium was only one among the many. Certain of these, such as mesothorium, had been given names suggestive of a paleontological bestiary. Others, more numerous, had received only common, cataloging letters, after the fashion of uranium X. Whatever their names were, the natural question that arose was how to organize all these new arrivals among the eighty-some elements that were known before the Curies began their first experiments.

Chemists had tried over the ages to classify chemicals, not necessarily with any excessive concern for order, but because a well-executed classification would cast new light on the origins of the elements and their mutual relations. The famous *periodic table*, now familiar to high school pupils, had been devised between 1869 and 1871 by the Russian chemist Dimitri Mendeleev (1834–1907). The columns of this table, in its definitive form, group the elements in order of increasing atomic mass and into families defined by similarities of physical and chemical properties. The alkalis (sodium, potassium, etc.) and the halogens (fluorine, chlorine, etc.) are two of the families. Even more important, the presence of empty cells led Mendeleev to predict the masses and properties of elements that remained to be discovered. The great usefulness of the periodic table was thus definitively asserted when new elements were isolated with the principal attributes that Mendeleev had predicted for them. In 1898 and 1899 the resounding trio of actinium, radium, and polonium came to occupy three of the still-vacant cells. The next problem that arose was that the "seats" in Mendeleev's table had been practically all taken when along came the radioelements, marching in by droves.

Radiochemists were thus confronted by difficulties that called into question the very concept of a chemical element. Ever since Lavoisier, after the end of the eighteenth century, the term *element* was used to refer to "a body into which other bodies may be analyzed," one that is not "itself divisible into other bodies different in form," according to a definition of Aristotle (which he had disputably followed, himself, with his four sublunary elements). A negative definition is not a very good one, but Aristotle's criteria had undergone repeated proofs. Independently of what one might think about atoms, the existence of which was still being denied by some at the end of the nineteenth century, it had been observed that elements have a "fingerprint": when they are heated to very high temperatures, they emit a light spectrum composed of lines having characteristic wavelengths (the yellow light emitted by sodium lamps, for example). These lines permit even extremely small quantities of a given element to be detected, and count has been lost of the elements, such as helium, that were first identified using *spectral analysis*.

Finally, the atomic mass completes the identity card of an element. Although chemists were still far away from being able to weigh atoms individually, through long effort they had learned how to compare the masses of samples containing a given number of atoms of different elements. Hydrogen is by far the lightest of the elements, and it is convenient to take its mass as a unit of measure. Helium comes just after hydrogen; since it is four times heavier than hydrogen, its mass is four units. Thorium and uranium were the heaviest known elements, with masses 232.2 and 238.5 times heavier, respectively, than that of hydrogen.

Because radioelements were produced in infinitesimal quantities, determinations of atomic masses presented considerable difficulties. After the existence of radium had been established on the basis of its radiations and spectral lines, Marie Curie took enormous pains to isolate a large enough quantity to determine the mass of the element. With Rutherford and Soddy's theory of radioactive disintegrations, this difficult work proved to be even more important, because atomic masses lent themselves well to a highly profitable arithmetic. If an element disintegrates by emitting a beta ray (an electron), its mass barely changes at all; but if it emits an alpha ray, its mass diminishes by 4 units, which are

State of development of Mendeleev's table by 1910

1	2	3	4	5	6	7	8	9	10	11	12	13	14	15	16	17	18
1H 1.0																	2He 4.0
3Li 7.0	4Be 9.1											5B 11.0	6C 12.0	7N 14.0	8O 16.0	9F 19.0	10Ne 20.2
11Na 23.0	12Mg 24.4											13Al 27.1	14Si 28.4	15P 31.0	16S 32.1	17Cl 35.5	18Ar 39.9
19K 39.1	20Ca 40.1	21Sc 44.1	22Ti 48.1	23V 51.2	24Cr 52.1	25Mn 55.0	26Fe 55.9	27Co 59.0	28Ni 58.7	29Cu 63.6	30Zn 65.4	31Ga 70.0	32Ge 72.0	33As 75.0	34Se 79.1	35Br 80.0	36Kr 81.8
37Rb 85.4	38Sr 87.6	39Y 89.0	40Zr 90.7	41Nb 94.0	42Mo 96.0	? ?	44Ru 101.7	45Rh 103.0	46Pd 106.0	47Ag 107.9	48Cd 112.4	49In 114.0	50Sn 118.5	51Sb 120.0	52Te 127.6	53I 126.8	54Xe 128.0
55Cs 133.0	56Ba 137.4	57La 138.0	? ?	73Ta 183.0	74W 184.0	? ?	76Os 191.0	77Ir 193.0	78Pt 194.8	79Au 197.2	80Hg 200.3	81Tl 204.1	82Pb 206.9	83Bi 208.5	84**Po** ?	?	86**Rn**
?	88**Ra** 226.0	89**Ac** 227.0	90Th 232.2	?	92U 238.5												

Notes: The superscripts are atomic numbers; beneath each symbol is its atomic mass; radioactive elements are indicated by bold type.

The fourteen rare earths, which follow Lanthanum (La), are not included. The atomic masses of the elements increase from left to right along the rows and from top to bottom down the columns, with a few inversions in order of mass—potassium (K) and argon (Ar), for example—which the discovery of the atomic number, the bona fide organizer of the table, accounted for. The discoveries of argon and helium (He) in 1895, which led to the creation of the last column, prompted the prediction of the existence of other noble gases; Ramsay and his collaborators isolated neon (Ne), krypton (Kr), and xenon (Xe) in 1898; radon (Rn), the mysterious emanation from radium, did not follow until 1910. The emission of an alpha ray, which diminishes the atomic mass by four units, displaces an element two cells to the left, in accordance with a rule formulated by Soddy and Kasimir Fajans even before the discovery of isotopes. The emission of a beta ray displaces an element one cell to the right without changing the atomic mass. Masses from Moissan, *Traité de chimie minérale*, except for the radioactive elements of the last row.

carried away as a helium nucleus. Between uranium (mass 238.5) and radium (226.4), the difference in mass is approximately 12: three alpha rays must therefore be ejected; between radium and lead (207.2), the difference is about 20, the equivalent of five alpha rays. Without knowing the details of the radioactive chain that leads from uranium to lead, one can still assert that eight alpha rays have been emitted along the chain.

This atomic arithmetic of childlike simplicity therefore allows one to know the quantity of helium formed throughout a chain of disintegration. And especially, it allows the end product of the radioactive chain to be determined because the difference in mass between the starting element and the end product of a chain must be some multiple of four, although this relationship seemed to hold true in a manner that was only approximate. The difference of mass between uranium and lead is 31.3, rather than 32. Holmes was hardly worried about this discrepancy. In 1913 he attributed the difference to the various experimental errors that sullied the atomic masses—but he had to change his opinion quickly: new tremors were about to shake radiochemistry, and new light was about to shine on the origins of lead.

Some Well-Filled Cells

These upheavals were triggered by the curious "genetic" problems that were posed more and more frequently by radioelements. The first was presented by ionium, a radioelement isolated in 1906 by Boltwood. Even though it had been clearly distinguished from thorium by its radiations, ionium appeared to have chemical properties that were remarkably close to those of thorium. So close, in fact, that if one blended ionium and thorium, it was impossible to separate them again. Nevertheless, chemists had enormous experience in the separation of elements; great storehouses of ingenuity had been utilized in getting to the bottom of the wiliest of families, that of the rare earths, whose numerous members all had chemical properties that were extremely similar but not identical.

History repeated itself after a discovery by Otto Hahn in 1907. He had already discovered two radioelements during a fruitful but fortuitous early stage of his career, which deserves mentioning. Hahn was the son of a prosperous glazier of Frankfurt, who had aspirations for his son to become an architect. After uneventful studies, which culminated in his doctorate in organic chemistry, Hahn decided to apply himself to industry at large. Toward this end, he left for London in 1904 in order to perfect his English. He was to have worked there for six months, but a recommendation led him to Ramsay. There, as an apprentice radiochemist (which he had become by sheer chance), Hahn isolated an element within a preparation of radium salts, which he called radiothorium. The great specialist Boltwood—who, like his friend Rutherford, held Ramsay's radiochemical activities in meager esteem—was so surprised about the discovery that he bluntly referred to radiothorium as the "compound of thorium X and stupidity." Hahn's discovery had given him the taste for research, and so this hasty slur did not trouble him. He later went to Montreal, to continue his studies with Rutherford, and there he discovered radioactinium. When he returned to Berlin in 1906, Hahn isolated yet another new element, mesothorium. More than one chemist found his head spinning when mesothorium was revealed to be chemically indistinguishable from radium, while radiothorium proved to be a twin brother to thorium.

Soddy did not lose his direction in this vortex. He was not the only one, but he was the first, to have the "courage" to take the plunge and draw forth all of the consequences of the situation. We encounter him again as a professor at Glasgow, where he observed that the atomic masses of these chemical look-alikes are never different by more than a few units of mass. He inquired in 1910, then, whether "any atomic weight is not merely a mean number," with the elements simply being mixtures of various constituents having different atomic masses. Shortly afterward, the importance of this point of view was confirmed when it was understood that twin element pairs, such as ionium and thorium, shared the same spectral lines: they had the same fingerprint, and these elements differed only in their masses and their radioactive properties.

Weary of describing them as "elements chemically identical and non-separable by chemical methods," in 1913 Soddy named them *isotopes* (*isotope* means "the same place"), indicating that they were to be found in the same cell in Mendeleev's table.[1] Although it did not provide the key to this table, the concept of the isotope clarified the sense of radioactive chains: like radiothorium and others, such as mesothorium, the majority of the radioelements were revealed to be isotopes of well-known, stable elements. With this surprising notion, isotopes of different elements could have the same mass: the well-established role of mass as a distinctive character for an element was thus bluntly challenged. Even

1. Until the discovery of the neutron in 1932, it was thought that the positive and negative electric charges coexisted in the atomic nucleus. In his original definition, Soddy thus called isotopes those elements whose nuclei have the same excess of positive charges.

though it made many problems disappear, as if by waving a magic wand, this concept, which upset a long century of physics and chemistry, was far too shocking to be quickly adopted.

For Crookes in his old age, the new understanding of isotopes was approaching the verification of one of his deepest conjectures, developed in 1886 in response to a remark by Faraday from twenty years earlier: "To discover a new element is a very fine thing, but if you could decompose an element and tell us what it is made of—that would be a discovery indeed worth making." According to the English chemist William Prout (1785–1850), the fact that all atomic masses are practically integral multiples of hydrogen's mass indicated that hydrogen was *protyle*, the prime matter from which the other elements were constituted. According to Crookes, protyle agglomerated in the stars, and in such a manner that the heavier elements were formed there at lower temperatures. Crookes thought that the amount of protyle within a given element could vary, and he deduced from this view that "probably, our atomic weights merely represent a mean value around which the actual atomic weights of the atoms vary within certain narrow limits." The majority of calcium atoms, thus, had a mass of 40, but Crookes supposed the existence of calcium atoms having masses of 39 and 41 and, even less frequently, masses of 38 and 42. Inspired by the intermediate character (between energy and matter), manifested by radiant matter, Crookes was finally led by this process of inorganic evolution to assert that "it is equally impossible to conceive of matter without energy, as of energy without matter; from one point of view the two are convertible terms." Twenty years before Einstein, Crookes had indeed announced the celebrated formula $E = mc^2$.

Reliable Chemical Ages

Now, after this introduction, it will be easier to understand why the lead in minerals was so significant in the new debate. The difference in mass between uranium (238.2) and thorium (232.2) is not a multiple of four. If the chains of these elements both terminate with lead, this means that lead must have at least two isotopes. In 1913 Soddy and the German radiochemist Kasimir Fajans (1887–1975) thus independently predicted that the atomic mass of lead ought to vary according to its origin, with the lowest (206) for lead produced exclusively from uranium and the highest (208) for that coming from thorium. Common lead, which is found elsewhere than among radioactive, uranium-bearing minerals, has an intermediate mass of 207.2, which indicates that it represents a blend of these two isotopes.

Such variations in mass constituted the first indubitable proof of the existence of isotopes. For chemists, thus, it was a question of primary importance. In order to resolve it, they turned to the old specimens, in which geological time had worked in their favor by accumulating these different isotopes. Previously, however they had not had the means to separate the isotopes in their laboratories. Now attempts at the purification of leads quickly became a feverish activity. The first results came from Glasgow in May of 1914; for the lead of a mineral from Ceylon that was rich in thorium, Soddy measured the high mass of 208.5. This result was joined in the following weeks by measurements made in Paris, Prague, and Boston: Maurice Curie, who was Pierre's nephew, Otto Höningschmid (1878–1945), and Theodore W. Richards (1868–1928) found abnormally light masses—as low as 206.4—for the lead of uranium-bearing minerals. The predictions made by Soddy and Fajans were thus confirmed remarkably well.

Because of the reputation of their author, Richards's analyses appeared to be the most conclusive. Richards, the son of a seascape painter and a Quaker poetess, was a professor at Harvard University. He succeeded in bringing the delicate art of the determination of atomic masses to new heights: as a result he became in 1914 the first American to receive the Nobel Prize for chemistry. During that year, the crowning of his career, Richards was the first to be surprised by the "amazing" result that the lead of uranium-bearing minerals had a mass that varied from one sample to another. The consistency of these masses had even become a dogma, with which more than a single chemist had vainly attempted to find fault. It was by confronting the constancy of masses in the case of nitrogen that Rayleigh had discovered argon. Again in 1907, Richards himself had not given up hope of refuting the dogma, noting that "the idea that the supposed constants may possibly be variable, adds to the interest which one may reasonably take in their accurate determination, and enlarges the possible field of investigation instead of contracting it." The irony of history is that it was Richards who had demonstrated, for dozens of elements, the rigorous constancy of atomic masses. This would not be the last time that a law collapsed just at the moment when it appeared to have been best established.

To be sure, the geologist Holmes did not hesitate to take advantage of this new incursion of chemistry into his home territory, as he sought to "expand" his own field of study. With these results, the question regarding the end product of the chain of thorium had finally been settled: it is indeed lead, in the form of its isotope of mass 208. An even simpler recipe allowed one to deduce the age of a mineral based on its quantities of lead, uranium, and thorium.[2] Owing to this formula, the edifice that Holmes had undertaken to construct was about to be reinforced. And above all, the existence of the two isotopes, lead 206 and lead 208, was going to open new perspectives: even if researchers were still quite far from being able to measure the respective quantities of these two isotopes, it was becoming possible to distinguish the fractions of lead produced through the disintegration of uranium from those that came from thorium over the course of geological time. But while the panorama of physics was brightening, the Western world was undergoing tragic cataclysms.

1914–1918

Of course, the Great War marked a deep rift. For the first time in any conflict, scientists were mobilized in their official capacities to contribute to the military effort of each side. As the belligerents became tired of the sterile, bloody trench combat, they realized that only new sorts of weapons would decide the victory. So research was amplified so as to permit killing more and more quickly, or to better protect the troops of one's own side. Alfred Nobel, a philanthropist, chemist, and spare-time idealist, had expected that his dynamite would put an end to the armed conflicts: it would cause too much damage, he thought, and the fear of formidable, mutual destructions would incite the nations to prudence. Nobel was unfortunately two wars too early; Big Bertha, the colossal cannon that bombarded Paris from afar, became the symbol of the alliance contracted between metallurgy and chemistry. Even before it entered on the scene, airplanes, submarines, and tanks had taken their places in the battle, united with command centers by means of radio or telephone, and each was equipped with its own appropriate form of Nobel's precious invention.

One particular death demonstrated, in its absurdity, that physicists could serve their countries in better ways than by being killed in action. In 1913 the electromagnetic nature of X-rays had just been demonstrated.

2. In one year 10^6 grams of uranium and thorium produce $1/7.400$ gram and $1/19.500$ gram of lead, respectively. The age in millions of years can be given simply as $7,400 \, Pb/(U + 0.38 \, Th)$, where Pb, U, and Th are the contents in the samples of lead, uranium, and thorium.

Henry Moseley (1887–1915), who was the son of one of Darwin's pupils and a student of Rutherford at Manchester, had thought that certainly these rays could "tell one much about the nature of an atom." In 1913 and 1914, first at Manchester then at Oxford, he had been able to measure the frequency of the X-rays emitted by a large number of elements, when these were bombarded by cathode rays. Moseley observed that the frequencies vary in a regular manner, from aluminum to gold, in accordance with the position of the elements within the periodic table, and even more especially, he noted that those positions are a simple function of the atomic number. (The atomic number, which goes from 1 for hydrogen to 92 for uranium, increases by one for each successive element in the periodic table.) Thus, the X-ray spectrum was found to be more fundamental even than the light spectrum, because it allowed not only the identification of an element, but also its proper placement in the element in the periodic table.

Rutherford's idea of the "planetary model" of atoms formed of electrons in revolution about a positively charged nucleus had not been taken seriously at all until that point. But it was justified at once when Moseley announced that the atomic number represented the positive electrical charge borne by the nucleus. The war exploded shortly after these great discoveries. Moseley was twenty-seven years old. His place was at the front; he had had to insist to be admitted into the engineering corps. His unit ended up being sent to the Dardanelles in the Franco-British expedition against Turkey. The terrain was trying and the expedition poorly prepared. On 10 August 1915, an implacable counterattack directed by Mustafa Kemal, the future Atatürk, decided the fate of the battle of Gallipoli. Moseley was killed, along with the majority of his unit, in the brave struggle.

During this time, Marie Curie, accompanied by her daughter Irène, had created a mobile radiology unit near the front. Crookes, J. J. Thomson, Rutherford, and Strutt participated in a committee of experts that counseled the British Navy. The ability to detect submarines took on considerable importance; Rutherford had to abandon the structure of atoms and convert a part of his laboratory for the study of submarine acoustics. The chemists, as well, were mobilized. In England, Soddy was occupied by technical problems, such as the extraction of ethylene from coal gas. In Germany, the chemist Fritz Haber had discovered how to synthesize ammonia. His discovery made it possible to manufacture fertilizer, without which the central powers would have been incapable of sustaining a war of attrition. He was named head of the chemical section of the Ministry of War in 1914, and it was he who invented poison combat gas. At first it was thought that the gas would make enemy soldiers flee from their trenches; experience proved that they could even more conveniently be buried there. Lieutenant Otto Hahn was among those who had been charged with spreading poison combat gas on the front. During the battle of Verdun, it was only the heavy French bombardment that prevented him from reaching the fort at Douaumont where he had been sent, and from being killed in an explosion along with the seven hundred German soldiers who were there. Once again, chance had served him well.

Gas and antisubmarine warfare had demonstrated (as if any proof were necessary) the importance of scientific research to the military. After the war, in Germany as elsewhere, science resumed its course, aware of its powers, advised of its responsibilities. The conflict nonetheless left painful scars on many scientists, as well as on many other people who did not even experience the front. Soddy, who before the war had been inclined to challenge the role of money in society, applied himself to understanding "why so far the progress of science has proved as much a curse as a blessing to humanity." At the age of forty, his career as a chemist was practically finished. He was individualistic and rebellious, and although he had become a professor at Oxford, he abandoned experimental science for economics and mathematics. For Soddy, the major problem of modern society was the fact that a "fictitious money system," which originated at the beginning of scientific civilization, "was being purposefully and consciously used to frustrate it, and to preserve the earlier civilizations founded on slavery." In *Cartesian Economics* and *A Physical Theory of Money*, Soddy pleaded for a reform of the monetary system conceived upon physical bases. These pleas were not heeded, any more than his further warnings in *The Wrecking of a Scientific Age*.

The First Ages of the Earth

In the United States, which was still in a state of neutrality, a voluminous work was published in 1917 by a professor at Yale University, Joseph Barrell (1869–1919). Barrell had studied engineering and taught astronomy as well as zoology; he contributed great syntheses, whose common theme was the effects of various physical agents upon the evolution of the earth and the biosphere. In particular, in order to account for the mechanics of the earth's crust, he defined two concepts that were to play key roles in plate tectonics a half century later: the *lithosphere*, which was the thick, rigid layer of about one hundred kilometers in thickness and which rested upon a much thicker, deformable substratum called the *asthenosphere*, of which Reverend Osmond Fisher had earlier been the advocate.

To use terms appropriate to the times, Barrell's article about geological time played the role of the preparation of heavy artillery in the debate that resumed after the war. In his preamble, Barrell conceded that the "skepticism of geologists was a correct mental attitude" in the face of the immense ages announced by physicists, and that they could properly be indignant at the "new promulgations of this dictatorial hierarchy of exact scientists." As Huxley had previously emphasized, it was necessary to use prudence, because "the exact formulas of a mathematical science often conceal the uncertain foundations on which reasoning rests and may give a false appearance of precise demonstration to highly erroneous results." The physicists' arrogance over their extremely long durations was as blatant as Tait's had been with his 15 million years. After a decade of trials, the time had come to correlate objectively the radiometric datings with the great geological observations.

Well supported by Holmes, Barrell confronted the new datings with all the terms of the old debate—sedimentation, erosion, salinity of the oceans, and heat flux. For Barrell, this confrontation required a change in perspective. Because of the innumerable variations in the climate or in the levels of the seas, sedimentation was irregular and discontinuous: the records that it registered gave a fragmentary view of the past. The pertinent notion was that of the "geological rhythm." The rhythms represented cycles of highly variable amplitude that integrated the irregularities of sedimentation. The greatest of the rhythms had seen the sea levels rising and falling, the sea bottoms subsiding and filling in with sedimentation, and the mountains rising up or disappearing. They had been useful in defining eras and periods, and their interest here is that their global durations could be established more precisely than could the durations of their various episodes. Using stratigraphy to interpolate the ages too infrequently available from radiometric methods, Barrell constructed the first detailed timescale. He correctly pushed back the start of the Cambrian to between 550 million and 700 million years ago and underscored the richness of the long anterior periods, even asserting that "four Precambrian eras" could have succeeded an "unknown Primordial time."

Shortly thereafter Chamberlin, another geologist, defended the extremely long durations, but he did so on a completely different basis. Since refuting Kelvin's arguments, Chamberlin had not concerned himself with chronology, not even in the voluminous treatises he had published in the meantime. He was the son of a Wisconsin farmer who was a part-time Methodist preacher. He had been attracted by fossils and geology since his childhood. Before becoming a professor at the University of Chicago, he had long studied the vast glacial terranes of Wisconsin. From there he had become interested in the causes of glaciations, in climate changes, and finally, as we will now see, in the great cosmological problems that were to make him a pioneer in two fields. Chamberlin was the first to draw attention in 1897—at the same time as the chemist Arrhenius did—to the effects on the climate caused by variations of the atmospheric content of carbon dioxide. He contested the solar nebula theory, according to which the planets had formed from the hot condensation of gaseous matter, claiming that instead they had been formed under conditions of cold by the slow gravitational attraction of small, meteoroidal bodies, the *planetesimals*. This mechanism had the great advantage that it made possible "the acquisition of an atmosphere and hydrosphere at a moderate temperature when the growing earth reached a medial size and introduced conditions congenial to life at a stage sufficiently anterior to the Cambrian period to satisfy the most strenuous demands of theoretical biology." Chamberlin waited until 1920 to outfit his scenario with a chronology, however. He proposed that biological evolution during the Precambrian had required a duration nearly ten times as long as what came after that, meaning that the 400 million years that had elapsed since the Cambrian indicated that the earth was 3 billion or 4 billion years old. In passing, Chamberlin attributed to the stars ages three or four times longer.

Chamberlin's estimations were remarkable, but they were based too much on his intuition to be convincing. Barrell's influence was thus fundamental, as the great master of French geology, Pierre Termier (1859–1930), for example, affirmed. After reading Barrell, he recognized in 1919 that "the general feeling among geologists is that a hundred million years is a minimum" for the total of the geological durations. "Among the results provided by the three processes of the radiometric method, there is an agreement that, even though not perfect, does not fail to make a strong impression." But, he added, "This is all according to appearance, and still quite uncertain."

In England, Holmes was suffering from malaria and had been judged unsuitable for military service during the war. His later financial difficulties led him to leave the university to prospect for petroleum in Burma (now Myanmar). A conference held in London in 1921 during his absence, under the auspices of the British Association for the Advancement of Science, proved to be a turning point. Sollas noted there that, with radioactivity, "the age of the earth was thus increased from a mere score of millions to a thousand million and more" and that "the geologist who before had been bankrupt in time now found himself suddenly transformed into a capitalist, with more millions in the bank that he knew how to dispose of." With perplexity, marked by weariness, he finally concluded that "geologists are not greatly concerned over the period which physicists may concede to them; they do not care much whether it is long or—within moderation—short, but they do desire to make reasonably certain that it is one which they can safely trust before committing themselves to the reconstruction of their science, should that prove necessary." What was especially novel here was that now even the old opponents of long geological timescales were beginning to have their own doubts.

In the year 1921 the first attempt was made at using radiometric dating methods for the great envelopes of the earth, not merely for isolated minerals. The method was proposed by the well-known Princeton astronomer Henry N. Russell (1877–1957). He supposed that all the lead present in the earth's crust had been formed from the disintegration of uranium or thorium. After innumerable chemical analyses of rocks, the average quantities of uranium, thorium, and lead were rather well known. From these quantities and the half-lives of uranium and thorium, Russell concluded that 8 billion years would have been required to form all the lead in the crust. Had he revealed the best guarded of secrets? No, certainly not. If the crust had effectively acquired its individual characteristic at an early moment in the history of the earth, it would have incorporated a certain quantity of lead, and the age given by Russell would have been only an upper limit. The oldest minerals dated reflected evidence of a billion years, so Russell had to conclude that the age of the crust was between 1 billion and 8 billion years. The question had not really been answered.

The Atoms' Assay Balance

In truth, chemical analysis appeared to have contributed everything that it could. Those analyses had the awkward inconvenience not only of requiring long and delicate work; more than that, it required significant quantities of matter. Separating and determining quantities of lead could be done only for relatively rare minerals that were rich in uranium or thorium; thus many rocks could not be dated because methods were lacking for analyzing the traces of lead present in ordinary minerals. Lead isotopes represented a memory of the elements that had produced them; a precise measurement of these became a prerequisite for finer datings. The solutions to these two problems came from experiments carried out at Cavendish Laboratory in Cambridge. Like many other solutions, these had their source in the instrumentation that initially had been devised for the study of cathode rays.

In addition to the cathode rays, rays with positive electrical charges had been observed in discharge tubes. The electrons of the cathode rays, when drawn out from the rarefied gases of the tube, left behind positive ions that also were accelerated by the electrical field of the tube, but in the direction opposite from that of the negative ions. In order to study these particles, J. J. Thomson began in 1906 utilizing the method of deflection of electrically charged particles by means of the electrical and magnetic fields that he had previously employed in his discovery of the electron. Because the ions that differed either in mass or in charge were deflected in distinct manners, Thomson thought of utilizing his *positive ray apparatus* for the purpose of chemical analysis. During the course of a study of noble gases in 1913, he was surprised to observe that, regardless of its origin, neon always behaved as a mixture of two different gases, one with an atomic mass of 20, the other with a mass of 22, which he thought to be a compound of neon. Without much success, his colleague Francis Aston (1877–1945) attempted to separate these two compounds using laborious methods of gaseous diffusion. Aston finally considered that neon might itself be formed of two distinct isotopes, which would also explain the nonintegral value (20.2) of its atomic mass. Isotopy had been revealed through the radioelements— but could it be a general property of matter, rather than the exclusive attribute of elements that were produced from radioactive disintegrations? In his search for the answer to this important question, Aston conceived "one of the most brilliant combinations of mathematical analysis and experimental skill this century has produced," according to Soddy's expression, by metamorphosing the primitive *positive ray apparatus* into a remarkable measuring instrument.

Aston had studied chemistry and was an expert in alcoholic fermentation. He worked several years in a brewery before turning to physics and finally being recruited by Thomson at Cambridge in 1910. He was an amateur photographer, a musician, and a sportsman. He was also an inveterate bachelor who preferred working alone, especially at night.

He single-handedly authored 137 of his 143 publications. He had barely begun work on neon when he was interrupted by the war. He had to wait until peace returned to complete his construction of the first *mass spectrograph,* an instrument that separates the isotopes of an element with the same clear simplicity with which a prism decomposes light into its constituent components. And isotopes 20 and 22 of neon were the first to be separated by the mass spectrograph. In relation to Thomson's first apparatus, the essential improvement that Aston incorporated was an adjustment of the electrical and magnetic fields, in such a manner that the ions would have a trajectory independent of their initial velocity. After passing into the ionization chamber of the spectrograph, the ions bearing the different isotopes of the element under study each follow trajectories characteristic of the individual isotopes. The ions are then intercepted by a photographic plate positioned along their course. After developing the photograph, one sees a series of bands, each of which distinguishes a particular isotope, and the intensity of each band is proportional to the number of ions. Because the ions have the same electrical charge, the differences in the positions of these bands depends solely upon their masses: with the mass spectrograph, it became possible to weigh the atoms individually, and even to count them. Aston could not possibly have imagined the magnitude of the repercussions that his new type of balance would prove to have, both for physics and for chemistry, not to mention geology and astronomy.

Aston's motto was "make more, more, and yet more measurements." He never tired of weighing atoms, using ever-more-improved versions of his apparatus: he discovered three-quarters of the 281 isotopes that can be found naturally on the earth. Isotopy appeared clearly as a fundamental attribute of matter, and elements such as fluorine or aluminum that consist naturally of one single isotope are rare. Many elements, such as lead, uranium, and thorium, still posed serious difficulties. For these it was necessary to introduce a gas or a liquid into the apparatus, appropriately heated, to vaporize them, but without allowing them to be polluted by impurities. For the heavy elements it was also difficult to keep high magnetic fields constant so as to deflect the ions correctly. Thus Aston could advance only by carefully measured steps. The situation was particularly irritating for lead, the first element for which variations in atomic mass had been observed, as early as 1913: its isotopes were not distinguished until the end of the 1920s, too late to contribute to the settling of the old debate initiated by Kelvin. The imposing antiquity of the earth was recognized without their help, but later it was their privilege to establish with precision the chronology of its evolution.

1926–1931: Consensus, Finally!

After his disappointment in prospecting for petroleum in Burma, Holmes returned to England in 1924. The economic situation was difficult. As it happened, the University of Durham was recruiting a professor. Holmes landed the job and quickly returned to his favorite research. Two years later, he made use of new analyses to refine Russell's model and to reduce to 3.2 billion years the time required to form all the radiogenic lead of the crust (evaluated as half the total lead). He also published a second edition, entirely rewritten, of his *Age of the Earth*. It was an indisputable sign of the shift of scientific opinion that Holmes did not even judge it worthy to dwell upon the old controversies that he had described at great length in his 1913 edition. He dedicated more space to the pioneers of geology than to Helmholtz or Kelvin, whose works he mentioned only as a sort of historical background note. Holmes followed Barrell in retaining the great rhythms as the only pertinent geological criterion. These cycles, some twenty in number, observed in Europe and in the Americas, were attributed durations of about 30 million years each by Holmes; and so on the whole, their total appeared to be coherent with the half billion years given by the radiometric datings after the Cambrian.

An important advance developed in 1926, when the National Research Council of the American Academy of Sciences formed a committee charged with comparing the age of the earth as determined from the geological point of view with the age derived from perspectives of the other disciplines, beginning with paleontology and astronomy. The final report was nearly five hundred pages long and was published in 1931. Holmes's contribution to this was of principal importance and was supported by a solid bibliography including 650 references. Holmes had reviewed everything available for consideration in this territory, by now so familiar to him: the physical bases of radioactivity, colored halos, the radioactive contents of minerals and meteorites, atomic masses, and, of course, datings and their geological implications. Even though these datings could still be counted by the tens, their correlations with geological ages were clear. These results seemed less convincing when taken individually, as Holmes had remarked, but he concluded that "they are nevertheless so consistently compatible with each other from end to end that, as a whole, they provide a most convincing demonstration of the method."

Charles Schuchert (1858–1942), one of Barrell's former colleagues from Yale, stood up for stratigraphy. Schuchert admitted in his contribution to the report that, with him, "one stratigrapher at least has wholly gone over into the camp of the radioactive workers." Like many of his colleagues, he did not actually dare to bridge the entire gap between the limited prescriptions written by the late physicists of the nineteenth century and the interminably long eras deduced from radioactivity. Initially favoring an overall geological history on the scale of 500 million years, Schuchert was the first to be "surprised over his own results, for he started with the idea that he could not find enough thickness of strata or enough breaks to meet the demands of time indicated by the radioactive minerals." With the hundred-odd kilometers of thickness accumulated in the sediments that he had accounted for in North America, Schuchert had "however, found easily enough marine strata since the beginning of Paleozoic time to call for 500 million years." When he added to this duration more than a billion years for the Precambrian, Schuchert had really taken the plunge.

Evaluations of the accrued thicknesses of sedimentation had still varied only little. Schuchert confirmed that because of poor knowledge of the parameters, stratigraphy had lent itself to highly divergent interpretations. The half billion years elapsed since the Cambrian corresponded to the ages given by the lowest rates of sedimentation. By contrast with the diminution of geological intensity postulated by Kelvin, the current rates appeared to be much higher than the mean of the past. The reason for this difference is that the mean altitude of continents is now higher, providing vaster fields of exercise for erosion: it was thus clear that strict uniformitarianism had to be abandoned. At the same time, however, through the use of radiometric chronological bases, it had finally become possible to determine the rates of sedimentation, as well as their variations over time. For the first time, these famous factors characterizing the dynamics of the earth over long periods had begun to take on values that were recognizably reliable. The new correlations between ages and thicknesses of sedimentary layers, which Holmes would continue to improve until his last days, were largely accepted.

As for biology, it wanted only to follow along with the new discoveries. John M. Clarke (1857–1925), one of the most eminent American specialists in the field, had pointed out in 1922 that paleontologists are not "the makers, but the users of time," and after having been "treated gingerly by astronomers and physicists in the allotment of time," they were content "that our colleagues in celestial mechanics are heaping upon us their munificence in their prescription of this heavenly commodity." But these "users of time" also found themselves standing on the wrong foot: they now had so much time that needed to be filled out, that the concepts of evolution were obviously going to require a complete

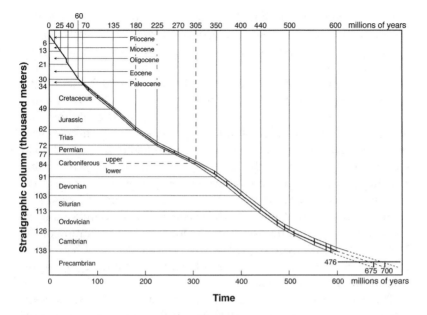

Figure 11.1 The most intense geological action over the course of the most recent periods, as revealed by thicknesses accrued in the stratigraphic column over the course of time. From Holmes, "Revised Geological Timescale," 205.

overhaul. By the end of the 1920s, the position of the last opponents had become untenable. The discovery of isotopy had, nonetheless, provided Joly with a new, ingenious argument in favor of the shorter ages: perhaps what had occurred was that a portion of the lead had been formed by a shorter-lived isotope of uranium, one that no longer existed. These final salvos had no effect. While still continuing to criticize the radiometric ages, Joly passed discreetly from 100 million years, which he still had not relinquished by 1925, to 600 million years in 1930. Thirty-five years after the discovery of radioactivity, and twenty-five years after the first radiometric datings, a consensus had finally been reached. By contrast with the conclusions that Kelvin had formed a good half century earlier, the radiometric timescales would no longer be contested.

Astronomy Left Aside

In the report of the National Research Council, astronomy had accepted the short end of the stick. Its contribution was reduced to a brief seven pages, in accord with an already well-established mutual lack of interest.

The stars, their timescales, and their role as mysterious crucibles in which the elements were formed before making up the various celestial bodies—none of that preoccupied geologists more than remotely. Certainly, one might expect astronomy to be able to specify the initial conditions of the earth's history, but that history was itself long enough that the geologists could not be bothered, in practice, by the veils covering the origins of their object. With radiometric dating, geological timescales were gradually founded on a solid basis, solid enough to make the astronomers jealous. Earlier, for example, Barrell had only touched upon astronomical questions in the last pages of his long paper, concluding that the nature of the great cosmic cycle whereby matter and energy were mutually transformed into each other "is still far beyond the bounds of scientific imagination." In brief, while geologists were feeling confident enough of themselves not to worry any longer about the future of the heavens, the astrophysicists had slowly realized that they had a new mission, which was to keep the age of the universe from lagging too far behind the age of their own planet.

There remained three great questions to resolve. The first was ancient. After the early infatuation with the role of radioactivity in the activity of the sun, the source of the energy of the stars still remained to be discovered. The second, intimately related to the first, concerned the formation of the elements within the stars, but it was difficult to answer that before elucidating the atomic nucleus. And the third became all the more urgent as the bounds of the universe appeared to recede farther and farther away. Ever more numerous galaxies were being discovered through the use of the new giant telescopes, raising the question of the dynamics of the entire universe. Although they relied upon the most fundamental theories of the times, astrophysicists had no cause to envy the humble geologists in regard to brusque turnabouts: the timescales deduced from astronomical observations fluctuated themselves—and with surprising amplitude and frequency. In truth, the road leading to solid solutions was marked with numerous pitfalls.

The Element Factory

Astrophysics had taken its very first steps in the 1860s. Shortly after the philosopher Auguste Comte (1798–1857) had proclaimed that "the chemical or mineralogical structure" and the temperatures of stars would remain "ever excluded from our recognition," William Huggins (1824–1910), near London, and the Jesuit Angelo Secchi (1818–1878), at Rome, used spectral analysis with marvelous success in determining the chemical composition of the stars. From spectral analyses, it was shown in 1864 that stars are formed of "matter identical with that upon the earth," which justified Descartes' old hypothesis of a common origin for the sun and the planets. Huggins, who saw this as an argument in favor of other inhabited worlds, added further, "It is remarkable that the elements most widely diffused through the host of stars are some of those most closely connected with the constitution of the living organisms of our globe." Huggins also proved that the nebulae were indeed constituted of hot gases, and not of clusters of stars, and he discovered two green lines in their spectra, which he attributed to *nebulium*; afterward, a yellow line, detected under particularly favorable conditions for observation during an eclipse in 1868, led Lockyer and Janssen to announce, at the same time, the presence in the sun's atmosphere of an element that was later so aptly named *helium*. (Helium continued on in its successful career, as we know, but nebulium was officially pronounced dead in 1928, when its lines were finally attributed to a positive ion of oxygen.)

It was Lockyer who most thoroughly exploited the resources of spectral analysis to place stellar history in an evolutionary perspective sustained by geological theories. Lockyer was a friend of Tennyson, then the greatest living English poet, and he authored a book titled *The Rules of Golf*. He was also a prehistorian, through his astronomical and religious theories regarding the alignments of megaliths. He founded the prestigious magazine *Nature* in 1869 and directed it for fifty years. Lockyer, more than anyone else, had a mind whose curiosity was equaled by its audacity. In the various types of stellar spectra, he saw the signature of the changes of temperature that chronicled the life of a star. At the very beginning of condensation from the nebula stage, the temperature increased and then began diminishing, until the point of complete extinction, when the energy of contraction could no longer compensate the heat that was radiated. Lockyer observed that, from blue stars to red stars, the number of spectral lines—and therefore the number of chemical elements—increased gradually as temperature decreased. Certain spectral lines were common to groups of elements, which he saw as an indication of subatomic entities from which matter was being constituted. Without taking up the concept of protyle, Lockyer suggested that the highest temperatures would dissociate matter into elementary fragments.

The stars thus revealed a progression of chemical forms that were analogous to those of the organic forms in geological terranes, to such an extent that one could even "define the various star-stages by means

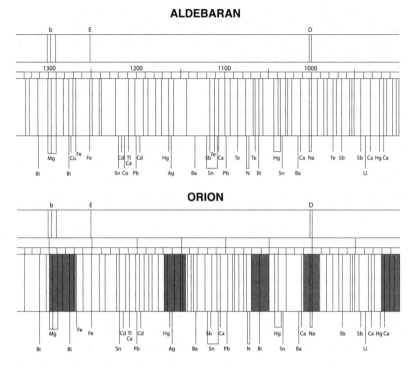

Figure 11.2 The identity of matter in the universe, as demonstrated by the light spectra of Aldebaran (α Tauri), of the Orion nebula, and of chemical elements. The wavelengths are in relative scale. From Huggins, "On the Spectra of Fixed Stars," plate 11.

of the chemical forms which they reveal to us in exactly the same way as the geologists have done in regard to organic forms; so that we may treat these stellar strata, so to speak, as the equivalent of the geological strata." From the hottest to the coldest star, one found "ten groups so distinct from each other chemically that they require being dealt with completely separately, just as do the Cambrian and the Silurian formations." And "imitating the geologists still further," Lockyer attributed names to the groups of stars, in decreasing order of stellar heat: Argonian, Alnitamian, Achernian, Algonian, Markabian, Sirian, Procyonian, Arcturian (solar), and Piscian. The analogy with geology was complete with the "break" that was introduced between the Markabian and Sirian "stellar strata."

Unfortunately, the great chemical complexity of the cooler stars was only illusory: the variations of the intensities of the spectral lines were due to differences in surface temperature, not to differences in chemical composition. But the concept of chemical evolution still benefited from new interest when radioactivity provided a solid experimental foundation. The first perceptions of the elementary constituents of atoms were with the alpha rays that are emitted all along the chains of disintegration. The discoveries of atomic numbers and, later, of isotopes then reinforced Prout's old hypothesis, according to which the elementary constituent of matter was instead the hydrogen atom.

Upon these foundations, the first attempts at a simple atomic "Erector Set" were made in 1915 by two chemists at the University of Chicago, William D. Harkins (1873–1951) and his colleague E. D. Wilson. Actually the atomic mass of helium is not exactly four times that of hydrogen, they noted, but rather 3.97 times its mass. It follows that the formation of a helium atom from four hydrogen atoms must be accompanied by a diminution in mass of 0.77 percent. The celebrated relation between mass and energy ($E = mc^2$) established by Einstein in 1905 gave a direct evaluation of the energy of atomic reactions from mass-balance equations. Even if the fraction appeared to be modest in magnitude, the loss of 0.77 percent of the mass liberated such a considerable energy as to give ample account for the heat of the sun. Harkins and Wilson further demonstrated how the elements of the periodic table could be produced from hydrogen (for odd atomic numbers) or helium (for even numbers), by gradually adding helium atoms. An interesting fact was that the variations of mass caused by these reactions in the synthesis of elements were very low in relation to that produced in the formation of helium from hydrogen. The helium atom thus proved itself to be a particularly stable atomic entity: it was helium that was emitted, not hydrogen, during the course of radioactive disintegration.

Harkins's model was worth much more than providing a general connection between stellar heat and the synthesis of elements. With such notions as the evolution of elements, chemistry took on a historical dimension; the composition of a star became a reflection of its past history. As a cosmological and chemical implication of his model, Harkins could even explain the greater stability (and subsequently, the greater abundance) of the elements having even atomic numbers as compared to elements having odd atomic numbers; this idea was experimentally confirmed by Rutherford in 1919, in his first experiments of artificial transmutations. At the same time, Aston had begun to measure the masses of the isotopes of many elements; it had become possible to make a precise assessment of the energy of atomic reactions. But by 1920, astronomers were hardly interested in the possible mechanisms behind stellar reactions.

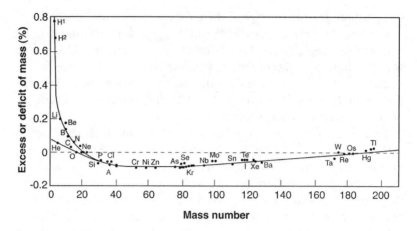

Figure 11.3 Excess or deficit of atomic mass in relation to the atoms' constituents, indicating the great loss of mass, dissipated in the form of energy, when helium is formed from hydrogen, and the greater stability of the atoms between silicon (Si) and tin (Sn). From Aston, *Mass Spectra and Isotopes* (1933), 167.

Stellar Dynamics

Astronomers assigned high priority to cataloging the myriad distant objects they discovered with their ever-more-powerful telescopes. Helmholtz's old theory still had adherents, even if "only the inertia of tradition keeps the contraction hypothesis alive—or rather, not alive, but an unburied corpse," as the great astrophysicist Eddington noted in 1920. Soon he himself rallied to Harkins's ideas (though without crediting them). The positions and the movements of stars were two parameters among others, such as color, temperature, and brightness, that could be determined through the use of spectral analysis and photographic observation. Between 1905 and 1907, the Dane Ejnar Hertzsprung (1873–1967), who was a graduate in chemical engineering from the Polytechnic Institute of Copenhagen, noticed that stars of the same spectral type could be quite different in brightness. This observation led him to distinguish the brilliant giants from the dense pale dwarfs. Russell finally realized in 1913 that he could report the correspondences between brightness and spectral type more exhaustively by using a diagram to display the brightnesses of stars as a function of their spectral types (and thus, implicitly, as a function of their surface temperatures, as is demonstrated in the following). The classification of spectral types was immediately justified by the simplicity of the correlations shown by the

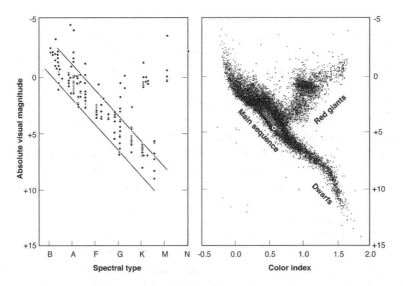

Figure 11.4 Hertzsprung and Russell's original diagram (from Russell, "Relation between the Spectra," 254) and the diagram obtained from observations of 41,453 stars taken from the satellite *Hipparcos* (M. A. C. Perryman, European Space Agency, pers. comm., 1998). In the diagram on the right, the position of the sun is indicated by the black dot in a white circle. It has long been thought that after its birth as a red giant, a star would follow the spectral scale toward the white stars before descending again toward increasingly colorful dwarfs.

diagram above: the great majority of the stars were placed on the *main sequence*, the diagonal that shunted the white dwarfs to one side and the red giants to the other. And our sun was revealed to be an ordinary star, representative of the main sequence in terms of both temperature and brightness.

The main sequence in this diagram, which soon became the Rosetta Stone of contemporary astronomy, was full of promise. Starting with Russell, there were some who viewed the diagram as showing the path followed by stars through the course of their long lives. Stellar evolution came back to the foreground, rekindling the ideas of Lockyer, though these had never been firmly established. One obvious question was how fast a star might travel through the main sequence. The first responses came from English astronomy's unfriendly rivals James Jeans (1877–1946) and Arthur S. Eddington (1882–1944). Both obtained ages of about a trillion years using different methods during the 1920s. Such ages overturned all the old concepts—and would even have made Kelvin, as well, turn over in his grave.

Jeans was a former pupil of J. J. Thomson who had started off in molecular physics. He was therefore familiar with the classical model established by Maxwell to calculate the time required for the molecules of a gas to achieve an equilibrium in terms of distribution of energy through their mutual interactions. The analogy that could be drawn between molecules moving within a gas and stars shifting in space gave him the idea of adapting this model to the distribution of energy between objects having masses and mutual distances analogous to those of the stars: one needed only replace the molecular force of attraction by that of gravitational attraction. Jeans found in 1924 that the distribution of the energy of the stars in our galaxy represented nearly a state of equilibrium and, more importantly, that several trillion years had been necessary to arrive at such a state.

An old problem now seemed to have become worse: it called for determining a mechanism of prodigious enough efficacy that it could account for stellar radiation over such long durations. Jeans was late in abandoning the idea of gravitational contraction, and he turned his thought to the annihilation of electrons and protons. This idea, which he communicated in 1904 in order to explain radioactivity, also led to his explanation of the dynamics of the formation of the elements in the stars. He claimed that at extreme pressures and temperatures, "matter in its earliest state consists of a mixture of elements of different atomic weights, those elements the atomic weights of which are highest having the greatest capacity for the spontaneous generation of radiation by annihilating themselves, and, in consequence having the shortest lives." The radioactive elements present on the earth thus played a pivotal role between the stable elements and the elements heavier than uranium, which were intrinsically unstable and, according to Jeans's hypothesis, the major constituents of the stars.

Eddington is considered to be the founder of modern astrophysics. He was the successor to George Darwin at Cambridge in 1913 and was a principal proponent of general relativity, which he was the first to put to the test with his experiment in 1919 measuring the deflection of light by a gravitational field during an eclipse. Through his research on the structure and the dynamics of the stars, he determined the equilibrium conditions between gravitation and the pressure of radiation. Eddington deduced from the fundamental relationship that thus resulted between mass and luminosity that the massive stars extinguished themselves more rapidly than others, even though these times were still to be measured in the trillions of years. Accordingly, he concluded in 1926 that "a star of mass 2 cannot be older than 10^{12} years, however great the mass with which it originally started," whereas a star half the size of the sun "must be at least 40×10^{12} years old."

In passing from the hands of Helmholtz into those of Jeans and Eddington, the stars had suddenly become ten thousand times older. First it was their distances, and now it was their ages that had become astronomical. In a time that seemed to be marked by many discontinuities, this new break was nonetheless remarkable. And as a painful, posthumous irony for Kelvin, it indicated a convergence of the new astrophysics with the worst uniformitarian excesses of the geologists. Jeans's model was generally accepted by physicists at the end of the 1920s thanks to the rather complete description it gave of the physics of the universe. But the success of his model was fleeting, because his conclusions were soon attacked on two fronts, namely, stellar dynamics and stellar energy.

Ever since Huggins and the very beginnings of spectral analysis, the Doppler effect had been used to measure the velocities of stars along the line of sight, that is, their *radial velocities*. An observer will notice that the wavelengths from a source of light appear to shift toward red for a light source traveling away from himself or herself. Very interesting measurements of this effect were made between 1912 and 1924 by one of America's most productive astronomers, Vesto M. Slipher (1875–1969). Slipher was hired by Percival Lowell, at the beginning of the creation of his observatory; later Slipher succeeded Lowell. Slipher began his career quite naturally with studies of Mars and the other planets, and he moved on to measuring the velocities of spiral nebulae, which were only poorly understood at the time. His first observations made of the Andromeda nebula indicated that it was approaching the sun at a speed of 300 kilometers per second, by far the most rapid velocity ever detected. Later Slipher found radial velocities as high as 1,800 kilometers per second for more distant nebula. An object moving at such a velocity would traverse the Milky Way in somewhat less than 20 million years. This conclusion was unacceptable in terms of the new timescales, and so the nebulae would have to be outside of the Milky Way. And the frontiers truly receded in 1924 when another American astronomer, Edwin P. Hubble (1889–1953), demonstrated that we are separated from the Andromeda nebula by a distance much greater than the Milky Way's diameter. The existence of other galaxies besides our own could no longer be doubted.

Hubble was an athlete; in fact he was a talented amateur boxer. He opted for law after his scientific studies, and even opened his own law firm. His attraction to astronomy finally prevailed, but his lawyer's training has often been linked to the capacity Hubble demonstrated to identify the fundamental aspects of complex problems. Just after completing his doctoral dissertation, he was engaged by the American forces in 1917. There was thus a two-year delay in his great beginnings at Mount Wilson Observatory, near Los Angeles, where the Carnegie Institution of Washington was preparing to put a very large telescope into operation. Hubble made optimal use of this instrument to explore the galaxies that by then were known to populate the universe and to investigate their structure, their movement, and their distribution throughout space.

In order to complete Slipher's measurements, Hubble undertook the task of systematically measuring the distances of the spiral galaxies. In taking into account the velocity of our own galaxy and the radial velocities of the others, he was forced to conclude in 1929 that galaxies moved away from each other at velocities proportional to their distances from each other. Relativistic theories for an expanding universe had been proposed shortly before; it is well known that this first observation in their favor marked the beginning of a new era. So much has since been written on the Creation, on the infinite, and on the nature of space and time that here we shall pass directly to the age of the universe, as it was estimated on the basis of the constant proportionality between the radial velocity and the distance of a star, which is referred to as Hubble's constant. With a constant rate of 0.15 meters per second per light-year, this expansion would have begun about 2 billion years ago. In becoming comparable to the age of our tiny earth, the age of the universe was becoming negligible in relation to the trillions of years presumed for the age of the Milky Way. Jean's model was also contradicted by other astronomical observations; it did not last long.

Jeans's speculations regarding the sources of stellar energy did not survive, either. Observations in the 1920s showed that the stars are essentially constituted of hydrogen and helium. There is not even the slightest trace of the super-heavy elements hypothesized by Jeans, and the "terrestrial" elements of heavier masses were found to be concentrated in the solar corona. The possibility of hydrogen's combustion to form other elements became full of relevance again. The mechanism of a *thermonuclear* reaction was first proposed in 1929 in Berlin. Its authors were Friedrich (Fritz) Houtermans (1903–1966), a young German physicist whose verve made a lasting impact on his contemporaries, and Robert d'Escourt Atkinson (1893–1981), an English astronomer familiar with the methods used by Eddington to estimate the internal temperatures of stars. Inspired by the theory of alpha disintegrations, Atkinson and Houtermans suggested that at extremely high temperatures, four hydrogen nuclei are first captured by the nucleus of a light element before forming a helium nucleus; two errors involving a factor of one thousand compensated each other exactly in these calculations—so the conclusion came out right.

On the same day the young authors finished writing their manuscript, Houtermans went out with a charming young lady. During their walk, the stars glistened above in the firmament. "Don't they shine beautifully?" his companion exclaimed. "I've known since yesterday why it is that they shine!" the love-struck physicist proudly replied. Physics can be romantic once in a while. Was that the end of an era?

TWELVE

From the Atomic Bomb
to the Age of the Earth

THE TRIUMPH OF IMMENSE TIMESCALES   Following the discovery of isotopes, radiometric dating methods became reliable and precise. This advance took place in the context of World War II. With these methods, Patterson demonstrated in the 1950s that the earth, meteoroids, and the sun all share the same age of 4.5 billion years. His work showed that the geological time recorded in the rocks dating from the Cambrian represents only 10 percent of the earth's history. A controversy lasting twenty-five hundred years was thus concluded; the solution had required contributions from most branches of physics, astronomy, and natural sciences.

On the Trail of Isotopes

By the beginning of the 1930s, the basic outlines had been sketched for the terrestrial and solar timescales. Now that terrestrial heat sources had begun to be understood, the sources of energy providing for stellar dynamics had finally been glimpsed. And even though the germs of the *Big Bang* theory had already appeared, the conflict of an earth that could prove older than the universe would threaten once again. When the neutron was discovered at Cambridge by James Chadwick in 1932, scientists were able to discard the theory that electrons were located within the atomic nucleus to bind the protons. At the same time, the new quantum mechanics gave a rigorous theoretical framework for establishing models of thermonuclear reactions. In 1938 Carl Friedrich von Weizsäcker (1912–) in Germany and Hans Bethe (1906–2005) in the United States improved the Atkinson-Houtermans model by showing that carbon and nitrogen isotopes act as catalysts in the cycle of the transformation of hydrogen into helium, which had provided for the heat of the stars since time immemorial. It is a star's initial mass alone that determines its evolution, and the path of its evolution on the Hertzsprung-Russell diagram intersects the main sequence rather than follows it. Nonetheless, it was necessary to wait until the end of the 1950s to understand how the heavier elements were formed, and this clarification began with carbon, the element consubstantial with life. Even though the first three minutes of the universe have already been described, with a wealth of details, there still remain some great questions without responses, such as whether the universe is in a state of perpetual expansion or whether it will at some point change to contraction. Hubble's constant and the age of the universe have fluctuated by a factor of seven since 1929, and neither is well established yet. Consensus for the age, though, has been converging recently to between 10 billion and 12 billion years.

None of these uncertainties any longer affect what we know of the age of the earth. Obviously that question was much less difficult, and resolving it was not merely for an anthropocentric interest. Naturally, the age of the earth represents a solid reference point. As given by an astrophysical model, the age of the universe ought not to come too close to that of the earth. And for the biologists, one must recall that the period separating the formation of the earth from the beginning of the Cambrian is equally crucial, as are also the durations of time that were required for life to appear and then for the first germs of organization to transform into trees or fish. From this point of view, the matter of whether our planet is 1 billion years old or 8 billion years was certainly not indifferent. This notion of the age of the earth had to be defined more precisely, but is it the date of its conception we are seeking? Or of its birth? Or the date of some catastrophic event, such as the birth of the moon, wrested from the earth after impact with an enormous meteorite, as the distant followers of George Darwin assert today? Whatever the sense we give it, this age represents a necessary point of departure for describing the formation and the dynamics of the planets, or the origin of life.

At the end of the 1920s, there was no choice but to recognize uranium as the best geological timepiece ever. While research continued on the subject of isotopes, the progress of analytical chemistry allowed the dating of more and more minerals. Friedrich Paneth (1887–1958), the Austrian radiochemist, made a specialty of analyzing, first in Berlin, then in Königsberg, ever smaller quantities of helium. He had the pleasant surprise of observing that, when produced in small quantities within uranium-poor minerals, helium is more likely to remain there. Because it then was no longer the exclusive prerogative of uranium-bearing minerals, the helium method of dating was approached with renewed interest: it completed and corroborated the lead-based ages in the determination of timescales. Beginning in 1929, Paneth even succeeded in dating some meteorites.

Nonetheless, new methods were required for dating the earth. It was not sufficient for different methods to give similar results in order for the result to be considered reliable. Otherwise, at the end of the nineteenth century, the convergence at that point toward an age of 100 million years would have closed the case. Until the late 1920s, the existence of isotopes had not been utilized, except to distinguish between the lead produced from uranium and that which came from thorium. In order to deduce *isotopic* ages, it was first necessary to identify the isotopes of lead and uranium. In the case of uranium, Rutherford's first attempt to use isotopic dating led him predict the existence of a rare isotope having mass 235, which was later confirmed, shortly before the discovery of nuclear fission. Because of the vicissitudes of history, this fissionable isotope immediately developed an extreme value when the possibility of exploiting atomic energy was discovered during World War II: uranium 235 became a military stake of strategic importance. As chance would have it, the first path leading to the correct age of the earth was glimpsed within the context of this conflict, and it started from the isotopic compositions of the lead in different minerals. The question was not completely resolved for another ten years, after the analyses of the Canyon Diablo meteorite. And the efforts made to provide a date for the earth had far-reaching consequences, well beyond the strictly geological or astrophysical framework. For example, as we shall see, gasoline containing lead was banned because of these efforts.

Rutherford's Adieus

In 1927 Aston put the second model of his mass spectrograph into service; this model had four times better resolution than the first and ten times greater precision. Lead could no longer resist it: its isotopic composition could now be measured directly, whereas before it could be evaluated only by measuring atomic masses. The proper approach was discovered in 1927 by Charles S. Piggot (1892–1973), a geochemist from the Carnegie Institution of Washington, D.C., during his attempts to determine isotopic ages for the very first time. Piggot's intention was to date some old minerals on the basis of their contents of lead 206 and 208, the isotopes of lead formed from the disintegration of uranium and thorium, respectively, whose existence had been postulated for fifteen years on the basis of atomic mass measurements. For this purpose, he suggested to Aston that he analyze samples of organic compounds of lead. He thus sent him some tetramethyl lead, a compound produced from common lead by the American Chemical Warfare Service. The idea proved to be excellent, because Aston finally succeeded in observing the two isotopes, 206 and 208, in this compound with his mass spectrometer. He was especially surprised to detect an additional isotope of mass 207 appearing in an abundance comparable to that of lead 206 (about 25%). Thus the atomic mass of a lead sample was shown to be a poor indicator of the lead's origin, because it can remain quite close overall to that of common lead (207.2) even with considerable variations of the relative proportions of the three isotopes.

Although the existence of a new isotope seemed to complicate the situation a bit, Piggot had foundations solid enough to pass to the following stage, the determination of the age of an old mineral. He and his colleague C. N. Fenner had some new tetramethyl lead prepared from a Norwegian sample of a uranium-bearing ore. Unfortunately the flask containing the precious liquid broke during its journey between Washington and Cambridge. In the new sample that reached him six months later, Aston observed very little lead 208 (about 5%), which conformed well with the fact that the mineral being examined was poor in thorium. By contrast, he found twice as much lead 207. This higher content could not have been the result of contamination by common lead; otherwise the lead 208 content would not have been so low. The logical conclusion was that lead 207 was also a radiogenic isotope, even though its parent was still unknown.

Because the chains of both uranium and thorium have an isotope of lead as the end product, Aston supposed by analogy that lead 207 would prove to be the end product of the third known radioactive chain, that of actinium. Rutherford was not the last to learn of the idea. Ten years earlier Rutherford had left Manchester to succeed his former teacher, J. J. Thomson, as the director of Cavendish Laboratory at Cambridge. At nearly sixty years of age, he ran the laboratory with a firm hand, all the while deploring the nearly industrial scale that experimentation was beginning to take on in his discipline. He became a monument to English science in his turn: the "Newton of atomic physics"—a Newton who had blossomed with honors and who had kept his distance from the bitter disputes and sterile sorts of research. Soon afterward, he was made baron, Lord Rutherford of Nelson (in honor of the town near where he was born and had gone to school). He even surpassed the future Soviet marshals in the number of medals he had amassed. His colleague A. S. Eve, who was also his first biographer, remarked to him on the day of one of his discoveries: "You are a happy man, Rutherford, always on the crest of the wave!" Rutherford replied, smiling: "Well! I made the crest, didn't I?" adding soberly, "At least to some extent."

In his only publication mentioning the age of the earth in the title, which was a note of less than one page, Rutherford made use of Aston's analyses to resume the task of geological dating. Taking it one step further, Rutherford returned to an idea then current by proposing that actinium belonged to the chain of some isotope of uranium that remained to be identified, for which he fixed the mass as 235 (lead 207 + 7 alpha rays). Some rather simple considerations led him to evaluate the abundance of this isotope and its half-life, which appeared to be around one-tenth the half-life of uranium 238. Rutherford now had available two different radioactive chains for the disintegration of uranium into lead, each beginning and ending with different isotopes of the two elements (uranium 238 / lead 206 and uranium 235 / lead 207). He also knew that the isotopes with even-numbered masses are produced in greater quantity in the stars than isotopes of odd-numbered masses. When the earth had separated from the sun, as he supposed, uranium 235 thus could not have been more abundant than uranium 238. The time required for the ratio of these two isotopes to proceed from unity, the highest possible value, to their current ratio would therefore mark the upper limit for the age of the earth, which Rutherford calculated as 3.4 billion years at the most. This age, twice that of the oldest minerals that had been dated until that point, represented the time of the earth's separation from the sun. Rutherford was not concerned about giving it a lower limit; what interested him was to compare this with the duration of the life of the sun. He adopted the seven trillion years recommended by Jeans and concluded that chemical elements such as uranium were formed in the very last period before their expulsion from the star. Naturally, interest in this conclusion diminished after the major downward revisions for the age of the sun.

Piggot and Fenner's work was directly at the roots of both Aston's discovery and Rutherford's speculations. The English physicists had left to the American geochemists the task of determining the age of their Norwegian mineral. The result obtained by Piggot and Fenner based on the lead 206 and lead 207 produced from uranium (908 million years) was in respectable accord with the chemical age (920 million years) but

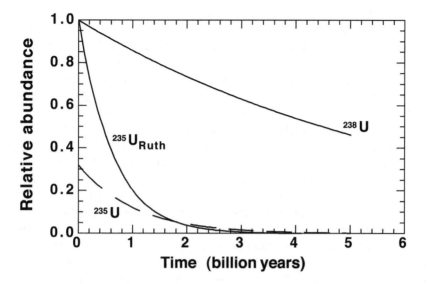

Figure 12.1 Rutherford's correct estimates. Actual ^{235}U and supposed ^{235}U$_{Ruth}$ variations of relative quantities of uranium 235 and uranium 238.

unfortunately quite different from that given by the lead 208 originating from thorium (1.313 billion years). Piggot and Fenner were somewhat disappointed. Was it possible that even the mass spectrographic analyses gave insufficient precision? To say the least, isotopic ages would not quickly supersede the chemical ages.

A New Form of Lead

Nonetheless, uranium was quite interesting, as Rutherford had foreseen, because it gives two different chronometers for dating the same phenomena. And since the half-life of uranium 238 is about ten times as long as that of uranium 235, any alteration or contamination of a mineral that modified its original uranium content will be revealed as a disagreement between the ages given by each of the two chains. Isotopic methods, because they are more complex, ought to have a much better resolution than chemical methods. But many important questions remained to be resolved, beginning with uranium 235, whose existence had yet to be demonstrated.

On his part, Aston was aware of the mediocre quality of his lead analyses. He began them anew under better experimental conditions. In 1933 he thought the tetramethyl lead from a series of old minerals was much more eloquent: five additional isotopes were detected. Unfortunately, further measurements confirmed the existence of only one of these, with mass of 204, which represented about 1.5 percent of the common lead. Even for an expert like Aston, it was not always easy to distinguish the main signal from background noise on a photo when a given isotope has a low abundance. The discovery of the new isotope 204, which is not radiogenic, revealed that not all of the lead on the earth comes from radioactive disintegrations, demonstrating furthermore that the age of the earth could not be determined on the basis of the quantity of lead present in the crust. Although this implication may have appeared to be a small step backward, actually it was not. Isotope 204 proved quite important because its quantity had not increased since the time of the earth's formation. It therefore represented an unchanging factor in the primordial isotopic composition of the earth's lead. This seemingly incidental fact would soon lead to very rigorous methods of dating.

Uranium is like lead in that it is not prone to forming gaseous compounds, and it consents unwillingly to mass spectrography. Even though uranium hexafluoride and uranium tetrabromide proved well suited to this kind of measurement, Aston had observed only the single isotope of mass 238 with his apparatus in 1931. At the University of Chicago in 1935, Arthur J. Dempster (1886–1950) was finally able to detect uranium 235, whose existence had been predicted by Rutherford, but he could not measure its quantity. More sensitive apparatuses were required, and they would have to be constructed by a new generation of physicists. Mass spectrography was going to become mass spectrometry. The photographic plates that were exposed to currents of ions deflected by magnetic fields were to be replaced by electric amplifiers. The sensitivity of instrumentation would be transformed.

Alfred Nier, Uranium, and Lead

In the forefront of the experimentation leading to the new mass spectrometry we find Alfred Nier (1911–1994), of Minnesota. Nier had been attracted to machines in childhood, after which his attention turned to electricity and radio. These interests led him into studies of electronics and physics. During the preparation of his PhD dissertation, he constructed a highly sensitive mass spectrometer. This instrument allowed him to determine the isotopic composition of elements, such as argon and potassium, that had previously been somewhat resistant to analysis.

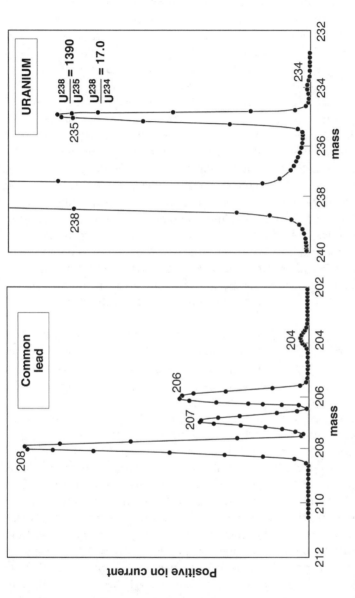

Figure 12.2 Uranium and lead isotopes, revealed by mass spectra. Quantities of isotopes are proportional to areas of respective peaks. From Nier, "Variations," 1572 (*chart on left*); Nier, "Isotopic Constitution," 151 (*chart on right*).

As he would later note, his own research could not be "considered to be of much practical importance in the 'real' world." A few years later, though, there was no doubt that the world's fate depended on it, at least in part.

After his thesis, Nier went to Harvard University as a postdoctoral fellow in 1936. There he constructed a new, even more sensitive apparatus, making use of an extremely powerful magnetic field created by a two-ton electromagnet. In addition, he was fortunate enough to find at Harvard a number of geologists genuinely interested in dating methods. He also found the lead collections that had been utilized by Richards and his colleagues for their atomic mass measurements. So Nier had arrived in the right place at the right time to pursue an interest in geological time. Beginning in 1938, he dedicated himself totally to this matter. In the following year he succeeded in measuring the abundances of uranium 235 and uranium 238 (which he found to be in a 1-to-139 ratio); from his results he deduced the half-life of uranium 235 much more precisely than Rutherford had been able to estimate it ten years earlier. And so, all the necessary elements had finally been assembled to complete the genealogy of lead. As illustrated in figure 12.3 by lead 206, radioactive chains can be rather complicated. Without leaving aside thorium, which has a single radioactive isotope, one can summarize the three radioactive chains that have lead isotopes as end products as follows:

	nonradiogenic		^{204}Pb
^{238}U	(4.56 billion years)	→.....→	^{206}Pb
^{235}U	(713 billion years)	→.....→	^{207}Pb
^{232}Th	(13.9 billion years)	→.....→	^{208}Pb

In this way Nier made use of solid foundations to pursue his isotopic analyses of lead, an effort that continued even after he went back to Minnesota, where he constructed a new mass spectrometer and where he remained until the end of his career. His results were published between 1938 and 1941, and for nearly fifteen years they formed the basis of all discussions regarding the age of the earth. Nier was advancing into virgin territory, and primary among his geological interests were precisely determining isotopic ages and identifying the metamorphoses undergone by minerals throughout their long history. Nier was also interested in primordial lead, which was incorporated into the earth at the time of its formation. This lead had progressively been diluted by radiogenic lead, which is produced in all places where uranium and thorium are present, even in trace quantities. Thus common lead appeared to him to be a blend in which the proportion of radiogenic lead ought to

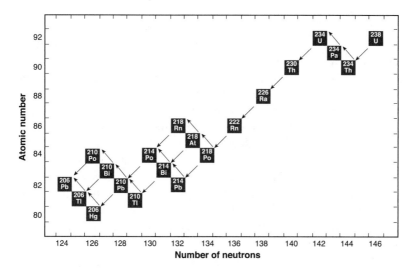

Figure 12.3 The prolific lineage of uranium 238.

increase with time. In order to distinguish these two components, Nier analyzed two quite different types of minerals: uranium-bearing samples were one type, and lead ores that were poor in uranium and thorium were the other.

For the first group, Nier observed that the quantity of lead 204 amounts to only about one-ten-thousandth of the quantity of the radiogenic isotopes: after thirty years of radiometric dating, the most important finding was that lead 204 at last justified the hypothesis that the main portion of the lead present in an unaltered mineral is indeed radiogenic. Furthermore, the quality of mass spectrometric analyses could now permit precise datings: based on three different chronometers (the two uranium-lead pairs and the thorium-lead chain), Nier obtained consistent ages that ranged, according to the samples, from the hundreds of millions to a billion years. The highest age that Nier found for a mineral was 2.2 billion years, for a sample from Manitoba. He nearly began to worry that this age seemed older than the earth. But the minerals provided consistent datings; the isotopes of lead were found to be reliable chronometers. What remained was to improve the sensitivity of the techniques, to allow dating minerals containing even lower quantities of uranium or thorium.

The question of the earth's age was far from being resolved, though, because doing so required first determining the isotopic signature of primordial lead. With this goal in mind, Nier analyzed lead ores that were scarce in uranium and in thorium, for which contamination by radiogenic lead would therefore be very low. The isotopic signature of these ores ought to be that of common lead, captured at the time of their formation. The oldest ores, in which the proportion of radiogenic lead was lowest, would then be recognizable by their high content of lead 204. A sample of galena (a form of lead sulfide) coming from Ivigtut (Ivittuut), Greenland, proved to be especially interesting from this point of view, but its content of isotope 204 was so high that Nier wondered whether this sample would actually be a good reference. At the time when he was ready to analyze more minerals in order to elucidate the point, the United States was entering the war. Nier then became distracted by other requirements, tasks that would make use of those very same instruments and methods he had engaged in the exploration of geological time, but this time for quite different purposes.

Uranium 235 and the Bomb

Nier's involvement in the war effort was related to the capital importance that uranium 235 was very soon to take on. In 1941 Nier was the only man capable of precisely measuring the abundance of this isotope. If we go back a few years to examine briefly one of the most curious discoveries of physics, that of uranium fission, we will better understand how this competence of his could have become so crucial.

Neutrons were discovered in 1932; they are relatively heavy particles devoid of electric charge that are expelled at great velocities from atoms. In Rome, Enrico Fermi (1901–1954) realized that neutrons constituted the projectiles of choice for the bombardment of atomic nuclei, which are positive in charge, because their use would avoid any sort of interaction resulting from electrostatic forces. Nearly all the elements of the periodic table became radioactive when Fermi and his colleagues submitted them to this treatment of bombardment. After absorbing one neutron and rejecting one electron, an element advanced forward one cell along its row in the periodic table. Using this sort of "alchemy made to measure," the possibility had thus been developed for producing the transuranic elements, those man-made elements that are even heavier than the already massive uranium atom.

The strange part of this episode began in 1934 when the Italian physicists attacked uranium with neutrons. In spite of the difficulties arising from the infinitesimal quantities produced, Fermi's team had good reasons for their claims of having identified new radioelements as the first transuranic elements ever formed. Further and further experiments were performed, and heavier and heavier transuranics were recognized. For more than four years, these super-heavy elements were the objects of numerous studies. In Berlin and in Paris, Hahn, the Joliot-Curies, and their respective collaborators were brought into the game. Even though the increasing complexity of this new radiochemistry caused some wonder, it was real consternation that developed one day in December 1938, when one of the products from the reaction undergone by uranium could not be identified by Hahn's team as anything other than an isotope of barium. How was it possible for uranium to give rise, under neutron bombardment, to an element that was so much lighter than itself? Hahn's discovery escaped any explanation. Once the commotion had subsided, the team resigned itself to admitting that the man-made transuranics did not yet exist. The correct explanation, though, was found by Lise Meitner (1878–1968), a close colleague of Hahn until her abrupt departure from Germany six months earlier to escape the Nazi threat. The encounter between a slow neutron and a uranium nucleus did indeed provoke the fragmentation of the latter: the process of *fission*, presumed impossible by the theoreticians, had been produced experimentally.

In the following months, it was observed that for each neutron absorbed, an average of between 2 and 3.5 neutrons were emitted, and these in turn went to attack other uranium atoms, causing each new target to emit the same number of new neutrons, and so triggering a chain reaction. And by contrast with all other known nuclear reactions, atomic fission liberated more energy than was required to be introduced in order to stimulate the reaction. The quantity of energy released thus increased exponentially through the course of development of the chain reaction. Thirty-six years after it had been envisaged by Rutherford, nuclear energy seemed to be utilizable. "One might ask whether it is an advantage for humanity to know the secrets of nature, whether it is mature enough to profit from this knowledge, or whether it will be harmful." Pierre Curie, who pronounced this opinion in 1905, concluded nonetheless that "the powerful explosives have allowed men to make admirable progress. They are also a terrible means of destruction in the hands of the great criminals who lead peoples toward war. I am among those who believe, with Nobel, that humanity will draw more good than evil from these new discoveries."

Perhaps Curie's son-in-law, Joliot, shared the same optimism. He described with his collaborators the controlling principle of chain reactions in what would later be called a nuclear reactor, and he hurried to obtain the first patents for the instrument in the autumn of 1939. But the concept of constructing a reactor was purely hypothetical; it was not even known which uranium isotope underwent fission. Pressed by theoreticians such as Fermi, Nier made recourse to his mass spectrometer to purify a quantity of uranium 235. From a sample used for his studies of geological time, he isolated some few billionths of a gram of the isotope of mass 235, but this was enough to establish that it was the only fissionable isotope. Nevertheless, the prospect of concentrating significant amounts (even though nobody really understood how much was needed to make a bomb) seemed practically impossible. Nier's result was no defense secret: it was published in March 1940.

But Nazi domination was not long in extending over Europe, nor that of Japan over Asia. Just when the possible military implications of nuclear fission were becoming more clear, Joliot's experiments in France were interrupted in June 1940 by the German invasion. At the initiative of the exiled European physicists, American research regarding uranium fission increased in pace. From this point, the need for discretion was understood: no further results would be made public before 1946. The United States entered the war in December 1941. Many physicists considered the fabrication of an atomic bomb to be impossible; even so, in November 1941, an American committee was created to study it. One year later, the Manhattan Project had been launched; the young physicist Robert Oppenheimer (1904–1967) and the energetic General Leslie R. Groves (1896–1970) provided the direction for the project.

From Chicago to Los Alamos, from one end of the country to the other, unlimited means were made use of in almost total secrecy. Nearly a half million persons were participating in the technical and logistical phases of the project, without even knowing what its purpose was. Entire industrial complexes sprang up out of the ground in a matter of just a few months. One of these, constructed at Oak Ridge, Tennessee, comprised a pharaonic battery of mass spectrometers; using them, hundreds of kilograms of uranium 235 were isolated, atom by atom. Isotopic separation was but one of the innumerable problems to be resolved, but the technical difficulties of this separation were immense. Nier played a crucial role more than once in resolving these before supervising, from beginning to end, the processes of enrichment and separation of uranium 235.

Never before had such a stunning constellation of scientific talents been summoned together. Theoreticians, experimenters, physicists, chemists, metallurgists—all the trades were mobilized. The dreamers of pleasant dreams, who previously had formalized the interiors of stars with their equations, now had the practical application, under the direction of the refugee Bethe, of producing models of chain reactions that could be activated in nuclear explosions. A large contingent of European scientists who had fled Hitlerism participated in the work. The result of these efforts is all too well known. On 6 August 1945, the uranium 235 in the bomb called *Little Boy* wiped Hiroshima off the map. Less than four years after the beginning of the Manhattan Project, this isotope, whose existence had been postulated on the basis of an old Norwegian mineral, passed from the status of laboratory curiosity to that of a terrifying weapon of destruction.

In case the industrial purification of uranium 235 had resulted in failure, the Manhattan Project harbored a second option, which was based on plutonium. The existence of this element had been kept secret after it was first produced in February 1941 by rapid neutron bombardment of uranium 238. This was the second artificial transuranic element to be artificially produced, following shortly after neptunium. One month later, plutonium's capacity for fission under the effect of a neutron flux was demonstrated. In September 1942, only three-millionths of a gram of plutonium had been painstakingly purified. In December this new element was produced more conveniently in the first nuclear reactor constructed by Fermi and his team, underneath the stadium bleachers at the University of Chicago. In passing to the industrial scale, it was produced by the tons in another nuclear reactor, before being separated from the initial uranium using more classical procedures of chemical purification. Three days after Hiroshima, Nagasaki experienced the fact that plutonium also was able to trigger chain reactions.

The ferociousness of the war left hardly any place for pure speculation. Yet it was within this context that productive ideas had begun to germinate regarding the age of the earth. Three important persons played key roles in the penultimate episode of our story. The first is a newcomer, a Russian geochemist, Eric Carlovich Gerling (1905–1985). The second is the German theoretician Fritz Houtermans, whom we have already met. And, as one might expect, the third is the geologist Arthur Holmes. Although each used a slightly different form and method, Gerling, Houtermans, and Holmes all independently analyzed the measurements made by Nier between 1938 and 1941 regarding the isotopic composition of lead.

Gerling and the Russian Campaign

Germany attacked the USSR in June 1941. The Soviet physicists were pressing toward the bomb by spring of 1943, but because Stalin did not comprehend the stakes of the American effort until then, the study did not receive increased support until after Hiroshima. Four years later, in August 1949, the first Soviet bomb exploded; despite Stalin's purges, not all of the Soviet scientific competence had been eliminated. In fact, it was at the Institute of Physics and Technology at Leningrad (now St. Petersburg) that the spontaneous fission of uranium, without any external intervention, had been discovered by Georgii Flerov and K. A. Petrjak in 1940: when present in sufficient quantities, uranium 235 could initiate and develop chain reactions on its own.

The Institute of Radiochemistry of the Soviet Academy of Sciences was also located at Leningrad. Natural radioactivity and uranium were the focus of the institute's research. Gerling worked there starting in 1928. He was born into a family of German origin from the Baltic, who had settled generations before in St. Petersburg. Although he was not a theoretician, he had great stores of intuition. He made numerous discoveries, some of which still have not been recognized in the West, and he was the founder of the methods of isotopic geochemistry used in Russia. Although he was not a dissident, Gerling maintained a constant aversion for the regime: his tendency for sharp reactions before the scientific apparatchiks brought him embarrassment on more than one occasion. He was in serious trouble several times under Stalin, and he owed his life to protection by his influential colleagues. He directed a laboratory that had the distinction of not including even a single member of the Communist Party; nonetheless, Gerling was awarded the Lenin Prize in 1961.

In 1937, during the worst of Stalin's purges, Gerling had been studying the behavior of noble gases in the context of geological research when he was threatened for the first time because of his distant foreign origins. He was protected from arrest by V. G. Khlopin, another specialist in the study of these gases. When the Germans invaded the USSR, the Institute of Radiochemistry was moved to Kazan, in the Urals. Unlike his friends Flerov and Petrjak, Gerling was not admitted into the Soviet atomic program because of his foreign ancestry. He was the director of the laboratory of gases in charge of the search for natural gas and petroleum. During these duties, he analyzed the hydrocarbon content of subterranean waters. Russia's war and his activities in regard to it did not prevent him from becoming interested in the age of the earth.

Gerling supposed, as did Nier, that minerals having the highest content of lead 204 were the most reliable depositories of primordial lead. In the vast ensemble of isotopic measurements made by Nier, it had been found that the galena of Ivigtut, Greenland, best satisfied this criterion, and it additionally yielded the primeval relative quantities of

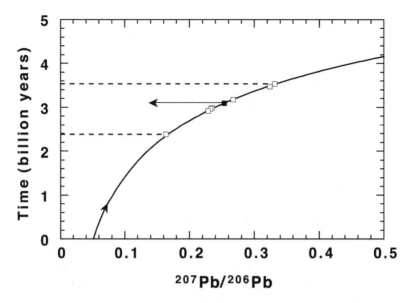

Figure 12.4 Increase over time of the ratio $^{207}Pb/^{206}Pb$ for exclusively radiogenic lead. The average value of the seven sulfides considered by Gerling, which corresponded to the age of 3.1 billion years, is indicated by the arrow. Sulfide by sulfide, the wide range of ages, between 2.4 and 3.5 billion years (white squares), still demonstrates the approximative nature of the method.

leads 204, 206, and 207. Compared to this galena, a twofold excess of lead 206 and lead 207 in a mineral with low uranium content would represent radiogenic lead, and their ratio was a simple function of the time elapsed before the formation of the said mineral (see appendix). In order to calculate an age, Gerling started with the average value (0.253) of the ratios of radiogenic lead ($^{207}Pb/^{206}Pb$) from seven sulfides, which he deduced from Nier's measurements. He obtained 3.1 billion years, to which had to be added the 130 million years elapsed since the formation of these sulfides. The earth thus appeared to Gerling to be at least 3.23 billion years old. Repeating the calculation for another of the oldest of the minerals dated by Nier, he obtained an age of 3.92 billion years. The consistency between the two results was reasonable.

By contrast with all preceding estimates, these elevated values represented lower limits for the age of the earth, because primordial lead could have been richer in lead 204 than was the galena of Ivigtut. Gerling concluded that "the age of the earth is not under 3 to 4 billion years" and added that "this is certainly not too much, since the age of certain minerals . . . was put at 2.2 and even 2.5 billion years." These results were published in September 1942, in three brief pages in the proceedings of the USSR Academy of Sciences, at the time when the German troops of Marshal Paulus penetrated Stalingrad. During the heat of the battle, few copies of the proceedings were printed, and they did not circulate abroad. Gerling's publication became known in the West only after the war, when Houtermans and Holmes had both published similar ideas. And so the first isotopic dating of the earth passed practically unnoticed: a fine way for such an old controversy to begin to give way!

Fritz Houtermans: After the NKVD and the Gestapo, on to the Age of the Earth

Houtermans had lived several lives since the time we left him in 1929, when he was contemplating the stars. While he was traveling in the Caucasus in 1930, he married his walking companion, who was also a physicist. Settling in Berlin, he continued his work in physics, something that his vitality and his inexhaustible repertory of amusing stories had not caused him to lose sight of. After Hitler came to power, Houtermans decided to leave Germany; he was a Communist and his mother was half Jewish. His retort to the anti-Semites: "When your ancestors were still living in the trees, mine were already forging checks," obviously had not engendered any new sympathies. He arrived in England in 1933; he did not like the country and the cuisine disgusted him: he left the following year for the USSR. The future of humanity there appeared glorious to him.

Houtermans accepted a position at the Institute of Physics and Technology at Kharkov, in the Ukraine. As he engaged in nuclear physics research at this renowned center, he was enthusiastic about belonging to "a little family of bright people occupying themselves with the problems of the universe as a sort of hobby," according to the expression of one of his colleagues. Nonetheless, Houtermans was not long in awakening from his illusions about the regime. The great purges came soon afterward; physicists were not spared. In 1937 Houtermans was fired. Before he could leave the USSR, he was arrested as a German spy. His wife had been able to flee to the United States and enlist some committed eminent scientists, such as Jean Perrin and the Joliot-Curies, to intervene on his behalf. Their efforts were in vain; Houtermans remained a prisoner in the NKVD (People's Commissariat of Internal Affairs) jail. To avoid madness, he devoted himself to mathematics, even using soap to write on his cell walls his demonstrations of theorems relating to number theory. During an interrogation that lasted eleven days, he admitted having transmitted to the Germans the secrets of an apparatus of his own invention, which allowed measuring from the ground the speed of aircraft. He even produced plans for his imaginary invention to support his confession. The names he gave for his supposed contacts were two German officers, Scharnhorst and Gneisenau—both of whom were Prussian generals during the Napoleonic wars—in honor of whom two ships of the ill-fated German navy were christened. The farce did not make it past a certain judge. Houtermans recanted, claiming that he had confessed under duress, but he still, of course, remained in prison.

Two and a half years later, the German-Soviet pact was signed. Houtermans was finally expelled, but not as far west as he would have hoped: the Gestapo were awaiting him at the German border. New interrogations ensued, though conducted this time without the professionalism of the NKVD, as Houtermans, by now an expert on the subject, noted. He escaped the worst through the courageous interventions of the physicist Max von Laue. Under the auspices of a program financed by the German postal service, Houtermans dedicated himself, even with his now-heretical reputation, to the theory of chain reactions. He came to learn that a new element called plutonium by the Americans was more easily induced to chain reaction than uranium was, and he managed to send a cable to the United States that the Germans were on the way to producing a bomb. Near the end of the war, Houtermans, who was an inveterate smoker, found himself short of tobacco. He managed to obtain some by claiming that he would be able to use it to produce heavy water. The stratagem worked twice before being discovered by the Gestapo, and once again Houtermans was in an uncomfortable position. Protected again by von Laue, he settled at Göttingen. There, after peace had returned, the age of the earth began to intrigue him.

Houtermans's method made use of a diagram that has now become classic, called "lead-lead" because it makes use of the two uranium-lead chains. Following the disintegration of uranium 238 and 235, the quantities of lead 206 and 207 increase, while that of lead 204 remains constant: the ratio of lead 206 or lead 207 to lead 204 thus increases regularly over time. Obviously, the point of departure for this evolution is fixed by the isotopic signature of primordial lead. If that were known, the time elapsed since the formation of the mineral could be obtained directly on the basis of the distance traversed along the curve depicted in a "lead-lead" diagram. In this diagram, the path followed by an isotopic signature depends additionally upon the uranium content of the mineral. The higher the uranium content, the faster lead 204 is diluted by radiogenic isotopes. Beginning from the same primordial composition, minerals of increasing contents thus define evolutions that are more and

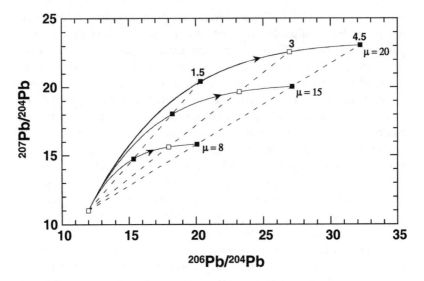

Figure 12.5 Evolution through time of the isotopic composition of lead (solid curves represent initial ratios μ = U/Pb 8, 15, and 20) and corresponding isochrons, depicted for 1.5, 3, and 4.5 billion years (broken lines). The half-life of uranium 235 is about one-tenth that of uranium 238, so the ratio $^{207}Pb/^{204}Pb$ increases more rapidly than the ratio $^{206}Pb/^{204}Pb$. This fact is responsible for the downward curvature in this "lead-lead" diagram.

more marked (see diagram below). Houtermans at last noticed that these paths have one very interesting mathematical feature: minerals of the same age, even those with different uranium contents, are all found on the same straight line that starts at the point representing the primordial composition. This age is then provided, quite simply, by the slope of the given straight line, which Houtermans called the *isochron* (of the same time). However, minerals that have different points of departure are not defined by a straight line. Houtermans thus had developed a much finer method than Gerling's, because not only did it offer the possibility of finding an age, but also of checking that this age was sound.

With the purpose of obtaining points along a given isochron that were more distant from one another, Houtermans researched Nier's observations of minerals that had the same age but quite different uranium contents. In so doing, he optimized the precision provided by the slope (age) and determined the ordinate of the origin for the given isochron (which represented the primordial composition of the lead isotope). Without giving further details about the analyses he selected to arrive at his result, Houtermans announced that 2.9 ($\pm$ 0.3) billion years had elapsed since the time when primordial lead began mixing with radiogenic lead. Houtermans cautiously supposed that this duration ought to have been intermediary between the solar system's age, that is, the age of the formation of terrestrial uranium, and that of the formation of "the solid crust of the earth." Using additional hypotheses, Houtermans succeeded in drawing from Nier's measurements an age that was consistent with the age obtained by Gerling. His result was imprecise, however, and its real meaning had to be clarified.

The Last Ages of Arthur Holmes

A few months before Houtermans, Holmes also had made use of Nier's measurements. In fact, no one else struggled so diligently to extract from them their entire import: he even used one of the very first computers to effect highly involved calculations. Holmes, still a professor at Durham University, was nearly fifty years old at the outbreak of World War II. He had been widowed in 1938 and remarried the following year. Because so many young people had left for the war, Holmes had fewer students, and he was less occupied with his teaching. He began writing his *Principles of Physical Geology* amid flashes from the incendiary bombs dropped by the Luftwaffe. This dynamic history of the earth—from the core to the crust, from the mountains to the oceans—was written in the form of a synthesis of geology and geophysics. For the stratigrapher D. V. Ager, it still represents "the great source-book of all original thought in geology." In 1943 Holmes left Durham for Edinburgh, where he continued revising his geological timescales even after his retirement.

In order to determine the age of the earth, Holmes sought the age and primordial composition that would allow him to account for the isotopic signatures of minerals taken by pairs.[1] In 1946 he used 72 pairs of minerals to obtain solutions ranging from 2.7 to 3.15 billion years. The following year, he made use of 1,419 pairs, leading to a much vaster range of ages, from 2 to 4 billion years, even though the distribution

---

1. As Gerling had done, Holmes was able to calculate the age of a mineral once he had fixed the composition of the earth's primordial lead. As he varied this composition, the age obtained also varied, and in a regular manner. By taking minerals in pairs, Holmes determined whether and how the "composition-age" curves of the two minerals coincided. Each time a correspondence was found, Holmes obtained a "solution" that gave the date at which the primordial lead had begun mixing with radiogenic lead. He also needed to combine successively, by pairs, the group of minerals studied by Nier in order to obtain all the possible "solutions."

within this range was marked by a peak around 3.35 billion years. After determining in 1946 that "the age of the earth is not far from 3,000 million years," Holmes persisted and finally concluded that on the basis of the available data, "the most probable age of the earth is about 3.350 million years." Shortly afterward, he further specified (somewhat recalling Leibniz) that "the age I have determined refers to the time when the granitic layer separated from average earth material during the consolidation of the globe." The ages obtained thus represented the time of the individualization of the crust, rather than the time of the formation of the earth, whose earliest moments have left no traces.

Nier's analyses and the 3 billion years concluded by Gerling, Houtermans, and Holmes suffered from the fact that, even in the absence of uranium and thorium, the lead of a given mineral could be partly radiogenic. The oldest of the samples analyzed by Nier thus would not allow establishing the true isotopic composition of primordial lead. Where could one find even older minerals, perhaps even as old as the earth? Harrison Brown (1917–1986) was a specialist in uranium and plutonium, working in Chicago near the end of the Manhattan Project. Both he and Houtermans gave the same response to this question independently in 1947: in the sky. Various geochemical arguments suggested a relationship between meteorites and the primitive earth. Because meteorites have undergone rather monotonous existences before falling to the earth, they might harbor minerals that could give clues to their formation. Some meteorites did indeed bear minerals that were found almost entirely free of uranium or thorium. Nier attempted to measure the isotopic composition of their lead. This was a failure: their lead contents were too low, and the samples had been contaminated during the course of analysis by traces of terrestrial lead. Nier abandoned his dating research, but the new developments that he brought to mass spectrometry were not lost. In the meantime, there were geochemists who had seized upon lead isotopes for their own research.

Patterson and Canyon Diablo

Like Nier and Brown, the other scientists of the Manhattan Project returned to their laboratories at the end of the war. A younger generation, people who had first served in the armed forces, had studied at the universities. The cold war was seen as a favorable climate for science. Each of the former allies, both east and west, thought the science that had helped them win the war would now provide for their own superiority. Budgets for science increased considerably. Space exploration and journeys to the moon and to the planets of the solar system were the most spectacular evidence of this situation.

Mass spectrometers, heavily utilized in the Manhattan Project, were no longer the prewar monsters; they had been transformed through the efforts of Nier and his physicist colleagues into laboratory instruments providing ever-greater possibilities. Their use expanded throughout the universities, where they were operated by researchers in many disciplines. Nier's laboratory manufactured these machines by the dozens. Isotopic geochemistry participated in the movement as a new discipline: with mass spectrometers, one could measure the temperatures of the oceans of times long past, or reconstitute the main stages of the life of the planets, or flush out the slightest traces of organic matter to be found in meteorites. Mass spectrometers are extremely sensitive analytical instruments; they are now used to detect slighter and slighter traces of pollutants of any kind. Miniature models are even produced today, to be installed in spacecraft. Nier analyzed the Martian atmosphere in 1976 using a spectrometer weighing less than 10 kilograms, borne by a Viking probe. Unfortunately for lovers of Mars, no biological activity was detected.

It is in this postwar context that we encounter Clair Patterson (1922–1995), an independent-minded geochemist. He spent his childhood in the Iowa countryside and then pursued chemistry studies. During the war, he passed a short time at Oak Ridge, where he undoubtedly encountered mass spectrometers for the first time. Afterward he went to Chicago, the cradle of isotopic geochemistry, for graduate work under the direction of Harrison Brown. During his first research, Patterson attempted to understand the origin of meteorites. Determining their ages had been helpful, and the isotopes of lead had served well to this effect. After completing his PhD dissertation, Patterson followed Brown to the West Coast. At Cal Tech, Brown established a laboratory offering extremely refined capabilities for the chemical preparation of samples: "ultra-clean" techniques were honed in order to avoid any contamination by traces of the lead omnipresent in the modern environment. With these improvements, Patterson made precise measurements of the isotopic composition of lead samples weighing only one-billionth of a gram. The major advantage of the new techniques combined with his previous experience with meteorites allowed him to achieve a decisive breakthrough.

In 1953 Patterson published the results of his analyses from the preceding five years. These results proved particularly interesting for the case of an iron meteorite that had fallen forty thousand years ago in Arizona, forming one of the most beautiful impact craters known. Aptly, it is called Meteor Crater. The meteorite fell relatively recently, out in the middle of the desert so that any possible contamination by lead dissolved in natural waters would have been inconsequential. The Canyon Diablo meteorite was well known. Crookes had dissolved five pounds of it in "an act of vandalism for the cause of science" (for which he begged the pardon of mineralogists). In so doing he discovered diamonds in the meteorite, but what Patterson was able to discover there was far more precious.

Uranium and thorium are present in very low quantities in Canyon Diablo, and sulfides in the form of inclusions are practically nonexistent. Patterson analyzed the infinitesimal traces of the lead contained in these inclusions and was not surprised to find the highest relative contents of lead 204 ever observed. It would have been tempting to regard this material as representing the isotopic signature of primordial lead. Patterson communicated his analyses to Houtermans, who used Gerling's simple method to conclude that it had required 4.5 (±0.3) billion years to pass from the composition of the "primordial" lead of Canyon Diablo to that of common lead, which he estimated on the basis of analyses of the galena from the Tertiary. In support of his reasoning, Houtermans underscored that all comparisons made for the heavy elements had demonstrated an "identity of isotopic constitution in meteoritic and terrestrial material." If the correspondence between the primordial lead and meteoritic lead was valid, the earth had just suddenly aged by a good billion years.

Patterson did not leave it at that. He wanted an even more rigorous demonstration because, although he accepted the figure of 4.5 billion years, it was "still impossible to defend the computation." With three iron and two stony meteorites, Patterson worked to determine whether the meteorites indeed formed a homogeneous system characterized by one same age. He observed in fact that all of these new results were positioned along one nice isochron of 4.55 billion years. Not only had he determined that the Canyon Diablo meteorite is not abnormal; in addition, the fact that these meteorites shared the same age as that found by Houtermans for the earth tended to confirm a common origin for the earth and the meteorites. It also confirmed the identity of the isotopic composition of their primordial lead. Furthermore, Houtermans's evaluation of the composition of common lead based on a galena from the Tertiary left much to be desired. Patterson considered the best samples of common lead to be not from isolated minerals, whose lead is an unlikely representative for the entire earth, but from the recent marine sediments. Because these sediments are produced by the erosion of

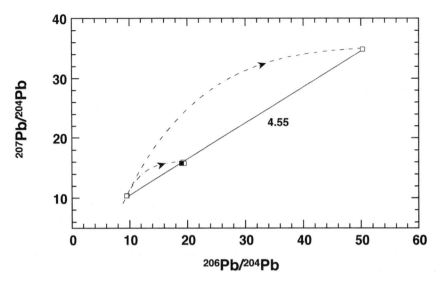

Figure 12.6 The mystery of the age of the earth finally solved. Isochron of the meteorites and the earth. The black square represents lead from oceanic sediments. The white squares (overlapping) represent meteorites. From Patterson, "Age of Meteorites," 232.

continents and are accumulated from rivers everywhere, they contain the most reliable possible blend of primordial and radiogenic lead. After his analyses, Patterson discovered that this common lead found its position perfectly along the isochron of the meteorites. Coincidence had to be excluded. The precise quality of the isochron method left no room for doubt. Patterson must have experienced a joy that was . . . ineffable!

Six points in alignment on the diagram finally revealed the "proverbially well-guarded secret of our Mother Earth," that Holmes had so eagerly sought. Centuries of interrogations came to an end when radiometric methods applied to lead isotopes bore their fruit, a half century after Rutherford's first speculations. The earth and the meteorites were shown to have the same age, namely, 4.55 billion years. Once again, a scientific controversy had been settled not by averaging out incoherent measurements or by combining inadequate tools, but by creating new methods, methods that reduced a problem whose complexity defied comprehension to the level of a college algebra exercise. "The whole object of science is to synthesize, and so simplify; and did we but know the uttermost of a subject we could make it singularly clear." For his presentation of the "lead-lead" diagram for the earth and meteorites, Patterson could even have attributed this maxim of Percival Lowell to himself.

Different Isotopes, Same Ages

The age determined by Patterson dated the last moments before the earth and the meteoroids were differentiated. As for the subsequent events, isotopic geochemistry, with its complete array of dating methods, has established that it all evolved quite quickly: the planets and the related bodies of the meteoroids had individualized completely and distinctly within about 100 million years. Such a short lapse of time thus would not have any real impact on "the age of the earth." Patterson's demonstration did have the weakness, however, of being based on only two chronometers, the two uranium-lead pairs. The figure of 4.55 billion years dates the last moment at which the uranium isotopes from the earth and from the meteoroids were homogeneous. Perhaps different elements might lead to a different conclusion.

Determining dates on the scale of billions of years requires radioactive elements with half-lives longer than 100 million years. There are eighteen radioactive isotopes of naturally occurring elements that satisfy this criterion. It would appear simple to make use of these, because the majority of the chains involve only the starting element and the end product. One of the main reasons such an exclusive interest has been accorded to uranium-lead chains thus far is that these chains have long been the only ones that could be considered workable. Uranium is a relatively abundant element, and all of its isotopes are radioactive. Consequently, the quantities of helium and lead formed from the chain are significant enough that they can be measured by means of the usual chemical methods. A secondary reason was simply that no other method would be applicable for determining the age of the earth. The key to the success of uranium comes from the fact that there are two different radioactive chains; this is what allows the initial quantities of lead to be determined. All the other elements besides uranium have only a single radioactive isotope, which does not occur abundantly. Accordingly, their primordial isotopic composition cannot be determined.[2]

With two other methods, which have become pet subjects of contemporary geochemistry, the first datings were attempted in 1913, as we have seen, when Hahn tried to make use of the radioactivity of rubidium and potassium. Despite his failure, Hahn did not discard geology, as

---

2. Here is a financial analogy: Without knowing the initial capital invested, it is impossible to determine the duration of an investment from the final capital and the rate of interest. But if two different amounts have matured through the same amount of time at different rates, one can determine the duration of the investments if one knows what the relative values of these amounts were initially.

evidenced by his short book published in 1926, *Was lehrt uns die Radioaktivität über die Geschichte der Erde?* (What Can Radioactivity Teach Us about the History of the Earth?) This interest continued even during his memorable research on nuclear fission; in 1937 and 1938 his measurements of strontium levels in rubidium-rich minerals gave the first (chemical) ages, based on the rubidium-strontium pair. The war did not cause his research to falter. Hahn kept his distance from Nazism and did not become involved in the fruitless German research into the atomic bomb. He declined his own government's invitation to go to Paris and work with the Joliot-Curies during the occupation. Instead he chose to pursue his own activities of isolating and identifying the ever-more-numerous artificial radioelements. At the same time he continued to date rubidium-rich minerals. His research was not secret: Hahn was one of the rare uranium specialists whose work continued to be published during the war. But the rubidium-strontium method did not provide decisive results until the traces of strontium 87 habitually produced in rocks could be measured precisely by means of mass spectrometry. These datings were initially restricted to the rare samples that were rich in rubidium, though they now can be made for samples bearing only one-billionth of a gram of the element.

Hahn had gotten nothing out of potassium, because the first product to be identified from its disintegration was calcium 40, which is the primary isotope of a very abundant element, and one with which potassium blends confusingly. Weizsäcker suggested in 1937 that potassium 40 must also form argon 40, concurrently, by means of another path.[3] It is, in fact, because of this transformation that argon is the third most abundant gas in our atmosphere. In 1948 Nier and his colleague L. T. Aldrich promoted the use of argon contents in minerals as a chronometer. The relationship with the good old uranium-helium method is obvious, but a very important difference is that the success of the potassium-argon pair was assured by the large size of argon atoms, which do not diffuse out of minerals unless heated.

These and various other methods give the same ages for minerals or rocks that have not been perturbed since their formation. The age of 4.55 billion years for meteorites, dated originally using lead isotopes, has been corroborated completely. The age of the earth as given by lead isotopes can be confirmed only indirectly; the arguments are numerous, and they agree with one another. We will cite only two of them. The first

3. To form calcium 40, the isotope potassium 40 transforms one neutron into an electron, which is ejected as a beta ray; one proton, which remains in the nucleus; and one neutrino. To produce argon 40, it captures one of its own electrons, which combines with a proton to make one neutron. The first type of disintegration is about ten times as frequent as the second.

is logical: since lead isotopes attribute the same age to the earth as they do to meteorites, it would be illogical for only the age of the meteorites to be correct. The second comes from the correspondence between the 4.5 billion years for the earth and datings performed on *shields*, whose antiquity has long been recognized by geologists. These are the roots of vanished continents, unsinkable rafts floating upon the surface of the earth since the dawn of time. They are found as vast fragments of crust outcroppings in Scandinavia, Greenland, Canada, Australia, Antarctica, and Ceylon. In accordance with the very first analyses made by Boltwood, Strutt, and Holmes, who indicated the extremely great antiquity of the Precambrian, ages surpassing 3 billion years have been established for some of the rocks in the shields. Certain minerals that are particularly resistant to alteration, such as the zircons, have already passed nearly 4 billion years near the surface of the globe. Since these were incorporated within younger rocks over the course of their history, a relatively brief interval of 500 million years separates their formation from the earliest moments of the earth.

The Invariability of Half-Lives

From the heights of its 4.55 billion years, the beginning of the Precambrian commands respect from the subsequent 540 million years during which life exploded. Erosion and the cooking of metamorphism have effaced very large sections of sedimentary history from the old shields. The accumulated thicknesses of terranes anterior to the Cambrian represent only a fraction of the stratigraphic column. Without radiometric datings, it would have been impossible to untangle the very long chronology of these rocks, which rarely bear fossils and whose relative ages estimated on the basis of mineralogy and geology have generally been incorrect. However, these divisions have provided a chronological framework for the traces of biological activity and for the primitive forms of life that have been detected. The beginning of an oxygenated atmosphere, due to the development of photosynthesis, also dates back 2 billion years. The first clues suggesting any organic activity whatsoever come from recently found rocks nearly 4 billion years old. During that period, when the friction from the tides had not yet pushed the moon very far away or slowed the rotation of the earth, the beings that were alive must have had days only one-third as long as our own.

As one recalls the disappointments of Kelvin and Tait, an obvious question that presents itself is, How can we be sure that some new phenomenon will not be discovered that will cause the structure of radiometric dating to collapse? And how can we assert that radioactive half-lives have not varied with time or that they have not been affected by the conditions of elevated pressure and temperature that reign within the depths of the earth? After all, what are the arguments that allow us to be more certain today of this age of 4.55 billion years than the Glasgow school was, yesterday, about its 25 or 100 million years?

A preliminary response to such questions is that the great enigmas that left Kelvin so bitter in 1896 have long been resolved. The atom is no longer an unknown; the relationship between matter and energy is now understood, particularly as regards conditions relevant to geology. The atomic bomb has quickly proved that the contemporary physical theories did correctly describe the forces of nature, under incomparably more severe conditions than those to be found at the interior of the earth. Isotopes by now have been used to describe radioactive chains completely, and the mechanisms of disintegration are now so well understood that it is possible even to predict the influence of pressure or temperature on half-lives; and indeed, these effects are negligible under geological conditions. Radioactivity is a property of atomic nuclei; in order for the electrons to be powerless to screen the nuclei from exterior perturbations, it would be necessary to place the atoms into the blast furnace of a star.

In short, a very simple proof of the immutability of half-lives comes from geology itself. The mechanisms of the different types of disintegration are quite different: the emission of an alpha ray or a beta ray, or the capturing of an electron do not depend upon external conditions in the same manner. If different chronometers give the same ages in spite of everything, it means that the possible effects of extreme pressures or temperatures really are negligible. The main errors made in radiometric dating actually come from the uncertainties with which half-lives are measured. Those errors, which are about 1 percent, are in fact far greater than all the other sources of error taken together. For the earth's 4.55 billion years, the absolute error is less than 50 million years.

The Hunt for Lead

After his work on the age of the earth, Patterson sought to utilize the variations of the quantities of lead isotopes as an indicator of the dynamics of the earth's mantle and crust. Few environments could escape his analyses, because his techniques allowed him to make measurements to a precision of one in a billion. Patterson discovered with surprise that the lead contents in the oceans were much higher than those predicted on the basis of deposits from rivers. This result, of apparently minor importance, caused Patterson to reorient his research completely, transforming him into the instigator of an all-out crusade against lead and its derivatives.

Patterson believed that this excess of lead in the oceans was produced from human activity: for example, from the tetraethyl lead added to gasoline for its antiknock properties. From that point, he began to inquire about the extent of the pollution caused by lead. New analyses demonstrated that the ocean surfaces were the most highly affected: there the lead had not yet had enough time to become diluted through mixing with the deeper waters. Patterson realized that what were considered "normal" lead contents were actually the results of industrial activity. Even the human body contains a thousand times more lead than it ought to, which conclusion was well confirmed by analyses of bones from 1,600 years ago. These "normal" levels are alarmingly close to the levels that are now thought to be toxic! In order to understand how pollution developed over the ages, Patterson analyzed core samples taken from glaciers in Antarctica and Greenland. The extreme sensitivity of his techniques allowed him to follow through the ancient ice column the inexorable increase of a pollution that began during the earliest days of metallurgy, ten thousand years ago, and diffused on the winds and the waters all the way down to the South Pole. As a secondary matter, archaeology and the history of metallurgy also benefited from these results.

It should not be surprising that Patterson's results did not arouse great enthusiasm, for considerable interests were at stake. Accordingly, questions were posed regarding the quality of his analyses and the pertinence of his conclusions. Industrial pressure was even exerted to have him fired from his institution. But Patterson was not the sort of scientist who could be satisfied, in the words of Crookes, "to acquire simply what others have already observed and discovered, with an eye directed mainly to medals, certificates, diplomas, and other honors recognized as the fruits of 'passing.'" Patterson had a mission; he had to fulfill it. And so he argued, he contradicted, and he held fast, using isotopes as his basis for revealing the origins of the lead, because the ore deposits exploited by humans have isotopic signatures differing from that of common lead. Based on the abundances of isotopes 206 and 207, it became possible to distinguish the proportion of the lead flux contributed by human activity in relation to that contributed by the forces of nature, and to demonstrate that industrial society contributes ten times as much as volcanism or the erosion of rocks. In 1970 the automobile traffic in the United States alone dispersed 370,000 tons of lead. Whether faster or slower, depending on the country involved, lead has now slowly been excluded from gasoline, from pipes, and from welding materials, which previously represented its principal uses.

With his customary intransigence, Patterson applied himself, at the end of his career, to denouncing the grip that utilitarianism holds on science. The utilitarian view has led to the atomic bomb, among other horrors, and the innumerable environmental alterations caused by human activities (and now accepted by too many as normal). He accused society of being utilitarian and incapable of emotion, of transforming knowledge into common "data," and of relegating to a subordinate role the old ideal of the pursuit of knowledge and beauty as ends in themselves. Although this theme was already classical, Patterson developed it further with his theory that the development of utilitarianism was a result of the differentiation, occurring along with the beginnings of agricultural capabilities, of the respective zones of the brain related to artistic reasoning and functional reasoning.

Our planet's 4.5 billion years provided Patterson with an unspeakable delight. Toward the end of his life it was, by contrast, his contemplation of the recent times of modest duration that caused him a painful dizziness. He was afraid that "a large part of the heritage of present-day knowledge provided by classical science will be perverted by the dominant breed of utilitarian scientists into new monstrosities such as genetically engineered production of cancer cures and a new species of hominid, *Homo sapiens android*, that will, besides being immune from cancer and devoid of humanity, live 500 years." But that would be another history of time, one that Patterson would not be on hand to narrate.

Epilogue

If one wonders whether it would be possible today to evaluate natural timescales through determinations of the age of the earth using Kelvin's methods, the response is negative. Neither the quantity nor the distribution of radioactive elements, nor the temperature profile nor the physical properties of terrestrial materials at great depths are known well enough; and other sources of heat must still be considered, such as the heat released from inside the earth through the crystallization of the very central part of the core. Furthermore, the slow movements by convection of materials between the crust and the core, expressed on the surface of the earth by plate tectonics, represent a very important manner of heat transfer that must be taken into account. These movements constantly modify the "layered" structure of the earth and even affect the shape of the earth and its period of rotation. Nevertheless, because the history of our planet is more complex on all scales than had ever been imagined, Kelvin's program to understand it is still completely relevant today. In fact, it has been necessary to wait until the dawning of the twenty-first century to be able to measure the melting temperatures of minerals at the pressure and temperature conditions prevailing in the interior of the earth.

At the same time, the array of techniques for exploiting this complexity has never ceased to expand. From the refined chronology of the solar system's formation to the timescales of the great cycles introduced by Hutton, there are few aspects of terrestrial activity for which radiometric methods have not been instrumental in improving our understanding. In just the few hundred millions of years since the end of the Precambrian, the displacements of the great plates have caused the oceans to open or close again and caused the mountains to rise up or be razed to plains. Over these periods, the salinity of the oceans has remained practically constant, with losses owing to sedimentation being compensated by the salts brought by the rivers. At the other extreme, the circulation of the oceans' waters and the cycles of glaciation and deglaciation have timescales in the thousands of years only, or some tens of thousands of years. Was the Flood—conserved in the memory of so many different peoples—actually the most recent of these rises of the seawaters? As for the first hominids, they have been dated, by means of the terranes from which their remains have been exhumed, back even beyond 2 million years.

"The heaven as a whole neither came into being nor admits of destruction," reiterated Aristotle, "but is one and eternal, with no end or beginning of its total duration." Perhaps Hutton winked an eye at Aristotle, when he also found "no vestige of a beginning, no prospect for an end" in the history of the earth. In any case, this matter would have satisfied Aristotle, whose mocking ghost perhaps keeps a finger on the passage from *On the Heavens* where he expounded another consequence of the cyclic movement of celestial bodies: "The same ideas, one must believe, recur in men's minds not once or twice, but again and again."

Many recurrences have thus been recognized in this tale; some are quite obvious, some rather subtle. Joly followed Halley in seeing the salts of the seas as a timepiece. Crookes rejoined Paracelsus in the search for the spiritual forces that govern the world. And the waning nineteenth century might recall the Renaissance, for some, as a period in which anything seemed to be possible, when the frontier between truth and fiction seemed to become blurred, when some believed they had found, in discovering the "planetary" structure of the atom, the last justifications of old analogies between the microcosm and the macrocosm. Even today, some of Kepler's influence remains, for example, in the contemporary thesis called the Gaia Theory, according to which the earth is a living, homeostatic being. Violent impacts of meteorites on it are invoked as causes for the great extinctions of species, for the wresting of the moon from the earth, and even, as Halley and Whiston conjectured, for the inclination of our planet on its axis. In a certain manner, geology has never ceased its wavering between uniformitarianism and catastrophism, between Buffon and Halley or between Lyell and Cuvier. As A. Koyré has pointed out, "Absurdity, as well as truth, is a daughter of time," and "time," occasionally, takes its amusement in being cyclical.

Nonetheless, these cycles and recurrences combine themselves with a sort of linear direction, and this latter tends to predominate in the spiral of knowledge that has sometimes been described. Thermodynamics is involved implicitly in it, because thermodynamics holds principles from which Nature never departs. Although we now know that the laws of thermodynamics impose an evolution, it is more difficult to persuade it to tell us the rate at which the evolution occurs. After the unequivocal evolution spanning all through the immense Precambrian, during which the earth lost its status as an incandescent star to become the haven for life we now enjoy, radioactive sources of heat triggered and maintained various terrestrial cycles, as proposed by Hutton. Kelvin was the best natural philosopher of his time, even though he placed too much faith in his solutions that, in the end, were irrelevant to questions of the ages of the earth and the sun. He repeated the same error that Newton had made two centuries earlier in his *Chronology*, where his concern for detail prompted him to annotate: "I do not pretend to be exact to a year: there may be errors of five or ten years, and sometimes twenty, and not much above."

In an age when modeling has had its influence on vast sectors of human activity, it might be useful to remember the "difficulties one finds when attempting to apply geometry or calculations to physical subjects that are too complicated, or to objects about whose properties we know too little to measure." Buffon, a mathematician himself, stressed this point in the *First Discourse* of his *Natural History*. "In any case, we are always obliged to make suppositions that contradict nature," he added, "to strip the subject from the majority of these qualities, to produce an abstract being that no longer resembles the actual being, and after we have reasoned much and calculated much about the relations and properties of this abstract being, and after we have arrived to a conclusion that, also, is completely abstract, we believe we have found something real, and we project this ideal result onto the real subject, and this is what produces countless falsehoods and errors." But Kelvin had never read Buffon; if he had, he would perhaps have heeded the advice of that great Newtonian. "This is the most delicate and most important point in scientific studies," Buffon concluded: "to know how to distinguish well between what there is of the real in a subject from that which we add to it arbitrarily as we consider it: to recognize clearly which properties belong to the subject and which properties we only imagine it to have."

The immensity of time, which Buffon worked harder to establish than anyone else, had not been entirely neglected by all others before him. Even more than the precession of the equinoxes, with its period of some tens of thousands of years, the much slower crisscrossings of the continents and the seas, attributed to the differences between the earth's center of mass and center of volume, had perpetuated this concept, during the Middle Ages. Certainly, medieval scholars were concerned about the future of the earth, and not its past, but the great scales of time brought into play well prepared the youthful geology to take on a historical dimension. Geology confirmed the distinction between human history and the history of the world (and in so doing, it preceded astronomy and biology) even before the history of the earth was divorced from that of the universe. According to the Big Bang model, the universe must have had a beginning, and it has not existed for all eternity: Aristotle was mistaken. But of course, he could not imagine that one movement could originate independently from another movement, or that the measure of time could be dissociated—as it is now, with radioactivity—from the movements of the stars. So then, did time start with the Big Bang? Would this be the end to the age-old question of time before the Creation, which long ago was solved by the Stoics, and then again later by Origen, in their own cyclical manner?

Appendix: Mathematical Complements

Heat Conduction

At any point within a body, indicated by Cartesian coordinates (x, y, z), Fourier's law states that the temperature (T) will vary with time (t) according to the equation

$$dT/dt = K(d^2T/dt^2 + d^2T/dy^2 + d^2T/dz^2),$$

where K is the ratio between thermal conductivity and specific heat.

Beginning from an initial state of elevated temperature (T_0), constant throughout the earth, Kelvin calculated the temperature profile at depths for different times after a lower temperature (T_s) had been reached at the surface. In assuming the earth to have the form of an infinite plane, he reduced the problem of cooling to that of heat conduction along a single dimension:

$$dT/dt = K d^2T/dx^2.$$

Integration of the equation with respect to x, the depth, gives the temperature equation:

$$T = T_s + 2(T_0 - T_s)\sqrt{\pi} \int \exp(-u^2)\, du,$$

where $u = x/\sqrt{Kt}$ and the integration is performed between 0 and $x/2\sqrt{(Kt)}$. The value Kelvin used for K was 400 in^2/year, or 0.0118 cm^2/s.

Radioactivity

The disintegration of a radioactive atom is described by the law $N = N_0 e^{-\lambda t}$, where N_0 is its initial quantity, N is its quantity at time t, and the constant of disintegration, λ, is related to the half-life by the equation $\lambda = \text{Ln } 2/T$. The values λ_{235} and λ_{238} are the disintegration constants of uranium 235 and uranium 238.

Within a closed system that has kept all its radiogenic lead throughout the ages, the ratio of the quantities of the two lead isotopes varies according to the equation

$$^{207}\text{Pb}/^{206}\text{Pb} = u_0(e^{\lambda_{235}t} - 1)/(e^{\lambda_{238}t} - 1),$$

where $u_0 = 1/137.88$ represents the current ratio $^{235}\text{U}/^{238}\text{U}$.

Furthermore, we have

$$(^{207}\text{Pb}/^{204}\text{Pb} - b_0)/(^{206}\text{Pb}/^{204}\text{Pb} - a_0) = u_0[(e^{\lambda_{235}t} - 1)/(e^{\lambda_{238}t} - 1)],$$

where a_0 and b_0 are the initial ratios $^{206}\text{Pb}/^{204}\text{Pb}$ and $^{207}\text{Pb}/^{204}\text{Pb}$, respectively. On a lead-lead diagram, an isochron passing through the point representing the quantity of primordial lead, a_0, b_0 is a straight line with slope

$$m = 1/137.88[(e^{\lambda_{235}t} - 1)/(e^{\lambda_{238}t} - 1)].$$

This was the method Patterson used to determine the age of 4.55 billion years for the earth and the five meteorites, whose isotopic measurements follow:

	$^{206}\text{Pb}/^{204}\text{Pb}$	$^{207}\text{Pb}/^{204}\text{Pb}$
Nuevo Laredo, New Mexico	50.28	34.86
Forest City, Iowa	19.27	15.95
Modoc, Kansas	19.48	15.76
Henbury, Australia	9.55	10.38
Canyon Diablo, Arizona	9.46	10.34
Ocean Sediments	19.00	15.80

Source Notes

19 *If heaven and earth*   Lucretius *On the Nature of Things* 5.324.
19 *The greater portion*   Herodotus *History* 2.10.83, 2.12.84.
20 *In Arabia*   Ibid., 2.11.84.
20 *by what means*   Polybius *Histories* 1.1.5.
20 *those commodities*   Ibid., 4.38.4.
20 *the Pontus has*   Ibid., 4.40.4–5.
21 *into closer connection*   Strabo *Geography* 1.3.10.
21 *Sea, what makes*   Psalm 114:5–6. All Bible quotations are from the New Jerusalem Bible.
21 *nothing comes from*   Quoted in Hippolytus Antipope *Refutatio Omnium Haeresium* 1.14.2.
21 *says further that*   Strabo *Geography* 1.3.4.
22 *little petrifications*   Ibid., 17.1.34.
22 *It is therefore clear*   Aristotle *Meteorology* 1.14.353a.
22–23 *Those whose vision*   Ibid., 1.14.352a.

CHAPTER 2

25 *if geology contradicted*   De Luc, *Traité élémentaire*, 3.
26 *has nothing to do*   Bottéro, *Birth of God*, 84.
26 *You must love*   Deut. 6:5.
26 *the most blessed*   Deut. 7:12–14.
26–27 *The faithful city*   Isa. 1:21, 11, 16–17.
27 *Who is there*   Jer. 15:5–6.
27 *no military disaster*   Eliade, *Myth*, 103.
27 *A son has been born*   Isa. 9:5, 6.
28 *uncovers depth*   Dan. 2:22, 44; 12:2.
28 *But you (Bethlehem)*   Mic. 5:1.
28 *to fulfill the Scriptures*   Mark 14:49.
29 *All clans on earth*   Gen. 12:3.
29 *the same sacrifices*   Heb. 10:1, 4, 14.
29 *the sheep on*   Matt. 25:33.
29 *eternal life*   Matt. 25:46.
30 *Their high antiquity*   Tertullian *Apology* 19.33.
30–31 *Moses the servant*   Theophilus *Theophilus to Autolycus* 3.23.
32 *Plato would have it*   Diderot and d'Alembert, *Encyclopédie*, s.v. "Création," 444.
32–33 *It is quite foolish*   Philo *Allegorical Interpretation* 1.2.
34 *What was, will be*   Eccles. 1:9.
34 *it is at once*   Origen *On First Principles* 3.5.3, 2.3.4.
34 *the effervescence*   Augustine *Confessions* 3.1.1.
35 *Do you not know*   Theophilus *Theophilus to Autolycus* 1.8.
35 *to gain over*   Celsus, quoted in Origen *Contra Celsus* 1.44.
35 *are good, and*   Augustine *Confessions* 7.c12.

35 *that which is*   Augustine *On Genesis* 52.c9.

35 *number six is*   Augustine *City of God* 11.30.6.

35 *the thoughts of men*   Ibid., 11.5.

36 *simultaneously with time*   Ibid., 11.6.

36 *Far be it*   Ibid., 12.14.2.

36 *Do not let us*   Basil *Homilies* 1.11.28b.

37 *Let there be*   Gen. 1:6, 7.

38 *a pious ignorance*   Allen, *Legend of Noah*, 4.

38 *for the honour*   Albertus *Physicorum* bk. 8, in *Opera omnia*, 3:1.

39 *there are two modes*   Bacon, *Opus majus*, vol. 2, part 6 of this plea, chap. 1, p. 583.

39 *one science is*   Ibid., vol. 1, part 2 of this plea, chap. 1, p. 36.

40 *another alchemy*   Bacon, *Opus tertium*, quoted in Crombie, *History of Science*, 69.

40 *the excommunications*   Duhem, *Système du monde*, 7:4.

41 *Since the creation*   Buridan, *Quaestiones*, quoted in Duhem, *Études sur Léonard*, 3:ix.

41 *by natural means*   Ibid., quoted in Duhem, *Système du monde*, 9:298.

42 *No one would say*   Oresme, *Livre*, 283.

42 *One could answer*   Ibid., 531.

43 *I also conjecture*   Buridan, *Quaestiones*, quoted in Duhem, *Système du monde*, 9:299.

44 *there was mention*   Luther, *Table Talk*, 358.

46 Lynn Thorndike quotes excerpts from John Ashenden's *Summa iudicialis de accidentibus mundi* in "John of Eschenden: Specialist in Conjunctions," in **his** *A History of Magic and Experimental Science* (New York: Columbia University Press, 1933), 3:325–346.

46 *I wish you would*   Augustine *Letter to Jerome* 70.2.3.

47 *the divine offspring*   Daniel Hensius, quoted in Scaliger, *Autobiography*, 77.

47–48 *Man feels within himself*   Chateaubriand, *Genius of Christianity*, chap. 3, p. 133.

48 *I myself have gathered*   Des Vignoles, *Chronologie*, 1:iii.

49 *To these deluges*   La Peyrère, *Theological Systeme*, 277.

50 *Here lies La Peyrère*   Moréri, *Grand dictionnaire*, 246.

51 *filled with curious*   Quoted in Lods, *Jean Astruc*, 32.

51 *everything we know*   Simon, *Histoire critique*, 229.

53 *there is no reason*   Cuvier, *Discourse*, 106.

CHAPTER 3

55 *Go forth my sons*   Severinus, *Idea Medecinæ Philosophicæ*, chap. 7, p. 73.

56 *the bosom of the earth*   Kepler *Harmony* book 4, chap. 7, p. 360.

56 *Heaven is man*   Paracelsus, *Samtliche Werke,* I, 8:100.

57 *Magic is natural*   Koyré, "Paracelse," 53.

59 *silence the Godless*   Quoted in Caspar, *Johannes Kepler* (1959), 228.

60 *since two truths*   Galileo to Piero Dini, May 1615, in *Opere,* 12:184.

60–61 *I confess that*   Descartes to Mersenne, 22 July 1633, in *Œuvres,* 6:238–39.

61 *everything that is*   Descartes, *Principles of Philosophy,* 2:64.

63 *in the great vortex*   Ibid., 3:33.

63 *for the occult qualities*   Laplace, *Système du monde,* 6th ed., chap. 5, p. 485.

63 *that the matter*   Descartes, *Principles of Philosophy,* 2:22.

64–65 *Nonetheless, mindful*   Ibid., 4:207.

65 *the single most important*   Heilbron, *Early Modern Physics,* 2.

66 *philosophy in a very*   Fontenelle, *Conversations on the Plurality,* 3.

66 *values a* Camelion   Astell, *Essay,* 91–92.

66 *the Infinity of the Sphere*   Halley, "Of the Infinity," 22.

67 *Ceiiinosssstuu*   Hooke, *Description of Helioscopes,* 151.

67 *bring in every Day*   Quoted in Drake, *Restless Genius,* 17.

68 *pleased the Almighty*   Cohen and Westfall, *Newton,* 141.

68–69 The anecdote is found in Voltaire, *Lettres philosophiques,* 15th letter, "Sur le système de l'attraction."

69 *the Planets in the Orbs*   Newton, MS Add. 3968.41, fol. 85r, quoted by I. B. Cohen in *Dictionary of Scientific Biography,* s.v. "Newton," 63.

69 *he seldom left*   Humphrey Newton quoted in More, *Isaac Newton,* 247.

70 *I want not*   Newton quoted in Manuel, *Portrait,* 303.

70 *hypotheses non fingo*   Newton, *Newton's Mathematical Principles* (1962), 546.

71 *verses and the wines*   Voltaire to M. Berger, 24 April 1735, in Voltaire, *Correspondance.*

71 *My serious occupation*   Voltaire to M. Thiriot, 4 February 1737, ibid.

71 *These vortices*   Voltaire, *Philosophie de Newton* (1992), 18.

72 *most inclined*   Newton to T. Burnett, 1 January 1681, in Newton, *Correspondence,* 2:329.

73 *the flattener of worlds*   Voltaire to Maupertuis, 6 October 1741, in Voltaire, *Correspondance.*

73 *You have confirmed*   Voltaire, *Quatrième discours.*

74 *a boldness*   Mairan, "Éloge de Halley," 26.

75 *he was generous*   Ibid., 27.

75 *the quantity*   Halley, "An Estimate," 866.

75 *true reason of*   Halley, "Cause of the Saltness," 296.

76 *push skepticism*   Arago, "Halley," 3:369.

76 *could not bear*   Conduitt quoted in Manuel, *Portrait,* 288.

76 *had no fear*   Mairan, "Éloge de Halley," 25.

76–77 *fully assured*   Halley, "Some Considerations," 118.

77 *considerations about the Cause*   Ibid.
77 *a person whose judgment*   Halley, "Some Further Thoughts," 123.
78 *He very rarely went*   Newton quoted in More, *Isaac Newton*, 247.
79 *Newton was piqued*   Granet in Newton, *Chronologie*, iii.
80 *I like these sorts*   Conti, *Lettre de M.*, 28.
80 *He lived in honor*   Voltaire, *Mélanges de philosophie*, 169.
81 *the real evidences*   Manuel, *Newton, Historian*, 9.
81 *Now since* Erathosthenes   Newton, *Chronology*, 7.
81 *the* Greek *Chronologers*   Ibid., 52.
82 *At least*   Mirabaud, *Le monde*, viii.
83 *I will dwell*   Fréret, *Défense*, 502.
83 *The idea of regulating*   Bailly, *Astronomie ancienne*, 509.
83 *I have drawn up*   Whiston, *Memoirs*, 1:39.
84 *I do not know*   Newton quoted in Turnor, *Collections*, 173.

CHAPTER 4

85 *was the fruit*   Newton, *Chronology*, v.
85–86 *this most beautiful*   Newton, *Newton's Mathematical Principles* (1962), 544.
86 *the light of*   Ibid.
87 *in general*   Dézallier d'Argenville, *Histoire*, 37.
89 *how the mountains*   Quoted in Ellenberger, *Histoire*, 1:79.
90 *Mountains have been*   Quoted in Duhem, *Système du monde*, 9:265. See also Ellenberger, *Histoire*, 1:82.
90 *it seems wonderful*   Albertus, *Book of Minerals*, 1.2.52.
91 *That it is*   Ibid., 1.1.14.
91–92 *That in the drifts*   Leonardo, *Notebooks*, no. 990, 2:215.
92–93 *practicing glassmaking*   Cap, in Palissy, *Discours admirable*.
93 *for some say*   Palissy, *Admirable Discourses*, 146.
93 *had not studied*   Ibid., 148.
93 *shellfish, which are*   Ibid., 158.
94 *Some are found*   Ibid., 164.
94 *Moses gives witness*   Ibid., 158.
94 *If you had thought*   Ibid., 157.
96–99 *The metallic materials*   Cardano, *Livres de Cardanus*, 106.
99–100 *if you contemplate*   Ibid., 131.
100 *For as the body*   Kepler, *Harmony*, book 4, chap. 7, p. 363.
100 *It is plain*   Aristotle *Physics* 2.8.199b; *Politics* 1.2.1253a.
101 *mixed with a sort*   Kircher, *Mundus subterraneus*, 2:53–59.
102 *yet did they*   Hooke, *Micrographia*, 107.
103 *except fleas*   Stimson, *Scientists and Amateurs*, 77.
104 *the World upside down*   Hooke, "Lectures," 29 February 1688, 411.

104 *these Shells*   Ibid., December 1686–January 1687, 335.
105–06 *so quick with*   Steno, *Discours sur l'anatomie du cerveau*, 1.
107 *given a substance*   Steno, *Prodromus*, 209.
107 *the strata of the earth*   Ibid., 227.
108 *There are those*   Ibid., 258.
109 *God saved me*   Steno to Leibniz, November 1677, quoted in Faller, "Niels Stensen," 160.
109–10 See Lhwyd, "Of the Origin of Marine Fossils."
110 *could he not*   Tournefort, "Description" (1720), 224.
111 *So if there be*   Burnet, *Sacred Theory* (1965), 48.
112 *the real Spoils*   Woodward, *Essay*, 15.
112 *the whole Terrestrial*   Ibid., ii.
112 *which yet is*   Ray, *Observations Topographical*, 8.
113 *I imagine a man*   Montaigne, *Complete Essays*, 2.37.594.
113 *What can we*   Ray, *Wisdom*, 26.

CHAPTER 5

114 *all of Europe*   Petit, *Dissertation*, 82.
114 *the heralds of*   Bayle, *Various Thoughts*, 18.
114–15 *It is in accordance*   Kepler quoted in Caspar, *Johannes Kepler* (1959), 302.
115 *to honor his*   Sévigné, *Lettres*, letter no. 686, 2 January 1681.
115 *While we were*   Molière, *Clever Women*, 4.3.1266–70.
116–17 *for a globe*   Newton, *Principles* (1962), 522.
117 *the Ancient chaos*   Whiston, *New Theory*, 4th ed., 73.
118 *well approved*   Newton quoted in Whiston, *Memoirs*, 1:43.
118 *one of those*   Locke, *Works*, 3:556.
118 *State of Darkness*   Whiston, *Astronomical Principles*, 155.
118 *These stars*   Maupertuis, "Lettre sur la comète," 26 March 1742 (1756), 3:210.
119 *It would take*   Voltaire, *Dissertation*.
119 *Let there be*   Gen. 1:6.
120 *that the globe*   Leibniz to M. Thévenot, 10 August 1691, in Leibniz, *Protogée*, 194.
120 *volcano*   Leibniz, *Protogée*, 17.
120 *even a slight*   Ibid., 12.
121 *Only a kind*   Dézallier d'Argenville, quoted by Carozzi in *Telliamed* (1968), 3.
121 *he knew very much*   Maillet, *Telliamed*, 2nd ed. (1755), 19.
121–22 *The transformation*   Maillet, *Telliamed* (1968), 188.
122 *before the invincible*   Maillet, MS quoted by Carozzi in *Telliamed* (1968), 46.
122 *system of a beginning*   Maillet, quoted by Carozzi in *Telliamed* (1968), 29.

122 *great views*   Buffon, *Histoire naturelle, Premier discours*, in *Œuvres philosophiques*, 7.

123 *to let the water*   Buffon to abbé Le Blanc, 22 February 1732, in Buffon, *Correspondance*.

124–25 *I would request*   Buffon to M. Hellot, 23 July 1739, ibid.

126 *until then*   Quoted by Piveteau in Buffon, *Œuvres philosophiques*, viii.

126 *The soul of*   Aude, *Vie privée du comte de Buffon*, 12.

126 *from sovereign*   Buffon to Catherine II, 14 December 1781, in Buffon, *Correspondance*.

126 *Daubenton is beautiful*   Buffon to Kingston, 26 September 1738, ibid.

126 *true happiness*   Buffon to Guyton de Morveau, March 1762, ibid.

126–27 *I lost a child*   Buffon to Ruffey, 21 November 1759, 2 April 1731, ibid.

127 *he was the delicate*   Aude, *Vie privée du comte de Buffon*, 41.

127 *To someone*   Hérault de Séchelles, *Buffon* (1979), 162.

127 *Style is the*   Buffon, *Discours sur le style*, in *Œuvres philosophiques*, 503.

127 *I am learning*   Hérault de Séchelles, *Buffon* (1979), 154.

127 *the general history*   Buffon, *Théorie de la Terre, Second discours*, in *Œuvres philosophiques*, 45.

127–28 Ideas summarized in Desmarest, *Géographie physique* (Paris: Agasse, 1794); and in Hampton, *Boulanger et la science de son temps*.

128 *The shock or*   Ibid., 56.

128 *Cannot one imagine*   Buffon, *Preuves de la théorie de la Terre*, in *Œuvres philosophiques*, 67.

129 *Each time one wants*   Ibid., 82.

129 *the waters*   MM. les députés et syndics de la faculté de théologie to Buffon, in *Œuvres philosophiques*, 106.

129 *The first religion*   Buffon to Mme de Staël, 22 March 1774, in Buffon, *Correspondance*.

129–30 *The deeper*   Buffon, *Époques*, cviii.

130 *I have abandoned*   Buffon to MM. les députés et syndics de la faculté de théologie, in *Œuvres philosophiques*, 106.

130 *The archbishop*   Quoted in Lavisse, *Histoire de France*, 157.

130 *The period was*   Hazard, *Pensée européenne*, 103.

130 *the unit of time*   Buffon, *Preuves de la théorie de la Terre*, in *Œuvres Philosophiques*, 87.

130 *A philosophical will*   Roger, *Buffon: A Life*, 115.

131 *various changes*   Buffon, *Époques*, 2.

131 *These are the changes*   Buffon, *Époques*, 40, 71, 93, 131, 139, 162, 205, for the 1st, 2nd, 3rd, 4th, 5th, 6th, and 7th epochs, respectively.

131 *It has been*   Buffon, *Introduction à l'histoire des minéraux. Partie expérimentale*, introduction.

131–32 *The earth is*   Buffon, *Époques*, 6.

132 *the primitive liquefaction*   Ibid., 12.

133 *A passage of* Buffon, *Introduction à l'histoire des minéraux, Partie expérimentale*, Expériences.

133 *organized nature* Buffon, *Introduction à l'histoire des minéraux, Partie expérimentale, Partie hypothétique, Premier mémoire: Recherches sur le re-froidissement de la Terre et des Planètes*, Conclusions.

133 *With the view* Buffon, *Époques*, 24.

133 *this new attack* Roger, in Buffon, *Époques*, cxxxv.

133 *Instead of going* Buffon, *Époques*, 70.

133 *the obstacles* Buffon, notebook quoted by Roger, in Buffon, *Époques*, 40.

133–34 *Are we not* Buffon, *Époques*, 115.

134 *Although it is* Buffon, notebook quoted in Roger, in Buffon, *Époques*, 40.

135 *Make this place* Aude, *Vie privée de Buffon*, 55.

135 *to strike down* Villars, secretary to the National Convention, to the city of Montbard, 4 Ventose year 2, in Buffon, *Correspondance*, 2:618.

135 *Still more* Guettard to Buffon, in *Époques*, cxxxix.

135 *interesting and little-known* Millin, *Discours*, 27.

136 *solar splatterings* Guettard to Buffon, in *Époques*, cxxxix.

136 *the action* Romé de l'Isle, *Action du feu*, vi.

136 *against 'faculties'* Roger, *Buffon: Époques*, cxlix.

136 *the constitution* Swedenborg, *The Principia*, 1:lxxix.

137 *sacred throne* Wright, unpublished manuscripts, quoted in Hoskin, "Cosmology of Wright," 45.

137 *cannot be made* Wright, *Original Theory*, 75.

137 *the Soul must be* Ibid., 74.

140 *Millions and entire* Kant, *Universal Theory*, 154, 184.

140 *become more excellent* Ibid., 189.

141 *From the sole* Laplace, *Système du monde*, 6th ed., chap. 6, p. 510.

141 *to consider* Laplace, *Philosophical Essay*, 2.

141 *veritable cause* Laplace, *Système du monde*, 6th ed., chap. 6, p. 510.

141 *construction* Herschel, "On the Construction of the Heavens," 213.

142 *the nebulae* Laplace, *Système du monde*, 6th ed., chap. 6, p. 520.

142 *Certain stars* Quoted in Arago, "Discours," 710.

142 *The state* Herschel, "Astronomical Observations," 284.

142 *I have looked* Herschel to Campbell, in Campbell, *Life and Letters*, 2:235–36.

CHAPTER 6

143 *created the earth* Cuvier, *Discourse*, 28.

144 *people in whose* Voltaire, *Dissertation* (conclusion).

144 *There are shells* Voltaire, 1768, *De la pierre*, in *Des singularités*.

144–45 *Sometimes one finds*   Romé de l'Isle, *Essai de cristallographie,* 246.

146 *preliminary mineralogical map*   Guettard, "Mémoire et carte," 363.

147 *crime of lese majesty*   Ruling of Parliament, 18 August 1770, quoted in Holbach, *Système de la nature,* i.

147 *reason guided*   Holbach, *Système de la nature,* 12.

147 *I still confess*   Voltaire, *Idées de Palissy sur les coquilles prétendues,* in *Des singularités.*

147–48 *The chronology*   Needham, *Nouvelles recherches,* 58.

148 *the future destruction*   Ibid., 102.

148 *One day*   Palassou, *Essai,* 87.

148–49 *since the time*   Soulavie, *Histoire naturelle,* 6:209.

149 *the friends*   Quoted in Aufrère, *De Thalès à Davis,* 17.

149 *the sea had*   Lavoisier, "Observations générales" (1789), 351.

149 *In the circle*   Toulmin, *Antiquity and Duration,* 185, 176, 200.

150 *One must not*   Leibniz, *Protogée,* 22.

151 *Each deposit forms*   Füchsel, *Historia terrae et maris,* 48.

151 *If there were*   Quoted in Keferstein, "Notice," 194.

151 *the series*   Arduino, *Due lettere,* second letter, clviii.

153 *centers of conflagration*   Kircher, *Mundus Subterraneus,* 2:184–85.

153 *earthquakes*   Hooke, "Lectures."

153 *From whence*   Maillet, *Telliamed* (1748), 2:101.

153 *quite capable*   Lémery, "Explication," 103.

153–54 *The best way*   Fontenelle, "Sur les feux," 51.

154 *Volvic!*   Condorcet, "Éloge de Guettard," 55.

154 *this realm has had*   Guettard, "Mémoire sur montagnes," 27.

156 *great oracle*   Cuvier, "Éloge de Werner," 2:116.

156 *went out to ask*   Aubuisson, *Traité de Géognosie,* 1:xxi.

157 *earth-onion*   See Ospovat, "Distortion of Werner," 192.

159 *one of those*   Playfair, "Biographical Account," 51.

159 *the globe*   Hutton, *Theory,* 1:17.

160 *objects which verified*   Playfair, *Illustrations,* 68.

160 *succession of worlds*   Hutton, "Theory" (1788), 304.

160–61 *destroyed in one part*   Hutton, *Theory,* 2:562.

161 *The result*   Hutton, "Theory" (1788), 304.

161 *judge of the great*   Hall, "Experiments on Whinstone," 44.

162 *common gun-barrel*   Hall, "Account of Experiments," 80.

162 *Go and see*   Cuvier, "Éloge de Desmarest," 156.

162 *Oh, how very*   Lamarck, *Hydrogeology,* 75.

164 *The great age*   Ibid., 77.

164 *worn and imperfect*   Arduino, quoted in Vaccari, *Giovanni Arduino,* 176.

164 *First Age*   Soulavie, *Histoire naturelle,* 1:317.

165 *The layers*   De Luc, "Dix-huitième lettre," 460.

165 *Alone I did it*   Quoted in Phillips, *Memoirs,* 133.

166 *furnish the best*   Smith, *Strata Identified,* i.

166 *He praised me*   Napoléon, quoted in Flourens, "Éloge de Cuvier," xxxvii.

166 *He is more liberal*   Lyell to his father, 8 July 1823, in *Life, Letters,* 1:128.

167 *the least prominence*   Cuvier, *Discourse,* 64.

167 *elephant*   Cuvier, "Mémoires," 440.

167 *A science that*   Ibid., 441.

168 *The method we*   Cuvier and Brongniart, "Essai," 306.

168 *I consider*   Brongniart, "Sur les caractères," 543.

169 *In order to unite*   Humboldt, *Geognostical Essay,* 475.

170–71 *a science which*   Buckland, *Geology and Mineralogy,* 1:593.

171 *We* Catholic *geologists*   Quoted by Lyell, *Life, Letters,* 1:276.

171 *Millions of millions*   Buckland, *Geology and Mineralogy,* 1:21.

172 *His theories*   Lyell, *Life, Letters,* 1:168.

173 *has been courageous*   Lyell to G. Mantell, Esq., 2 March 1827, in *Life, Letters,* 1:168.

173 *This method*   Aubuisson, *Traité de géognosie,* 1:xxiv.

CHAPTER 7

178 *If you were*   Carnot, *Réflexions,* 4.

179 *For us*   Fourier, "Remarques générales," 159.

179 *The profound study*   Fourier, *Théorie,* xiii.

180 *to doff*   Arago, "Joseph Fourier," 1:295.

180 *several sermons*   Ibid., 1:299.

180 *Fourier is not*   Ibid., 1:300.

181 *Yesterday I turned*   Fourier to Bonard, 22 March 1789, unpublished letter available at Municipal Library, Auxerre, France.

181 *You'll be able*   Cousin, *Œuvres* (Paris, 1849), 3:25.

182 *lessons on theology*   Quoted in Herivel, *Joseph Fourier,* 79.

182 *to that very*   Herivel, *Joseph Fourier,* 1.

182–83 *This heat*   Fourier, "Extrait," 420.

183 *different bodies*   Fourier, *Théorie,* iii–iv.

184 *exquisite mathematical poem*   "Fourier's Analytical Theory of Heat," 192.

185 *The analytical solution*   Fourier, "Extrait," 419.

186 *the lowering*   Ibid., 435.

187 *It is less*   Ibid., 430.

188 *God might have*   Chateaubriand, *Genius of Christianity,* 136.

188 *Without this original*   Ibid., 137.

189 *total quantity*   Pouillet, "Mémoire," 24.

189–90 *a source*   Ibid., 34.

191 *Respiration*   Lavoisier and Laplace, "Mémoire," 2:283.

191 *Nature, science*   Quoted in Caneva, *Robert Mayer,* 11.

192 *there are more*   Quoted in Mayer, "Celestial Dynamics. I," 388.

193 *if the diameter*   Helmholtz, "On the Interaction," 514.

193–94 *within a finite*   Thomson, "On a Universal Tendency," 306.

194 *For myself*   Thomson, "On Geological Dynamics," 2:113.

195 *to draw man*   Quoted in Smith and Wise, *Energy and Empire* 39.

195 *in his lectures*   Quoted in Sharlin, *Lord Kelvin*, 24.

195 *and he used*   J. J. Thomson, *Recollections and Reflections*, 423.

195 *combined*   Quoted in Eve, *Rutherford*, 355.

195 *what the telescope*   Sharlin, *Lord Kelvin*, 17.

196 *the minimum thickness*   Hopkins, "Researches in Physical Geology— Third Series," 51.

197 *a faultless technique*   Quoted in Thompson, *Life of William Thomson*, 1154.

197 *truly is the* Alma mater   Ibid., 948.

200 *The entire scientific*   Ibid., 986.

200 *Marvel of Science*   Dupuy, in Pasteur, "Discours," 10.

201 *1. Energy is constant*   Clausius, "Ueber verschiedene," 400.

201 *We would be*   Fourier, *Théorie*, 432.

202 *divine artificer*   Lyell, *Principles*, 10th ed. (1867), 213.

202 *For eighteen years*   Thomson, "On the Secular Cooling," 3:295.

203 *It seems, therefore*   Thomson, "On the Age of the Sun's Heat" (1862), in Thomson, *Popular Lectures*, 1:368.

203 *Although the mathematical*   Thomson, "On the Rigidity of the Earth; Shiftings of the Earth's Instantaneous Axis of Rotation; and Irregularities of the Earth as a Timekeeper" (1863), in Thomson, *Mathematical and Physical Papers*, 3:312, 329.

205 *with much probability*   Thomson, "On the Secular Cooling," 3:300.

205 *fitted as an abode*   Thomson, "Age of the Earth as an Abode," 665.

CHAPTER 8

207 *What is the cause*   Boubée, *Cours abrégé*, 2–6.

208 *A great reform*   Thomson, "On Geological Time," 2:10.

208 *I am as incapable*   Quoted in Thomson, "Age of the Earth as an Abode," 668.

208 *geology in much*   Thomson, "Age of the Earth as an Abode," 666.

209 *Our experiments*   Cordier, "Essai," 536.

210 *of the main*   Élie de Beaumont, *Notice*, 1222.

212 *He who can read*   Darwin, *Origin of Species*, 1st ed., 282.

212 *You care for nothing*   Quoted in Beer, *Charles Darwin*, 25.

214 *in all probability*   Darwin, *Origin of Species*, 1st ed., 287.

214 *overpoweringly strong proofs*   Thomson, "Presidential Address," 2:204–5.

214 *because he believed*   Burchfield, *Lord Kelvin*, 72.

215 *the* absolute *antiquity*   Phillips, *Life on Earth*, 121.

215 *The Geological Scale*   Ibid.

216 *too vast*   Ibid., 126.

217 *the present figure*   Thomson, "On Geological Time," 2:43.

217–18 *Recent research*   Geikie, "On Modern Denudation," 189.

218 *as the grandest mill*   Huxley, "Anniversary Address," l (p. ell = 50).

218 *British popular geology*   Thomson, "On Geological Time," 2:44.

218 *The critical examination*   Huxley, "Anniversary Address," liii.

218 *It is quite certain*   Thomson, "On Geological Time," 2:44.

218 *I do not suppose*   Huxley, "Anniversary Address," xlvii.

218 *The existing state*   Thomson, "On Geological Time," 2:64.

219 Some such period Huxley, "Anniversary Address," xlviii.

219 *Admission*   Thomson, "On Geological Dynamics," 2:80.

221 *to apply*   Becquerel and Becquerel, *Éléments*, 43.

222 *the forces*   Daubrée, *Etudes et expériences*, 3.

222 *When science is*   Launay, *Science géologique*, 722.

222 *become cold*   Flammarion, *Popular Astronomy*, 78.

222–23 *in these successive*   Ibid., 80.

225 *privilege of making*   King, "Age of the Earth," 2.

226 *affectionate pupil*   Perry, "On the Age of the Earth," 226.

226 *as extinct*   Perry to Tait, 29 November 1894, *Nature* 51 (1895): 227.

226 *I thought*   Kelvin (Thomson) to Perry, 13 December, *Nature* 51 (1895): 227.

227–28 *the sodium*   Joly, "Estimate," 48.

228 *sorest troubles*   Darwin to Wallace, quoted in Burchfield, *Lord Kelvin*, 75.

229 *High eccentricity*   Wallace, "Measurement II," 454.

229 *fully satisfy*   Ibid.

230 *In this most*   De Vries, "Evidence," 399.

230 *from the need*   Ibid., 396.

230 *Some thousand*   De Vries, *Species and Varieties*, 713.

232 *Thus we can*   Tait, *Lectures*, 169.

232–33 *Instead of seeking*   Fisher, "Depression of Ice-Loaded Lands," 526.

233 *(1) Time which*   Sollas, *Age of the Earth*, 16, 24, 25.

234 *Total since*   Goodchild, "Geological Evidence," 306.

234 *The amount*   Fisher, "Review of an Estimate," 130.

235 *extravagant*   Geikie, "Geological Change," 111.

235 *There is a*   Geikie, *Textbook of Geology*, 1:77.

235 *period of between*   Fisher, "Review of an Estimate," 132.

CHAPTER 9

237–38 *the concordance*   Thomson, "Age of the Earth," 440.

238–39 *the ideas*   Schuster, *Progress of Physics*, 82.

239 *sameness*   Franklin, *Experiments*, 334.

239–40 *Il a ravi*   Franklin, *Œuvres*, frontispiece.

240 *understood a lot*   Quoted by Heilbron in *Dictionary of Scientific Biography*, s.v. "Volta," 69.

241 *artificial electrical organ*   Volta, "On the Electricity," 429.

241 *heat and light*   Oersted, "Experiments," 27.

242 *brings together*   Fourier, *Théorie*, xiij.

243 *in a few years*   Maxwell, *Scientific Papers* (1890), 2:244.

244 *One word*   Quoted in J. J. Thomson, *Recollection and Reflections*, 425.

246 *alteration in the*   Quoted by Medhurst in Crookes, *Crookes and the Spirit World*, 23, 22, 35.

247 *wretched superstition*   Quoted in Thompson, *Life of William Thomson*, 2:1104.

247 *we know that*   Lodge, *The Case For and Against Psychical Belief*, 12.

248 *ectoplasmic formations*   Richet, *Traité de métapsychique*, 627.

250 *radiant matter*   Crookes, "Radiant Matter," in Fournier d'Albe, *Life of Crookes*, 290.

252 *second decimal*   G. P. Thomson, *J.J. Thomson and the Cavendish Laboratory*, 169.

254 *I shall admit*   Perrin, "Quelques propriétés," 186.

254 *X-ray apparatus*   Advertisement reproduced in *Physics Today*, November 1995, 30.

254 *I wish some*   *Life*, 27 April 1886, 27:259, quoted in Glasser, *Wilhelm Conrad Röntgen* (English ed.), 365.

256 *Jewish physics*   Lenard, *Deutsche Physik*, viii.

256 *probably enjoy*   Herschel, "On the Remarkable Appearances," 273.

256–57 *seas*   Schiaparelli quoted in Morse, *Mars*, 22–24.

257 *in advance of*   Lowell, *Mars*, 209.

258 *Mars, therefore*   Wallace, *Is Mars Habitable?* 110.

261 *make life a dream*   Marie Curie, in P. Curie, *Œuvres*, viii.

261 *a title that*   Quoted by Sainte-Claire Deville in Faraday, *Histoire*, 13.

262 *That's the last*   Quoted in Eve, *Rutherford*, 11.

262 *very pleasant*   Ibid., 15, 14.

263 *Thus, it is*   Poincaré, "Rayons cathodiques," 56.

CHAPTER 10

265 *invisible radiations*   Becquerel, "Sur les radiations invisibles," 501.

265–66 *the first example*   Becquerel, Émissions," 1088.

267 *whether bodies*   Sklodowska Curie, "Rayons," 126:1101.

267–68 *an asphalt-paved shed*   Marie Curie, in P. Curie, *Œuvres*, vii.

268 *the most wonderful*   Jauncey, "Early Years," 239.

268 *Were I younger*   Quoted in Fournier d'Albe, *Life of Crookes*, 386.

269 *what the internal*   Chamberlin, "Lord Kelvin's Address," 10:12.

269 *profound transformation*   Curie and Laborde, "Sur la chaleur," 675.
270 *he termed*   Rutherford, "Uranium Radiation," 1:175.
271 *the maintenance*   Rutherford and Soddy, "Radioactive Change," 591.
272 *It is not improbable*   Rutherford, *Radioactivity* (1904), 344.
272 *Science offers*   Rutherford, "Radium," in *Collected Papers*, 1:785.
272 *filthy habit*   Quoted in Eve, *Rutherford*, 91.
273 *centers of condensations*   P. Curie, Sur la radioactivité," 226.
275 *standard of time*   P. Curie, "Sur la mesure absolue,"
       439.
278 *great master*   Becker, "Age of a Cooling Globe," 228.
280 *if it were not*   Launay, *Science géologique*, 722.
280 *a few decigrams*   Quoted in Fournier d'Albe, *Life of Crookes*, 386.
281 *I came*   Quoted in Eve, *Rutherford*, 107.
281 *electro-ethereal hypothesis*   Thomson, "Contribution by Kelvin," 220.
281 *any serious probability*   Thomson to Orr, 29 January 1906, quoted in
       Smith and Wise, *Energy and Empire*, 550.
282 *that before*   Quoted in J. J. Thomson, *Recollections and Reflections*, 420.
283 *to avoid deception*   Röntgen, "On a New Kind of Rays," in Klickstein,
       Wilhelm Conrad Röntgen, 6.
283 *In fact, the observer*   Blondlot, *"N" Rays*, 83.
284 *N-rays have shown*   Piéron, "Grandeur et décadence," 159.
285 *We ask*   Basil *Homilies* 3.3.57b.
287 *to be an old man*   Quoted in Thompson and Thompson, *Silvanus
       Thompson*, 338.
287 *were left*   Thompson, "On Hyperphosphorescence," 104.

CHAPTER 11

289 *the surprises*   Holmes, *Age of the Earth* (1913), 167.
289 *radium will be*   Arrhenius, *Life of the Universe*, 237.
291 *is of great promise*   Quoted in Woodward and Watts, "William Johnson
       Sollas," 272.
292 *the origin*   Lindeman, "Age of the Earth," 372.
293 *a body*   Aristotle *On the Heavens* 3:3.302a.
296 *compound of thorium*   Boltwood to Rutherford, 22 September 1905,
       quoted in Eve, *Rutherford*, 132.
296 *any atomic weight*   Quoted in Aston, *Mass Spectra and Isotopes*
       (1933), 9.
296 *elements chemically identical*   Soddy, "Discussion on Isotopes," 98.
297 *To discover*   Quoted in Crookes, "Chemical Science," 424.
297 *probably, our*   Crookes, "Chemical Science," 429.
298 *amazing*   Richards and Lembert, "Atomic Weight of Lead," 1339.
298 *the idea that*   Richards, "New Outlook," 303.
300 *tell one much*   Quoted in Heilbron, *Moseley*, 81.

301 *why so far*   Fleck, "Frederick Soddy," 210.
302 *skepticism of geologists*   Barrell, "Rhythms," 749.
302 *geological rhythm*   Ibid., 745.
303 *the acquisition*   Chamberlin, "Group of Hypotheses," 676.
303 *the general feeling*   Termier, *Gloire de la terre*, 348.
304 *the age*   Sollas, "Age of the Earth" (1921), 282.
305 *one of the most*   Soddy, "Discussion on Isotopes," 100.
306 *make more*   Quoted by Brock in *Dictionary of Scientific Biography*, s.v. "Aston," 321.
307 *they are nevertheless*   Holmes, "Radioactivity," 439.
307–8 *one stratigrapher*   Schuchert, "Geochronology," 14.
308 *the makers*   Clarke, "Age of the Earth," 275.
310 *is still far beyond*   Barrell, "Rhythms," 904.
310 *the chemical*   Comte, *Positive Philosophy*, 1:148.
311 *matter identical*   Huggins, "On the Lines of the Spectra," 434.
311–12 *define the various*   Lockyer, *Inorganic Evolution*, 159.
314 *only the inertia*   Eddington, "Internal Constitution" (1920), 353.
316 *matter in its*   Jeans, "Recent Developments," 37.
316–17 *a star of mass*   Eddington, *Internal Constitution*, 310.
319 *Don't they shine*   Quoted in Khriplovich, "Life of Houtermans," 30.

CHAPTER 12

323 *Newton of atomic physics*   Quoted in Eve, *Rutherford*, 431.
324 *You are*   Ibid., 436.
328 *considered to be*   Nier, "Reminiscences of Mass Spectrometry," 388.
331 *One might ask*   P. Curie, *Prix Nobel*, 7.
334 The biographical information given for E. C. Gerling was provided in a personal communication to the author from Professor Igor Toltsikhin, a Russian isotope geochemist.
335 *the age*   Gerling, "Age of the Earth," 260.
336 *When your ancestors*   Quoted in Khriplovich, "Life of Houtermans," 29.
336 *a little family*   Ibid., 32.
339 *the solid crust*   Houtermans, "Das Alter des Urans," 327.
339 *the great source-book*   Ager, *Stratigraphical Record*, 85.
340 *the age*   Holmes, "Estimate of the Age" (1946), 684.
342 *an act*   Quoted in Fournier d'Albe, *Life of Crookes*, 376.
342 *identity of isotopic*   Houtermans, "Determination," 1625.
342 *still impossible*   Patterson, "Age of Meteorites," 230.
243 *proverbially well-guarded*   Holmes, *Age of the Earth* (1913), ix.
343 *The whole object*   Lowell, *Mars and Its Canals*, viii.
348 *to acquire simply*   Crookes, "Chemical Science," 423.
349 *a large part*   Patterson, "Acceptance Speech," 1387.

EPILOGUE

351 *The heaven*   Aristotle *On the Heavens* 2.1.283b.
351 *The same ideas*   Ibid., 1.3.270b.
351 *Absurdity, as well*   Koyré, "Unpublished Letter," 317.
352 *I do not pretend*   Newton, *Chronology*, 8.
352 *difficulties one finds*   Buffon, *Histoire naturelle, Premier discours,* in *Œuvres philosophiques,* 26.

Suggestions for Further Reading and Reference

See page 382 for a list of works that provide information about prominent persons discussed in this book.

Preface

Adams, Frank Dawson. *The Birth and Development of the Geological Sciences*. Baltimore: William and Wilkins, 1938. Reprint, New York: Dover, 1954. An early "modern" history of geology.

Albritton, Claude C., Jr. *The Abyss of Time: Changing Conceptions of the Earth's Antiquity after the Sixteenth Century*. San Francisco: Freeman, Cooper, 1980. A series of portraits from Steno to Holmes.

Bottéro, Jean. *Naissance de Dieu: La Bible et l'historien*. 2nd ed. Paris: Gallimard, 1992. Published in English as *The Birth of God: The Bible and the Historian*, translated by K. W. Bolle (University Park: Pennsylvania State University Press, 2000). An introduction to the Yawhist and Elohist accounts of Genesis.

Brush, Stephen. G. *Transmuted Past: The Age of the Earth and the Evolution of the Elements from Lyell to Patterson*. New York: Cambridge University Press, 1996. A view from the physical standpoint.

Diderot, Denis, and Jean le Rond d'Alembert, eds. *Encyclopédie, ou dictionnaire raisonné des sciences, des arts et des métiers*. 28 vols. Paris: Briasson, 1751–72.

Duhem, Pierre. *Le système du monde: Histoire des doctrines cosmologiques de Platon à Copernic*. 10 vols. Paris: Hermann, vols. 1–7, 1913–1919, reprinted in 1958; vols. 8–10, 1956–58.

The most comprehensive account of science from antiquity to the scientific revolution.

Ellenberger, François. *Histoire de la géologie*. 2 vols. Paris: Lavoisier, 1988–94. Published in English as *History of Geology,* translated by R. K. Kaula and M. Carozzi; vol. 1, *From Ancient Times to the First Half of the Seventeenth Century* and *History of Geology*; vol. 2, *The Great Awakening and Its First Fruits—1660– 1810* (Rotterdam, Netherlands: A. A. Balkema, 1996, 1999). A standard history.

Fraser, Julius Thomas, ed. *The Voices of Time: A Cooperative Survey of Man's View of Time as Expressed by the Sciences and the Humanities*. New York: G. Braziller, 1966. Time, thought, man, and matter.

Geikie, Archibald. *The Founders of Geology*. London: Macmillan, 1905. Reprint, New York: Dover, 1960. A well-written classic.

Gohau, Gabriel. *Histoire de la géologie*. Paris: La Découverte, 1987. Published in English as *A History of Geology*, translated by A. V. Carozzi and M. Carozzi (New Brunswick, NJ: Rutgers University Press, 1991).

Haber, Francis C. *The Age of the World: From Moses to Darwin*. Baltimore: Johns Hopkins Press, 1959. Reprint, Westport: Greenwood Press, 1978. A pioneering work on the age of the world.

Hawking, Stephen, and Roger Penrose. *The Nature of Space and Time*. Princeton, NJ: Princeton University Press, 1996. A viewpoint from contemporary cosmology.

Hooykaas, Reijer. *Religion and the Rise of Modern Science*. Grand Rapids, MI: Eerdmans, 1988. The question of Christianity as a prerequisite to modern science.

Kuhn, Thomas S. *The Structure of Scientific Revolutions*. 2nd ed. Chicago: University of Chicago Press, 1970. How great discoveries can initially raise more problems than they solve.

The New Jerusalem Bible.

Rossi, Paolo. *I segni del tempo: Storia delle terra e delle nazioni da Hooke a Vico*. Milan: Feltrinelli, 1979. Published in English as *The Dark Abyss of Time: The History of the Earth and the History of Nations from Hooke to Vico*, translated by L. G. Cochrane (Chicago: University of Chicago Press, 1987). Historical views of the earth and man.

Sorabji, Richard. *Time, Creation, and the Continuum*. London: Duckworth, 1983. The reality of time, from the Greeks to the fathers of the church and Islamic thinkers.

Toulmin, Stephen, and June Goodfield. *The Discovery of Time*. New York: Harper and Row, 1965. On history, geology, and biology.

White, Andrew Dickson. *A History of the Warfare of Science with Theology in Christendom*. 2 vols. New York: D. Appleton, 1901. Theology sentenced without remission.

Whitrow, Gerald J. *The Natural Philosophy of Time*. 2nd ed. Oxford: Clarendon Press, 1980.

Chapter 1

Bonnefoy, Yves, ed. *Dictionnaire des mythologies et des religions des sociétés traditionnelles et du monde antique*. Paris: Flammarion, 1994. A comprehensive dictionary of myths.

Bottéro, Jean. *Lorsque les dieux faisaient l'homme*. Paris: Gallimard, 1989. Mesopotamia and its legacy to history.

Cameron, Alan. "The Last Days of the Academy at Athens." *Cambridge Philos. Soc. Proc.* 15 (1969): 7–29. The life of the Academy after its official closure.

Cohen, Morris R., and Israel E. Drabkin. *A Source Book in Greek Science*. New York: McGraw-Hill, 1948. Selected readings in all branches of Greek science.

Dicks, David R. *Early Greek Astronomy to Aristotle*. Ithaca, NY: Cornell University Press, 1970. A review of ancient astronomy on the basis of primary texts.

Eliade, Mircea. *Le mythe de l'éternel retour*. Paris: Gallimard, 1949. Published in English as *The Myth of the Eternal Return*, translated by W. R. Trask (New York: Pantheon Books, 1954). A standard presentation of cycles.

Freund, Philip. *Myths of Creation*. New York: Washington Square Press, 1965. Includes the myths of the Blackfoot, among many others.

La Naissance du monde. Vol. 1, *Sources orientales*. Paris: Seuil, 1959. A compilation of cosmologies from the Middle East.

Puech, Henri-Charles. *En quête de la gnose*. Paris: Gallimard, 1978. The end of cycles.

Sarton, George. *Ancient Science through the Golden Age of Greece*. Cambridge, MA: Harvard University Press, 1952. Reprint, New York: Dover, 1980.

Whitrow, Gerald J. *Time in History: Views of Time from Prehistory to the Present Day*. Oxford: Oxford University Press, 1988.

Chapter 2

Allen, Don C. *The Legend of Noah: Renaissance Rationalism in Art, Science, and Letters*. Urbana: University of Illinois Press, 1963. The Flood and Noah's Ark in confrontations between faith, reason, and the Bible.

Bottéro, Jean. *Naissance de Dieu: La Bible et l'historien*. 2nd ed. Paris: Gallimard, 1992. Published in English as *The Birth of God, the Bible, and the Historian*, translated by K. W. Bolle (University Park, PA: Pennsylvania State University Press, 2000).

Chenu, Marie-Dominique. *La théologie comme une science au XIIIe siècle*. 3rd ed. Paris: Vrin, 1969. Published in English as *Is Theology a Science?* translated by A. H. N. Green-Armytage (New York: Hawthorn Books, 1959).

Crombie, Alistair C. *The History of Science from Augustine to Galileo*. New York: Dover, 1995.

Duhem, Pierre. *Le système du monde: Histoire des doctrines cosmologiques de Platon à Copernic*. Vol. 9. Paris: Hermann, 1958. The small-scale motions of the earth and the origins of geology (chap. 18); the earth's rotation (chap. 19).

Eliade, Mircea. *Le mythe de l'éternel retour*. Paris: Gallimard, 1949. Published in English as *The Myth of the Eternal Return*, translated by W. R. Trask (Princeton, NJ: Princeton University Press, 1954).

Gilson, Etienne. *La philosophie au moyen âge: Des origines patristiques à la fin du XIXe siècle*. Paris: Payot, 1944. Translated as *History of Christian Philosophy in the Middle Ages* (New York: Random House, 1955).

Grant, Edward. *Planets, Stars, and Orbs: The Medieval Cosmos, 1200–1687*. Cambridge: Cambridge University Press, 1994. A recent synthesis, following Duhem's.

Lindberg, David C. "Science and the Early Christian Church." *Isis* 74 (1983): 509–30.

Lods, Adolphe. *Histoire de la littérature hébraïque et juive*. Paris: Payot, 1950. Reprint, Geneva: Slatkine, 1982. A history of Hebrew and Jewish literature.

———. *Jean Astruc et la critique biblique au XVIIIe siècle*. Strasbourg: Librairie Istra, 1922. On Astruc and his contribution to biblical studies.

North, John D. "Chronology and the Age of the World." In *Cosmology: History and Theory*, edited by Wolfgang Yourgrau and Allen D. Breck. New York: Plenum Press, 1977. The great cycles of chronology.

Puech, Henri-Charles, ed. *Histoire des religions*. 3 vols. Paris: Gallimard, 1970–76.

Serres, Marcel de. *De la cosmogonie de Moïse comparée aux faits géologiques*. 2 vols. Paris: Lagny Frères, 1859. Geology and Mosaic cosmogony by a late upholder of biblical literalism in France.

———. "On the Physical Facts Cited in the Bible Compared with the Discoveries of the Modern Sciences." *Edinb. Philosoph. J.* 38 (1845): 239–71.

van Seters, John. *In Search of History: Historiography in the Ancient World and the Origins of Biblical History*. New Haven, CT: Yale University Press, 1983. An account of the debates over the beginnings of history.

Chapter 3

Bertrand, Joseph. *L'Académie des sciences et les académiciens de 1666 à 1793*. Paris: J. Hetzel, 1869. A history of the Paris Academy of Science.

Birch, Thomas. *The History of the Royal Society of London for Improving of Natural Knowledge, From Its First Rise. In which the most considerable of those papers communicated to the society, which have hitherto not been published, are inserted in their proper order, as a supplement to the Philosophical Transactions*. 4 vols. London: A. Millar, 1756–57.

Hall, Alfred Rupert. *Philosophers at War*. Cambridge: Cambridge University Press, 1980. Newton against Leibniz.

Heilbron, John L. *Elements of Early Modern Physics*. Berkeley: University of California Press, 1982. A sociology of science as a complement to the history of electricity in the seventeenth and eighteenth centuries.

Koyré, Alexandre. *From the Closed World to the Infinite Universe*. Baltimore: Johns Hopkins Press, 1957. The meaning of infinity, from Nicholas of Cusa to Newton and Leibniz.

Leibniz, Gottfried Wilhelm. *The Leibniz-Clarke Correspondence, Together with Extracts from Newton's Principia and Optics*, ed. H. G. Alexander. New York: Barnes and Noble, 1978. Against Newton's God, who has to rewind his clock once in a while.

Peignot, Gabriel. *Dictionnaire critique, littéraire et biographique des principaux livres condamnés au feu, supprimés ou censurés: Précédés d'un discours sur ces sortes d'ouvrages*. 2 vols. Paris: A. Renouard, 1806. Reprint, Amsterdam: P. Schippers N. V., 1966. The encounters with censorship of Bayle, Diderot, La Peyrère, Simon, Spinoza, and many others.

Todhunter, Isaac. *Theories of Attraction and the Figure of the Earth*. London: Macmillan, 1873. Reprint, New York: Dover, 1962. The figure of the Earth as a major problem in physics.

Chapter 4

Adams, Frank Dawson. *The Birth and Development of the Geological Sciences*. Baltimore: William and Wilkins, 1938. Reprint, New York: Dover, 1954. See in particular the chapters on generation of stones and on medieval mineralogy.

Dézallier d'Argenville, Antoine-Joseph. *Histoire naturelle éclaircie dans une de ses parties pricipales, la conchyologie, qui traite des coquillages de mer, de rivière et de terre; ouvrage dans lequel on trouve une nouvelle méthode Latine et Françoise de les diviser: Augmentée de la zoomorphose, ou représentation des animaux à coquille, avec leurs explications*. Paris: De Bure l'Aîné, 1757. A period summary of opinions about fossils.

Ellenberger, François. *Histoire de la géologie*. 2 vols. Paris: Lavoisier, 1988–94. Published in English as *History of Geology*. Vol. 1, *From Ancient Times to the First Half of the Seventeenth Century*, translated by R. K. Kaula. (Rotterdam, Netherlands: A. A. Balkema, 1996, 1999). Careful studies of Leonardo, Palissy, and Steno, among many others.

Pannier, Léopold. *Les lapidaires français du Moyen Age, des XIIe, XIIIe et XIVe siècles*. Paris: F. Vieweg, 1882. Valuable original versions of medieval lapidaries.

Rudwick, Martin. *The Meaning of Fossils: Episodes in the History of Paleontology*. London: Macdonald, 1972. A history of the understanding of fossils from 1585 on.

Scheuchzer, Johann Jacob. *Physique sacrée, ou histoire naturelle de la bible*. 8 vols. French translation from the Latin. Amsterdam: P. Schenk and P. Mortier, 1732–37. By one of the most illustrious naturalists of the times; the Bible justified and beautifully illustrated for the benefit of atheists.

Chapter 5

Desmarest, Nicolas. *Géographie physique: Résumé des opinions anciennes et des travaux antérieurs à 1789*. Vol. 1. In the series *Encyclopédie méthodique*. Paris: Agasse, 1794–95. See p. 34 for a biographical notice on Boulanger.

Hazard, Paul. *La pensée européenne au XVIIIᵉ siècle, de Montesquieu à Lessing*. Paris: Boivin, 1946. Published in English as *European Thought in the Eighteenth Century*, translated by J. Lewis May (Harmondsworth, Eng.: Penguin, 1965).

Pingré, Alexandre Guy. *Cométographie, ou traité historique et théorique des comètes*. 2 vols. Paris: Imprimerie Royale, 1783–84. The eighteenth-century reference work on comets.

Yeomans, Donald K. *Comets: A Chronological History of Observation, Science, Myth, and Folklore*. New York: Wiley, 1991.

Chapter 6

Buffetaut, Eric. *A Short History of Vertebrate Paleontology*. London: Rutledge and Kegan Paul, 1987.

Cannon, Walter F. "The Uniformitarian-Catastrophist Debate." *Isis* 51 (1960): 38–50.

Cohen, Claudine. *Le destin du mamouth*. Paris: Editions du Seuil, 1994. Published in English as *The Fate of the Mammoth*, translated by W. Rodarmor (Chicago: University of Chicago Press, 2002). Beyond the mammoth, a history of paleontology.

Dean, Dennis R. "On Early Uses of the Word 'Geology.'" *Ann. Sci.* 36 (1979): 35–43.

Desmarest, Nicolas. *Géographie physique: Résumé des opinions anciennes et des travaux antérieurs à 1789*. Vol. 1. In the series *Encyclopédie méthodique*. Paris: Agasse, 1794–95. A review of the works of Arduino, Buffon, Guettard, Halley, Lehmann, etc.

Figuier, Louis. *La terre avant le Déluge*. Paris: Hachette, 1862. The earth before the Flood as described and pictured in the middle of the nineteenth century.

Gillispie, Charles C. *Genesis and Geology: A Study in the Relations of Scientific Thought, Natural Theology, and Social Opinion in Great Britain*. 1951. Reprint, Cambridge, MA: Harvard University Press, 1996.

Gohau, Gabriel. *Les sciences de la terre aux XVIIᵉ et XVIIIᵉ siècles: Naissance de la géologie*. Paris: Albin Michel, 1990. The birth of geology.

Hallam, Anthony. *Great Geological Controversies*. Oxford: Oxford University Press, 1984. The great debates concisely told.

Laudan, Rachel. *From Mineralogy to Geology: The Foundations of a Science, 1650–1830*. Chicago: University of Chicago Press, 1987. An original and critical review of the foundations of geology.

Rappaport, Rhoda. *When Geologists Were Historians, 1665–1750*. Ithaca, NY: Cornell University Press, 1997.

Rudwick, Martin. *Scenes of Deep Time: Early Pictorial Representations of the Prehistoric World*. Chicago: University of Chicago Press, 1992. How the landscapes representing the Creation or the dinosaur backwaters came to be so beautifully pictured.

Rupke, Nicholas A. *The Great Chain of History: William Buckland and the English School of Geology (1814–1849)*. Oxford: Clarendon, 1983. British geology in the nineteenth century beyond Buckland.

Schneer, Cecil J., ed. *Toward a History of Geology*. Cambridge, MA: MIT Press, 1969. A varied collection of contributions to geology and its history.

Chapter 7

Brush, Stephen G. *The Kind of Motion We Call Heat*. 2 vols. Amsterdam: North Holland, 1976. A collection of fifteen papers devoted mainly to the history of heat and of the statistical theory of gases.

Cardwell, Donald S. L. *From Watt to Clausius: The Rise of Thermodynamics in the Early Industrial Age*. Ithaca, NY: Cornell University Press, 1971.

Elkana, Yehuda. *The Discovery of the Conservation of Energy*. London: Hutchinson Educational, 1974.

Kuhn, Thomas S. "Energy Conservation as an Example of Simultaneous Discovery." In *Critical Problems in the History of Science*, edited by M. Clagett. Madison: University of Wisconsin Press, 1959. The context within which the first law of thermodynamics was formulated.

Chapter 8

Bowler, Peter J. *Evolution: The History of an Idea*. 2nd ed. Berkeley: University of California Press, 1989. The starting point and synthesis of the enormous body of literature devoted to Darwinism.

Burchfield, Joe D. *Lord Kelvin and the Age of the Earth*. 2nd ed., with a new postface. Chicago: University of Chicago Press, 1990. A comprehensive account of the reactions of Anglo-American geologists to Kelvin's thesis.

Smith, Crosbie, and M. Norton Wise. *Energy and Empire: A Biographical Study of Lord Kelvin*. Cambridge: Cambridge University Press, 1989. An extensive physical analysis of Kelvin's positions, as a complement to Burchfield's discussion focused on geology.

With, Etienne. *L'écorce terrestre*. Paris: Plon, 1874.

Chapter 9

Badash, Lawrence. "The Completeness of Nineteenth-Century Science." *Isis* 63 (1972): 48–58.

Crawford, Elisabeth. *The Beginnings of the Nobel Institution: The Science Prizes, 1901–1915*. Cambridge: Cambridge University Press, 1985. Scientific activity, conflicts, and awards in physics and chemistry at the end of the nineteenth century.

Forman, Paul, John L. Heilbron, and Spencer Weart. "Physics circa 1900: Personnel, Funding, and Productivity of the Academic Establishments." *Hist. Stud. Phys. Sci.* 5 (1975): 1–185.

Grigg, Emanuel R. N. *The Trail of the Invisible Light*. Springfield, IL: C. C. Thomas, 1965. A comprehensive survey of X-rays, from their discovery to radiobiology.

Oppenheim, Janet. *The Other World: Spiritualism and Psychical Research in England, 1850–1914*. Cambridge: Cambridge University Press, 1985. The cloudy relationships of distinguished British physicists with spiritualism.

Schuster, Arthur. *The Progress of Physics during 33 Years (1875–1908): Four Lectures Delivered to the University of Calcutta during March 1908*. Cambridge: University Press, 1911. An on-the-spot account of the upheavals of physics at the end of the nineteenth century by an able witness.

Segré, Emilio. *From Falling Bodies to Radio Waves: Classical Physicists and Their Discoveries*. New York: W. H. Freeman, 1986. A widely accessible history.

Travers, Morris W. *The Discovery of the Rare Gases*. London: E. Arnold, 1928. The discovery of the noble gases, as related by a scientist working with Ramsay.

Chapter 10

Badash, Lawrence. "Radioactivity before the Curies." *Amer. J. Phys.* 33 (1965): 128–35.

———. "Rutherford, Boltwood, and the Age of the Earth: The Origin of Radioactive Dating Techniques." *Proc. Amer. Phil. Soc.* 112 (1968): 157–69.

Jauncey, George E. M. "The Early Years of Radioactivity." *Amer. J. Phys.* 14 (1946): 226–41.

Chapter 11

Hufbauer, Karl. "Astronomers Take up the Stellar Energy Problem, 1917–1920." *Hist. Studies Phys. Sci.* 11 (1981): 277–302.

Knopf, Adolph, E. W. Brown, Arthur Holmes, A. F. Kovarik, A. C. Lane, and Charles Schuchert. *The Age of the Earth*. Washington, DC: National Research Council, 1931. The general acceptance of radioactive dating methods.

North, John D. *The Measure of the Universe*. Oxford: Clarendon Press, 1965. Reprint, New York: Dover, 1990. A history of the physics of the universe.

van Spronsen, Jan W. *The Periodic System of Chemical Elements: A History of the First Hundred Years*. Amsterdam: Elsevier, 1969. The history of the classification of elements.

Williams, Henry S. *Super-Engines of War*. Vol. 3 of *The Story of Modern Science*. London: Funk and Wagnalls, 1923. The requisitioning of science during World War I.

Chapter 12

Allègre, Claude J., Gérard Manhès, and Christa Göpel. "The Age of the Earth." *Geochim. Cosmochim. Acta.* 59 (1995): 1445–56.

Dalrymple, G. Brent. *The Age of the Earth*. Stanford, CA: Stanford University Press, 1991. Radiogenic dating methods and their application to the ages of the moon, the earth, and meteorites.

Faure, Gunter. *Principles of Isotope Geology*. 2nd ed. New York: Wiley, 1986. The reference treatise on radiogenic dating methods.

Harper, Christopher T. "Geochronology: Radiometric Dating of Rocks and Minerals." In *Benchmark Papers in Geology*, edited by C. T. Harper, vol. 5. Stroudsburg, PA: Dowden, Hutchinson, and Ross, 1973. An annotated selection of important publications devoted to radiogenic dating.

Holloway, David. *Stalin and the Bomb*. New Haven, CT: Yale University Press, 1994. An account of the Soviet atomic program.

Lambeck, Kurt. *The Earth's Variable Rotation: Geophysical Causes and Consequences*. Cambridge: Cambridge University Press, 1980. The variations of the length of the day, from their causes to their astronomical and paleontological effects.

Lee, Der-Chuen, Alex N. Halliday, Gregory A. Snyder, and Lawrence A. Taylor. "Age and Origin of the Moon." *Science* 278 (1997): 1098–1103.

Mojzsis, Stephen J., Gustaf Arrhenius, Kevin D. McKeegan, T. Mark Harrison, Allen P. Nutman, and Clark R. L. Friend. "Evidence for Life on Earth Earlier than 3.8 Billion Years Ago." *Nature* 384 (1996): 55–59.

Nriagu, Jerome O. *The Biogeochemistry of Lead in the Environment*. 2 vols. New York: Elsevier, 1978. Lead pollution from the Roman empire to the contemporary world.

Odin, Gilles S., ed. *Numerical Dating in Stratigraphy*. 2 vols. New York: Wiley, 1982. A large variety of current radiogenic dating methods.

Rhodes, Richard. *The Making of the Atomic Bomb*. London: Simon and Schuster, 1986. A history of the first American atomic bombs.

Watson, Andrew. "The Universe Shows Its Age." *Science* (1997): 981–83.

Epilogue

Ager, Derek V. *The Nature of the Stratigraphical Record*. 3rd ed. New York: Wiley, 1993. A modern catastrophist standpoint.

Berggren, William A., and John A. Van Couvering, eds. *Catastrophes and Earth History: the New Uniformitarianism*. Princeton, NJ: Princeton University Press, 1984. Catastrophes as an agent of uniformitarianism.

Gradstein, Felix M., James G. Ogg, and A. Gilbert Smith. *A Geologic Time Scale, 2004*. Cambridge: Cambridge University Press, 2004. The new reference timescale.

Kagan, Boris A., and Jürgen Sündermann. "Dissipation of Tidal Energy, Paleotides, and Evolution of the Earth-Moon System." *Advances in Geophysics* 38 (1996): 179–266.

Knauth, L. Paul. "Salinity History of the Earth's Early Ocean." *Nature* 395 (1998): 554–55.

Laskar, Jacques. "The Chaotic Motion in the Solar System: A Numerical Estimate of the Size of the Chaotic Zones." *Icarus* 88 (1990): 266–91.

Lovelock, James. *The Ages of Gaia: A Biography of Our Living Earth*. New York: Norton, 1995.

Richter, Frank M. "Kelvin and the Age of the Earth." *J. Geol.* 94 (1986): 395–401. The current implications of terrestrial heat flux on calculation of the earth's age.

Trompf, Garry W. *The Idea of Historical Recurrence in Western Thought from Antiquity to Reformation.* Berkeley: University of California Press, 1979. History: cyclical or not?

Wasserburg, Gerald J. "Isotopic Abundances: Inferences on Solar System and Planetary Evolution." *Earth Planet. Sci. Lett.* 86 (1987): 129–73.

Influential Students of Time

Albertus Magnus

BIOGRAPHY

Duhem, Pierre. "Albert le Grand." In Duhem, *Le système du monde: Histoire des doctrines cosmologiques de Platon à Copernic,* 5:412–65. Paris: Hermann, 1917; reprinted in 1958.

Weisheipl, James A., ed. *Albertus Magnus and the Sciences: Commemorative Essays.* Toronto: Pontifical Institute of Mediæval Studies, 1980.

FURTHER READING

Wyckoff, Dorothy. "Albertus Magnus on Ore Deposits." *Isis* 49 (1958): 109–22.

Giovanni Arduino

BIOGRAPHY

Vaccari, Ezio. *Giovanni Arduino (1714–1795) Il contributo di uno scienzato veneto al dibattito settecentesco sulle scienze della Terra.* Florence: L. S. Olschki, 1993.

Aristotle

FURTHER READING

Duhem, Pierre. "La physique d'Aristote." In Duhem, *Le système du monde: Histoire des doctrines cosmologiques de Platon à Copernic,* 1:28–101. Paris: Hermann, 1913; reprinted in 1958.

Svante August Arrhenius

BIOGRAPHY

Crawford, Elisabeth. *Arrhenius: From Ionic Theory to the Greenhouse Effect.* Canton, MA: Watson, 1996.

FURTHER READING

Weart, Spencer R. "The Discovery of the Risk of Global Warming." *Physics Today*, January 1997, 34–40.

John Ashenden

FURTHER READING

Thorndike, Lynn. "John of Eschenden: specialist in conjunctions." *In A History of Magic and Experimental Science*. 3:325–346. New York: Columbia University Press, 1933.

Augustine

BIOGRAPHY

Marrou, Irénée. *Saint Augustin et l'augustinisme*. Paris: Editions du Seuil, 1956. Published in English as *St. Augustine and His Influence through the Ages*, translated by P. Hepburne-Scott (New York: Harper Touchbooks, 1957).

FURTHER READING

Duhem, Pierre. "La cosmologie des pères de l'Église." In Duhem, *Le système du monde: Histoire des doctrines cosmologiques de Platon à Copernic*, 2:394–494. Paris: Hermann, 1914; reprinted in 1958.

Avicenna

AUTOBIOGRAPHY

Avicenna [Ibn Sînâ]. *Avicenna's Treatise on Logic*. Translated by F. Zabeeh. The Hague: Nijoff, 1971.

FURTHER READING

Duhem, Pierre. "La géologie des Arabes." In Duhem, *Le système du monde: Histoire des doctrines cosmologiques de Platon à Copernic*, 9:257–66. Paris: Hermann, 1958.

Roger Bacon

BIOGRAPHY

Easton, Stewart C. *Roger Bacon and His Search for a Universal Science*. New York: Columbia University Press, 1952.

FURTHER READING

Carton, Raoul. *L'expérience physique chez Roger Bacon: Contribution à l'étude de la méthode et de la science expérimentales au XIIIe siècle*. Paris: Vrin, 1924.

Crombie, Alistair Cameron. *The History of Science from Augustine to Galileo.* New York: Dover, 1995.

Jean-Sylvain Bailly

BIOGRAPHY

Arago, François. "Bailly." In Arago, *Œuvres completes,* 2:247–426. Paris: Gide and Baudry, 1854.

Joseph Barrell

BIOGRAPHY

Schuchert, Charles. "Biographical Memoir of Joseph Barrell, 1869–1919." *Biog. Mem. Nat. Acad. Sci.* 12 (1929): 1–40.

Henri Becquerel

BIOGRAPHY

Crookes, William. "Antoine Henri Becquerel, 1852–1908." *Proc. Roy. Soc. London* A83 (1909): xx–xxiii.

Darboux, Gaston. "Discours prononcé aux funérailles de M. Henri Becquerel." *C. R. Acad. Sci.* 147 (1908): 143–51.

FURTHER READING

Lodge, Oliver. "Becquerel Memorial Lecture." *J. Chem. Soc.* 101 (1912): 2005–42.

René Blondlot

FURTHER READING

Piéron, Henri. "Grandeur et décadence des rayons N: Histoire d'une croyance." *Année Psychologique* 13 (1907): 143–69.

Toulouse, Dr. "Les rayons N existent-ils?" *Revue scientifique* 2 (1904): 545–52, 590–91, 620–23, 656–60, 682–86.

Wood, Robert W. "The N-Rays." *Nature* 71 (1904): 530–31.

Johann Friedrich Blumenbach

FURTHER READING

Héron de Villefosse, Antoine-Marie. "Considérations sur les fossiles, et particulièrement ceux que présente le pays de Hanovre; ou extrait raisonné d'un ouvrage de M. Blumembach, ayant pour titre: *Specimen Archeologia telluris, terrarum que imprimis Hannoveranarum.*" *J. Mines* 16 (1804): 5–36.

Bertram Boltwood

FURTHER READING

Badash, Lawrence. *Rutherford and Boltwood: Letters on Radioactivity.* New Haven, CT: Yale University Press, 1969.

Eve, Arthur Stewart. *Rutherford: Being the Life and Letters of the Rt Hon. Lord Rutherford, O. M.* New York: Macmillan, 1939.

Nicolas Boulanger

BIOGRAPHY

Hampton, John. *Nicolas-Antoine Boulanger et la science de son temps.* Geneva: Droz, 1955.

FURTHER READING

Ellenberger, François. "Nicolas Boulanger, le pionnier de génie méconnu." In *Histoire de la géologie,* 2:197–210. Paris: Lavoisier, 1988. Published in English as *History of Geology,* translated by R. K. Kaula and M. Carozzi (Rotterdam, Netherlands: A. A. Balkema, 1996).

Tycho Brahe

BIOGRAPHY

Thoren, Victor E., and John Robert Christianson, *The Lord of Uraniborg.* Cambridge: Cambridge University Press, 1990.

Alexandre Brongniart

BIOGRAPHY

Launay, Louis de. *Une grande famille de savants, les Brongniart.* Paris: G. Rapilly, 1940.

Brothers of Purity

FURTHER READING

Ellenberger, François. "Le miracle arabe." In Ellenberger, *Histoire de la géologie,* 1:77–81. Paris: Lavoisier, 1988. Published in English as *History of Geology,* translated by R. K. Kaula and M. Carozzi (Rotterdam, Netherlands: A. A. Balkema, 1996).

Harrison Brown

BIOGRAPHY

Revelle, Roger. "Harrison Brown, September 26, 1917–December 8, 1986." *Biog. Memoir Nat. Acad. Sci.* 65 (1994): 41–55.

FURTHER READING

Smith, Kirk R., Fereidun Fesharaki, and John P. Holdren, eds., *Earth and the Human Future: Essays in Honor of Harrison Brown*. Boulder: Westview Press, 1986.

William Buckland

BIOGRAPHY

Rupke, Nicholas A. *The Great Chain of History: William Buckland and the English School of Geology (1814–1849)*. Oxford: Clarendon, 1983.

Georges-Louis Leclerc Buffon

BIOGRAPHY

Aude, Chevalier d'. *Vie privée du comte de Buffon, suivie d'un recueil de poésies, dont quelques pièces sont relatives à ce grand homme*. Lausanne, 1788.

Roger, Jacques. *Buffon: Un philosophe au Jardin du Roi*. Paris: Fayard, 1989. Translated by S. L. Bonnefoi as *Buffon: A Life in Natural History* (Ithaca, NY: Cornell University Press, 1996).

FURTHER READING

Gayon, Jean, ed. *Buffon 88: Actes du colloque international pour le bicentenaire*. Paris: Vrin, 1992.

Heim, Roger, ed. *Buffon*. Paris: Muséum National d'Histoire Naturelle, 1952.

Hérault de Séchelles, Marie Jean. *Voyage à Montbar, contenant des détails très-intéressans sur le caractère, la personne et les écrits de Buffon*. Paris: Solvet, 1800. Reprint by Y. Gaillard, *Buffon: Biographie imaginaire et réelle*. Paris: Hermann, 1979.

Royou, Thomas Marie. *Le monde de verre réduit en poudre, ou analyse et réfutation des Epoques de la nature de M. le comte de Buffon* (Paris: Mérigot le jeune, 1780).

Thomas Burnet

FURTHER READING

Roger, Jacques. "The Cartesian Model and Its Role in Eighteenth-Century 'Theory of the Earth.'" In *Problems of Cartesianism*, ed. T. M. Lennon, J. M. Nicholas, and J. W. Davis, 95–112. Montreal: McGill–Queen's University Press, 1982.

———. "La théorie de la terre au XVIIe siècle." *Rev. Hist. Sci.* 26 (1973): 23–48.

Dominique Cassini

BIOGRAPHY

Fontenelle, Bernard le Bovier de. "Éloge de M. Cassini." In Fontenelle, *Éloges des Académiciens de l'Académie royale des sciences, morts depuis l'année 1699*, 2:296–334. New ed. Paris: Libraires associés, 1766.

Gabrielle-Émilie du Châtelet

FURTHER READING

Havard, Jean-Alexandre. *Voltaire et Mme du Châtelet: Révélations d'un serviteur attaché à leurs personnes.* Paris: E. Dentu, 1863.

Vaillot, René. *Voltaire et son temps: Avec Madame du Châtelet.* Oxford: Voltaire Foundation, 1988.

John Mason Clarke

BIOGRAPHY

Schuchert, Charles. "Biographical Memoir of John Mason Clarke." *Biog. Mem. Nat. Acad. Sci.* 12 (1929): 182–244.

Antonio Schinella Conti

BIOGRAPHY

Conti, Guinguéné, P.-L., "L'abbé Antoine Schniella Conti." In *Biographie universelle ancienne and moderne, ou histoire, par ordre alphabétique, de la vie publique and privée de tous les hommes qui se sont faits remarquer par leurs écrits, leurs actions, leurs talents, leurs vertus ou leurs crimes,* ed. M. Michaud, new ed., 9: 122–24. Paris: Mme. C. Desplaces, 1854–65.

Nicolaus Copernicus

FURTHER READING

Kuhn, Thomas S. *The Copernican Revolution.* Cambridge, MA: Harvard University Press, 1957.

James Croll

FURTHER READING

Imbrie, John, and Katharine P. Imbrie. *Ice Ages: Solving the Mystery.* London: Macmillan, 1979.

William Crookes

BIOGRAPHY

Fournier d'Albe, Edmund Edward. *The Life of Sir William Crookes.* London: T. F. Unwin, 1923.

Strutt, Robert John (fourth Baron Rayleigh). "Some Reminiscences of Scientific Workers of the Past Generation, and Their Surroundings." *Proc. Phys. Instit.* 48 (1936): 217–46.

FURTHER READING

Hall, Trevor. H. *The Spiritualists: The Story of Florence Cook and William Crookes.* New York: Helix Press, 1963. Reprinted as *The Medium and the Scientist* (Buffalo, NY: Prometheus Books, 1984).

Marie Curie

BIOGRAPHY

Curie, Eve. *Madame Curie.* Paris: Gallimard, 1938. Published in English as *Madame Curie: A Biography*, translated by V. Sheean (New York: Doubleday, 1939).
Curie, Marie. *Pierre Curie.* New York: Macmillan, 1923.
Reid, Robert. *Marie Curie.* New York: E. P. Dutton, 1974.

FURTHER READING

Davis, J. L., "The Research School of Marie Curie in the Paris Faculty, 1907–14." *Ann. Sci.* 52 (1995): 321–55.

Pierre Curie

BIOGRAPHY

Barbo, Loïc. *Curie: Le rêve scientifique.* Paris: Belin, 1999.
Curie, Marie. *Pierre Curie.* New York: Macmillan, 1923.
Hurwic, Anna. *Pierre Curie.* Paris: Flammarion, 1995.

Georges Cuvier

BIOGRAPHY

Coleman, William. *Georges Cuvier, Zoologist: A Study in the History of Evolution Theory.* Cambridge, MA: Harvard University Press, 1964.
Flourens, Pierre. "Éloge historique de G. Cuvier." In Cuvier, *Recueil des éloges historiques lus dans les séances publiques de l'Institut de France,* 1:i–liii. Paris: Firmin-Didot, 1861.
Outram, Dorinda. *Georges Cuvier: Vocation, Science, and Authority in Post-Revolutionary France.* Manchester, Eng.: Manchester University Press, 1984.

FURTHER READING

Rudwick, Martin J. S. *Georges Cuvier, Fossil Bones, and Geological Catastrophes: New Translations and Interpretations of Primary Texts.* Chicago: University of Chicago Press, 1997.

Charles Darwin

BIOGRAPHY

Beer, Gavin de. *Charles Darwin: Evolution by Natural Selection.* New York: Double-day, 1964.

Darwin, Francis. *Life and Letters of Charles Darwin.* London: John Murray, 1887. Reprint, New York: Basic Books, 1959.

Desmond, Adrian, and J. Moore. *Darwin.* New York: Warner Books, 1991.

FURTHER READING

Burchfield, Joe D. "Darwin and the Dilemma of Geological Time." *Isis* 65 (1974): 301–21.

Jean-André De Luc

FURTHER READING

Ellenberger, François, and Gabriel Gohau. "A l'aurore de la stratigraphie paléontologique: Jean-André De Luc, son influence sur Cuvier." *Rev. Hist. Sci.* 39 (1981): 217–57.

Arthur Jeffreys Dempster

BIOGRAPHY

Allison, Samuel King. "Arthur Jeffrey Dempster." *Biog. Mem. Nat. Acad. Sci.* 27 (1952): 318–29.

René du Perron Descartes

BIOGRAPHY

Gaukroger, Stephen. *Descartes: An Intellectual Biography.* Oxford: Clarendon, 1995.

FURTHER READING

Ellenberger, François. "Descartes and the First Theory of the Earth." In *History of Geology,* translated by R. K. Kaula and M. Carozzi (Rotterdam, Netherlands: A. A. Balkema, 1996).

Roger, Jacques. "The Cartesian Model and Its Role in Eighteenth-Century 'Theory of the Earth.'" In *Problems of Cartesianism,* ed. T. M. Lennon, J. M. Nicholas, and J. W. Davis, 95–112. Montreal: McGill–Queen's University Press, 1982.

Nicolas Desmarest

BIOGRAPHY

Cuvier, Georges. "Éloge historique de Desmarest." In Cuvier, *Recueil des éloges historiques lus dans les séances publiques de l'Institut de France,* 2:137–62. Paris: Firmin-Didot, 1861.

FURTHER READING

Taylor, Kenneth L. "Nicolas Desmarest and Geology in the Eighteenth Century." In *Toward a History of Geology,* edited by C. J. Schneer. Cambridge, MA: MIT Press, 1969.

Hugo De Vries

FURTHER READING

Bowler, P. J. "Hugo De Vries and Thomas Hunt Morgan: The Mutation Theory and the Spirit of Darwinism." *Ann. Sci.* 35 (1978): 55–73.

Arthur Stanley Eddington

BIOGRAPHY

Chandrasekhar, Subrahmanyan. *Eddington: The Most Distinguished Astrophysicist of His Time.* Cambridge: Cambridge University Press, 1983.

Douglas, Allie Vibert. *The Life of Arthur Stanley Eddington.* London: Thomas Nelson, 1956.

Enrico Fermi

BIOGRAPHY

Fermi, Laura. *Atoms in the Family.* Chicago: University of Chicago Press, 1954.

FURTHER READING

Noddack, Irma Z. "Über das Element 93." *Ang. Chem.* 47 (1934): 653.

Osmond Fisher

BIOGRAPHY

Davison, Charles. "Eminent Living Geologists: Rev. Osmond Fisher, M.A., F.G.S." *Geol. Mag.* 7 (1900): 49–54.

"Rev. Osmond Fisher, M.A., F.G.S." *Geol. Mag.* 1 (1914): 383–84.

Bernard le Bovier de Fontenelle

FURTHER READING

Niderst, Alain. *Fontenelle à la recherche de lui-même (1657–1702)*. Paris: Nizet, 1972.

Rappaport, Rhoda. "Fontenelle Interprets the Earth's History." *Rev. Hist. Sci.* 44 (1991): 281–300.

Joseph Fourier

BIOGRAPHY

Arago, François. "Joseph Fourier." In Arago, *Œuvres completes,* 1:295–369. Paris: Gide and Baudry, 1854.

Champollion-Figeac, Aimé-Louis. *Fourier, Champollion-Figeac: L'égypte et les cent jours.* Paris: Firmin Didot, 1844.

Cousin, Victor. "Discours de réception à l'Académie française et notes additionnelles." In Cousin, *Œuvres,* 4:1–101. Paris, 1849.

Dhombres, Jean, and Jean-Bernard Robert. *Fourier: Créateur de la physique mathématique.* Paris: Belin, 1998.

Herivel, John. *Joseph Fourier: The Man and the Physicist.* Oxford: Clarendon Press, 1975.

Nicolas Fréret

FURTHER READING

Wade, Ira O. *The Clandestine Organization and Diffusion of Philosophic Ideas in France from 1700 to 1750.* Reprint, New York: Octagon Books, 1967.

Georg Christian Füchsel

FURTHER READING

Keferstein, Christian. "Notice sur Füchsel et ses ouvrages." *J. Géologie* 2 (1830): 191–97.

Galileo Galilei

BIOGRAPHY

Drake, Stillman. *Galileo at Work: His Scientific Biography.* Chicago: University of Chicago Press, 1978.

FURTHER READING

Redondi, Pietro. *Galileo eretico.* Turin: G. Einaudi, 1983. Published in English as *Galileo Heretic,* translated by R. Rosenthal (Princeton, NJ: Princeton University Press, 1987).

Luigi Galvani

BIOGRAPHY
Pera, Marcello. *La rana ambigua.* Turin: G. Einaudi, 1986.

FURTHER READING
———. *Galvani-Volta.* Norwalk, CT: Burndy Library, 1952.

John Goodchild

BIOGRAPHY
"John George Goodchild, F.G.S." *Geol. Mag.* 33 (1906): 189–90.

Leslie Groves

BIOGRAPHY
Lawren, William. *The General and the Bomb.* New York: Dodd and Mead, 1988.

Jean-Etienne Guettard

BIOGRAPHY
Condorcet. "Éloge de M. Guettard." *Hist. Acad. Roy. Sci.* (1786): 47–62.

FURTHER READING
Beer, Gavin de. "The Volcanoes of Auvergne." *Ann. Sci.* 18 (1963): 49–61.
Cuvier, Georges. "Éloge historique de Desmarest." In Cuvier, *Recueil des éloges historiques lus dans les séances publiques de l'Institut de France,* 2:137–62. Paris: Firmin-Didot, 1861.
Ellenberger, François. "Précisions nouvelles sur la découverte des volcans de France: Guettard, ses prédécesseurs, ses émules clermontois." *Histoire et Nature* 12–13 (1978): 3–42.

Otto Hahn

BIOGRAPHY
Hoffmann, Klaus. *Schuld und Verantwortung: Otto Hahn, Konflikte eines Wissenschaftlers.* Berlin: Springer, 1993. Published in English as *Otto Hahn: Achievement and Responsibility,* translated by J. M. Cole (New York: Springer, 2001).

Edmond Halley

BIOGRAPHY
Armitage, Angus. *Edmond Halley.* London: Nelson, 1966.

Mairan, Jean-Jacques Dortous de. "Éloge de M. Halley." *Mém. Acad. Roy. Sci.* (1742): 172–88. Reprinted in *Correspondence and Papers of Edmond Halley,* edited by E. F. MacPike (Oxford: Clarendon, 1932). Page references are to the reprint.
Ronan, Colin A. *Edmond Halley: Genius in Eclipse.* New York: Doubleday, 1969.

FURTHER READING
Schaffer, Simon. "Halley's Atheism and the End of the World." *Notes Records Roy. Soc.* 32 (1978) 17–40.

Hermann von Helmholtz

BIOGRAPHY
Köningsberger, Leo. *Hermann von Helmholtz.* Brunswick, Germany: Vieweg, 1911.

William Herschel

BIOGRAPHY
Hoskin, Michael A. *William Herschel and the Construction of the Heavens.* With notes by D. W. Dewhirst. London: Oldbourne, 1963.

Heinrich Hertz

BIOGRAPHY
Fölsing, Albrecht. *Heinrich Hertz: Eine biographie.* Hamburg: Hoffmann and Campe, 1997.
Susskind, Charles. *Heinrich Hertz: A Short Life.* San Francisco: San Francisco Press, 1995.

FURTHER READING
O'Hara, James G., and Willibald Pricha. *Hertz and the Maxwellians: A Study and Documentation of the Discovery of Electromagnetic Wave Radiation, 1873–1894.* London: Peter Peregrinus, 1987.

Ejnar Hertzsprung

BIOGRAPHY
Herrmann, Dieter B. *Ejnar Hertzsprung: Pionier der Sternforschung.* Berlin: Springer, 1994.

William Francis Hillebrand

BIOGRAPHY AND FURTHER READING
Clarke, Frank Wigglesworth. "William Francis Hillebrand 1853–1925." *Biog. Mem. Nat. Acad. Sci.* 12 (1929): 42–70.

Arthur Holmes

BIOGRAPHY

Dunham, Kingsley Charles. "Arthur Holmes 1890–1965." *Biogr. Mem. Fellows Roy. Soc.* 12 (1966): 291–310.

Lewis, Cherry. *The Dating Game: One Man's Search for the Age of the Earth.* Cambridge: Cambridge University Press, 2000.

Robert Hooke

BIOGRAPHY

Andrade, Edward Neville Da Costa. "Robert Hooke." *Proc. Roy. Soc.* A201 (1950): 439–73.

Drake, Ellen T. *Restless Genius: Robert Hooke and His Earthly Thoughts.* Oxford: Oxford University Press, 1996.

FURTHER READING

Harwood, John T. "Rhetoric and Graphics in *Micrographia.*"In *Robert Hooke, New Studies,* edited by M. Hunter and S. Schaffer. Woodbridge, Eng.: Boydell Press, 1989.

Koyré, Alexandre. "An Unpublished Letter of Robert Hooke to Isaac Newton." *Isis* 43 (1952): 312–37. Reprinted in Koyré, *Newtonian Studies.*

Stimson, Dorothy. *Scientists and Amateurs: A History of the Royal Society.* New York: H. Schuman, 1948.

William Hopkins

FURTHER READING

Lyell, Charles. "Award of the Wollaston Medal and Donation Fund." *Quart. J. Geol. Soc.* 6 (1850): xxiii–xxvi.

Friedrich Georg Houtermans

BIOGRAPHY

Khriplovich, Iosif B. "The Eventful Life of Fritz Houtermans." *Physics Today,* July 1992, 29–37.

FURTHER READING

Weissberg, Alexander. *The Accused.* New York: Simon and Schuster, 1951.

Edwin Powell Hubble

BIOGRAPHY

Sharov Alexander S., and Igor D. Novikov. *Edwin Hubble: The Discoverer of the Big Bang Universe*. Translated from the Russian by V. Kisin. Cambridge: Cambridge University Press, 1993.

James Hutton

BIOGRAPHY

Playfair, John. "Biographical Account of the Late Dr James Hutton, F.R.S. Edin." *Trans. Roy. Soc. Edinb.* 5 (1805): 39–99.
Dean, Dennis R. *James Hutton and the History of Geology*. Ithaca, NY: Cornell University Press, 1992.

FURTHER READING

Craig, Gordon Y., Donald B. McIntyre, and Charles D. Waterston. *James Hutton's Theory of the Earth: The Lost Drawings*. Edinburgh: Scottish Academic Press, 1978.
De Luc, Jean-André. *Traité élémentaire de géologie* (Paris: Courcier, 1809).
Desmarest, Nicolas. "Notice sur plusieurs points importans de l'histoire naturelle de la terre." In *Géographie Physique*, by Desmarest, 1:409–31. Paris: Agasse, 1794–95.
Jones, Jean, Hugh S. Torrens, and Eric Robinson. "The Correspondence between James Hutton (1726–1797) and James Watt (1736–1819) with Two Letters from Hutton to George Clerk Maxwell (1715–1784): Part II." *Ann. Sci.* 52 (1995): 357–82.
MacCulloch, John. "A Geological Description of Glen Tilt." *Trans. Geol. Soc. London* 3 (1816): 259–337.

Thomas Huxley

BIOGRAPHY

Desmond, Adrian. *Huxley: Evolution's High Priest*. London: M. Joseph, 1997.

James Jeans

BIOGRAPHY

Milne, Edward Arthur. *Sir James Jeans*. Cambridge: Cambridge University Press, 1952.

John Joly

BIOGRAPHY

"John Joly 1857–1903." *Obit. Not. Roy. Soc. London* 1 (1934): 259–86.

James Prescott Joule

BIOGRAPHY

Cardwell, Donald S. L. *James Joule: A Biography*. Manchester, Eng.: Manchester University Press, 1989.

Lord Kelvin. See William Thomson.

Johannes Kepler

BIOGRAPHY

Caspar, Max. *Johannes Kepler*. Munich: V. Kohlhammer, 1948. Published in English as *Johannes Kepler*, translated by C. Doris-Hellman (New York: Abelard-Schuman, 1959).

FURTHER READING

Koyré, Alexandre. *La révolution astronomique*. Paris: Hermann, 1961. Published in English as *The Astronomical Revolution*, translated by R. E. W. Maddison (New York: Dover, 1992).
Simon, Gérard. *Kepler astronome astrologue*. Paris: Gallimard, 1979.

Clarence King

BIOGRAPHY

Emmons, Samuel Franklin. "Clarence King. 1842–1901." *Biog. Mem. Nat. Acad. Sci.* 6 (1906): 25–55.

Athanasius Kircher

BIOGRAPHY

Reilly, P. Connor. *Athanasius Kircher: Master of a Hundred Arts, 1602–1680*. Wiesbaden, Germany: Edizioni del Mondo, 1974.

FURTHER READING

Jahn, Melvin E., and Daniel J. Woolf. *The Lying stones of Dr. Johann Bartholomew Adam Beringer being his Lithographiæ Wirceburgensis*. Berkeley: University of California Press, 1963.
Oldenburg, Henry. "Of the *Mundus Subterraneus* of Athanasius Kircher." *Phil. Trans.* 6 (1665): 109–17.

Jean-Baptiste de Monet Lamarck

BIOGRAPHY

Burkhardt, Richard W., Jr. *The Spirit of System: Lamarck and Evolutionary Biology.* Cambridge, MA: Harvard University Press, 1977).

Cuvier, Georges. "Éloge historique de M. de Lamarck." In Cuvier, *Recueil des éloges historiques lus dans les séances publiques de l'Institut de France,* 3:179–210. Paris: Firmin-Didot, 1861.

FURTHER READING

Guédès, Michel. "Les revenus de Lamarck." *Histoire et Nature* 21 (1982): 49–60.

Isaac de La Peyrère

BIOGRAPHY

Moréri, Louis. *Grand dictionnaire historique de Moréri.* New ed. Vol. 8. Paris, 1749.

FURTHER READING

Lods, Adolphe. "Histoire de la critique du Pentateuque." In *Histoire de la littérature hébraïque et juive,* 83–118. Paris: Payot, 1950.

McKee, D. Rice. "Isaac de la Peyrère, a Precursor of Eighteenth-Century Critical Deists." *PMLA* 59 (1944): 456–85.

Rossi, Paolo. *I segni del tempo: Storia delle terra e delle nazioni da Hooke a Vico.* Milan: Feltrinelli, 1979.

Pierre-Simon Laplace

BIOGRAPHY

Arago, François. "Laplace." In *Œuvres complètes,* 3:456–515. Paris: Gide and Baudry, 1854.

Fourier, Joseph. "Éloge historique de M. le marquis de La Place." In Laplace, *L'Exposition du Système du monde,* 6th ed. Paris: Bachelier, 1836. Earlier edition published in English in two volumes as *The System of the World,* translated by H. H. Harte (London: Longmans, 1830).

Gillispie, Charles Coulson, Robert Fox, and Ian Grattan-Guinness. *Pierre-Simon Laplace (1749–1827): A Life in Exact Science.* Princeton, NJ: Princeton University Press, 1997.

Antoine-Laurent de Lavoisier

BIOGRAPHY

Bensaude-Vincent, Bernadette. *Lavoisier: Mémoires d'une révolution.* Paris: Flammarion, 1993.

Donovan, Arthur. *Antoine Lavoisier: Science, Administration, and Revolution.* Oxford: Blackwell, 1993.

Poirier, Jean-Pierre. *Antoine-Laurent de Lavoisier, 1743–1794.* Paris: Pygmalion, 1995. Published in English as *Lavoisier: Chemist, Biologist, Economist,* translated by R. Balinski (Philadelphia: University of Pennsylvania Press, 1996).

FURTHER READING

Rappaport, Rhoda. "Lavoisier's Geologic Activities, 1763–1792." *Isis* 58 (1968): 375–84.

Gustave Le Bon

FURTHER READING

Nye. Mary-Jo. "Gustave LeBon's Black Light: A Study in Physics and Philosophy in France at the Turn of the Century." *Hist. Stud. Phys. Sci.* 4 (1974): 163–95.

Gottfried Wilhelm Leibniz

BIOGRAPHY

Aiton, Eric J. *Leibniz: A Biography.* Bristol, Eng.: Adam Hilger, 1985.

Fontenelle, Bernard le Bovier de. "Éloge de Mr. G.G. Leibniz." In Fontenelle, *Éloges des Académiciens de l'Académie royale des sciences, morts depuis l'année 1699,* 1:447–506. New ed. Paris: Libraires associés, 1766.

FURTHER READING

Pécaut, Catherine. "L'œuvre géologique de Leibniz." *Rev. Gén. Sci.* 58 (1951): 282–96.

Nicolas Lémery

BIOGRAPHY

Fontenelle, Bernard le Bovier de. "Éloge de M. Lémery." In Fontenelle, *Éloges des Académiciens de l'Académie royale des sciences, morts depuis l'année 1699,* 1:356–71. New ed. Paris: Libraires associés, 1766.

Leonardo da Vinci

BIOGRAPHY

White, Michael. *Leonardo: The First Scientist.* London: Little, Brown, 2000.

FURTHER READING

Duhem, Pierre. *Études sur Léonard de Vinci, ceux qu'il a lus et ceux qui l'ont lu.* 3 vols. Paris: A. Hermann, 1906.

Norman Lockyer

BIOGRAPHY
Lockyer, Thomazine Mary, and Winifred L. Lockyer. *Life and Work of Sir Norman Lockyer.* London: Macmillan, 1928.

Percival Lowell

FURTHER READING
Hoyt, William Graves. *Lowell and Mars.* Tucson: University of Arizona Press, 1976.

Charles Lyell

BIOGRAPHY
Wilson, Leonard G. *Charles Lyell: The Years to 1841: A Revolution in Geology.* New Haven, CT: Yale University Press, 1972.

FURTHER READING
Gould, Stephen Jay. *Time's Arrow, Time's Cycle: Myth and Metaphor in the Discovery of Geological Time.* Cambridge, MA: Harvard University Press, 1987.
Ospovat, Alex M. "The Distortion of Werner in Lyell's Principles of Geology." *Brit. J. Hist. Sci.* 9 (1976): 190–99.

Guglielmo Marconi

BIOGRAPHY
Jolly, W. P. *Marconi.* New York: Stein and Day, 1972.

William Diller Matthew

BIOGRAPHY
Gregory, William K. "William Diller Matthew, Paleontologist (1871–1930)." *Science* 72 (1930): 642–45.

James Clerk Maxwell

BIOGRAPHY
Goldman, Martin. *The Demon in the Aether: The Story of James Clerk Maxwell.* Edinburgh: Paul Harris, 1983.

Julius Robert Mayer

BIOGRAPHY

Caneva, Kenneth L. *Robert Mayer and the Conservation of Energy*. Princeton, NJ: Princeton University Press, 1993.

Schmolz, Helmut, and Hubert Weckbach, *Robert Mayer: Sein Leben und Werk in Dokumenten*. Heilbronn, Germany: Anton H. Konrad, 1964.

Jean-Baptiste de Mirabaud

FURTHER READING

Wade, Ira O. *The Clandestine Organization and Diffusion of Philosophic Ideas in France from 1700 to 1750*. Reprint, New York: Octagon Books, 1967.

Henry Gwinn Jeffreys Moseley

BIOGRAPHY

Heilbron, John L. *H. G. J. Moseley: The Life and Letters of an English Physicist*. Berkeley: University of California Press, 1974.

Sarton, George. "Moseley. The Numbering of the Elements." *Isis* 9 (1927): 96–111.

Isaac Newton

BIOGRAPHY

Manuel, Frank E. *A Portrait of Isaac Newton*. Cambridge, MA: Belknap Press of Harvard University Press, 1968.

More, Louis Trenchard. *Isaac Newton: A Biography, 1642–1727*. London: Scribner's, 1934.

Turnor, Edmund. *Collections for the History of the Town and Soke of Grantham. Containing Authentic Memoirs of Sir Isaac Newton*. London: W. Miller, 1806.

Westfall, Richard. *Never at Rest*. Cambridge: Cambridge University Press, 1980.

FURTHER READING

Cohen, I. Bernard, and Richard Westfall, eds. *Newton*. New York: Norton, 1995.

Greenberg, John L. "Isaac Newton and the Problem of the Earth's Shape." *Arch. Hist. Exact Sci.* (1996): 371–91.

Koyré, Alexandre. *Newtonian Studies*. Cambridge, MA: Harvard University Press, 1965.

Manuel, Frank E. *Isaac Newton, Historian*. Cambridge, MA: Belknap Press of Harvard University Press, 1963.

Alfred O. C. Nier

BIOGRAPHY

Craig, Harmon. "Introduction of Alfred O.C. Nier for the V.M. Goldschmidt Award 1984." *Geochim. Cosmochim. Acta* 49 (1985): 1161–74.

De Laeter, John, and Mark D. Kurz. "Alfred Nier and the Sector Field Mass Spectrometer." *J. Mass. Spectrosc.* 41 (2006): 447–54.

Goldich, Samuel Steven. "Alfred O. Nier, 1911–1994." *EOS* 75 (1994): 66.

Robert Oppenheimer

BIOGRAPHY

Goodchild, Peter. *Oppenheimer: Shatterer of Worlds*. Boston: Houghton Mifflin, 1981.

Nicole Oresme

FURTHER READING

Duhem, Pierre. "Nicole Oresme expose l'hypothèse de la rotation terrestre." In Duhem, *Le système du monde*: *Histoire des doctrines cosmologiques de Platon à Copernic,* 9:329–44. Paris: Hermann, 1958.

———. "La géologie de Buridan et l'université de Paris. I. Un adversaire: Nicole Oresme." In Duhem, *Le système du monde*: *Histoire des doctrines cosmologiques de Platon à Copernic*, 9:306–9. Paris: Hermann, 1958.

Paracelsus

FURTHER READING

Koyré, Alexandre. "Paracelse." *Rev. Hist. Philos. Relig.* 13 (1933): 46–75, 145–63.

Clair C. Patterson

BIOGRAPHY

Church, Thomas M. "Clair C. Patterson (1922–1995)." *EOS* 77 (1996): 306.

Epstein, Samuel. "Introduction of Clair C. Patterson for the V.M. Goldschmidt Medal 1980." *Geochim. Cosmochim. Acta* 45 (1981): 1383–97.

Jean Perrin

BIOGRAPHY

Charpentier-Morize, Micheline. *Jean Perrin: Savant et homme politique, 1870–1942*. Paris: Belin, 1997.

Charles Snowden Piggot

BIOGRAPHY

Tilton, George R. "Charles Snowden Piggot, 1892–1973." *Biog. Memoirs Nat. Acad. Sci.* 66 (1995): 2–20.

John Phillips

BIOGRAPHY

Reyment, Richard. "John Phillips (1800–1874)." *Terra Nova* 8 (1996): 5–7.

John Philoponus

FURTHER READING

Duhem, Pierre. "La première tentative concordiste entre le récit de la Genèse et la physique: Jean Philopon." In Duhem, *Le système du monde: Histoire des doctrines cosmologiques de Platon à Copernic,* 2:494–501. Paris: Hermann, 1914; reprinted in 1958.

Sorabji, Richard. *Philoponus and the Rejection of Aristotelian Science.* London: Duckworth, 1987.

Siméon Denis Poisson

BIOGRAPHY

Arago, François. "Poisson." In Arago, *Œuvres completes,* 2:593–689. Paris: Gide and Baudry, 1854.

William Ramsay

BIOGRAPHY

Tilden, William A. *Sir William Ramsay: Memorials of His Life and Work.* London, 1918.

John Ray

BIOGRAPHY

Raven, Charles E. *John Ray Naturalist: His Life and Works.* 2nd ed. Cambridge: Cambridge University Press, 1950. Reprint, 1986.

Lord Rayleigh (John William Strutt)

BIOGRAPHY

Lindsay, Robert Bruce. *Lord Rayleigh—The Man and His Work.* Oxford: Pegamon, 1966.

Strutt, Robert John (fourth Baron Rayleigh). *John William Strutt, Third Baron Rayleigh*. London: E. Arnold, 1924. Reprinted as *Life of John William Strutt, Third Baron Rayleigh, O.M., F.R.S.* (Madison: University of Wisconsin Press, 1968).

———. "Some Reminiscences of Scientific Workers of the Past Generation, and Their Surroundings." *Proc. Phys. Instit.* 48 (1936): 217–46.

Jean Richer

FURTHER READING

Olmsted, John W. "The Scientific Expedition of Jean Richer to Cayenne." *Isis* 34 (1942–43): 117–28.

Ole Römer

FURTHER READING

Cohen, I. Bernard. "Roemer and the First Determination of the Velocity of Light" (1676). *Isis* (1939): 327–79.

Taton, René, ed. *Roemer et la vitesse de la lumière*. Paris: Vrin, 1978.

Wilhelm Conrad Röntgen

BIOGRAPHY

Fölsing, Albrecht. *Wilhelm Conrad Röntgen: Aufbruch ins Innere der Materie*. Munich: C. Hanser, 1995.

Glasser, Otto. *Wilhelm Conrad Röntgen und die Geschichte der Röntgenstrahlen*. Berlin, 1933. Published in English as *William Conrad Röntgen and the Early History of the Roentgen Rays* (Baltimore: Charles C. Thomas, 1934; reprint, San Francisco: Norman, 1993).

FURTHER READING

Klickstein, Herbert S. *Wilhelm Conrad Röntgen on a New Kind of Rays*. Vol. 1. Philadelphia: Mallinckrodt Classics of Radiology, 1966.

Guillaume François Rouelle

BIOGRAPHY

Rappaport, Rhoda. "G.F. Rouelle: An Eighteenth-Century Chemist and Teacher." *Chymia* 6 (1960): 68–102.

FURTHER READING

Desmarest, Nicolas. "Rouelle." In *Géographie Physique, Résumé des opinions anciennes et des travaux antérieurs à 1789. II Partie*, 1:409–31. Paris: Agasse, 1794–95.

Mayer, Jean. "Portrait d'un chimiste: Guillaume-François Rouelle." *Rev. Hist. Sci. Appl.* 23 (1970): 305–32.

Henry Norris Russell

BIOGRAPHY

DeVorkin, David H. *Henry Norris Russell: Dean of American Astronomers*. Princeton, NJ: Princeton University Press, 2000.

Ernest Rutherford

BIOGRAPHY

Eve, Arthur Stewart. *Rutherford: Being the Life and Letters of the Rt Hon. Lord Rutherford, O. M.* New York: Macmillan, 1939.

Oliphant, Mark. *Rutherford: Recollections of the Cambridge Days*. Amsterdam: Elsevier, 1972.

Wilson, David. *Rutherford: Simple Genius*. Cambridge, MA: MIT Press, 1983.

Joseph Scaliger

FURTHER READING

Grafton, Anthony T. *Joseph Scaliger: A Study in the History of Classical Scholarship*. Oxford: Oxford University Press, 1983.

———. "Joseph Scaliger and Historical Chronology: The Rise and Fall of a Discipline." *History and Theory* 14 (1975): 156–85.

Richard Simon

FURTHER READING

Auvray, Paul. *Richard Simon (1638–1712)*. Paris: PUF, 1974.

Vesto Melvin Slipher

BIOGRAPHY

Hoyt, William Graves. "Vesto M. Slipher, 1875–1969." *Biog. Mem. Nat. Acad. Sci.* 52 (1980): 410–49.

William Smith

BIOGRAPHY

Phillips, John. *Memoirs of William Smith, LL.D.* London, 1844. Reprint, New York: Arno Press, 1978.

Frederick Soddy

BIOGRAPHY

Fleck, Alexander. "Frederick Soddy." *Biogr. Mem. Fellows Roy. Soc.* 3 (1957): 203–16.

Merricks, Linda. *The World Made New: Frederick Soddy, Science, Politics, and Environment.* Oxford: Oxford University Press, 1997.

William Johnson Sollas

BIOGRAPHY

Woodward, A. Smith, and William Whitehead Watts. "William Johnson Sollas 1849–1936." *Obit. Not. Roy. Soc. London* 2 (1938): 265–81.

Jean-Louis Giraud Soulavie

FURTHER READING

Aufrère, Louis. *De Thalès à Davis: Le relief et la sculpture de la Terre (auteurs, textes, doctrines, ambiances). IV. La fin du XVIIIe siècle, I. Soulavie et son secret, un conflit entre l'actualisme et le créationisme, le temps géomorphologique.* Paris: Hermann, 1952.

Nicolaus Steno (Niels Stensen)

FURTHER READING

Faller, Adolf. "Niels Stensen und der Cartesianismus." In *Nicolaus Steno and His Indice,* ed. Gustav Scherz, 140–66. Copenhagen: Munksgaard, 1958.

Poulsen, Jacob E., and Egill Snorrason, eds. *Nicolaus Steno, 1638–1686: A Reconsideration by Danish Scientists.* Gentofte, Denmark: Nordisk Insulinlaboratorium, 1986.

Scherz, Gustav, ed. *Nicolaus Steno and His Indice.* Copenhagen: Munksgaard, 1958.

Robert John Strutt (fourth Baron Rayleigh)

BIOGRAPHY

Strutt, Guy. "Robert John Strutt, Fourth Baron Rayleigh." *Appl. Opt.* 3 (1964): 1105–12.

Emmanuel Swedenborg

BIOGRAPHY

Jonsson, Inge. *Emanuel Swedenborg.* Translated by C. Djurklou. New York: Twayne, 1971.

FURTHER READING

Arrhenius, Svante August. *Emmanuel Swedenborg as a Cosmologist*. Stockholm: Aftonbladets Tryckeri, 1908.

Peter Guthrie Tait

BIOGRAPHY

Knott, Cargill Gilston. *The Life and Scientific Work of Peter Guthrie Tait*. Cambridge: University Press, 1911.

Etienne Tempier

FURTHER READING

Duhem, Pierre. "Le nombre infini actuel et l'immortalité de l'âme." In Duhem, *Le système du monde: Histoire des doctrines cosmologiques de Platon à Copernic*, 7:4. Paris: Hermann, 1919; reprinted in 1956.

Mandonnet, P. *Siger de Brabant and l'Averroïsme latin au XIIIeme siècle*. Vol. 1, *Etude critique*, 2nd ed. (Louvain: Institut supérieur de philosophie de l'université, 1911). Vol. 2, *Textes inédits* (Louvain: Institut supérieur de philosophie de l'université, 1908).

Piché, David. *La condamnation parisienne de 1277*. Paris: Vrin, 1999.

Joseph John Thomson

BIOGRAPHY

Strutt, Robert John (fourth Baron Rayleigh). *The Life of Sir J. J. Thomson*. Cambridge: University Press, 1943.

Thomson, George Paget. *J. J. Thomson and the Cavendish Laboratory in His Day*. New York: Doubleday, 1965.

FURTHER READING

Davis, Edward Arthur, and Isabel J. Falconer. *J. J. Thomson and the Discovery of the Electron*. London: Taylor and Francis, 1997.

William Thomson (Lord Kelvin)

BIOGRAPHY

Gray, Andrew. *Lord Kelvin: An Account of His Scientific Life and Work*. London: J. M. Dent, 1908.

J. L. "Lord Kelvin." *Proc. Roy. Soc.* A81 (1908): iii–lxxvi.

King, Elizabeth. *Lord Kelvin's Early Home*. London: Macmillan, 1909.

Sharlin, Harold Issadore. *Lord Kelvin: The Dynamic Victorian*. University Park: Pennsylvania State University Press, 1979.

Smith, Crosbie, and M. Norton Wise. *Energy and Empire: A Biographical Study of Lord Kelvin.* Cambridge: Cambridge University Press, 1989.

Thompson, Silvanus P. *The Life of William Thomson, Baron Kelvin of Larg.* 2 vols. London: Macmillan, 1910.

FURTHER READING

Gray, Andrew. "Famous Scientific Workshops. I. Lord Kelvin's Laboratory in the University of Glasgow;" *Nature* 55 (1897): 486–92.

Thomson, Joseph John. *Recollections and Reflections.* New York: Macmillan, 1937.

Silvanus Thompson

BIOGRAPHY

Greig, James. *Silvanus P. Thompson Teacher.* London: H. M. Stationery Office, 1979.

Thompson, Jane Smeal, and Helen G. Thompson. *Silvanus Phillips Thompson: His Life and Letters.* London: T. F. Unwin, 1920.

George Hoggart Toulmin

FURTHER READING

Porter, Roy. "George Hoggart Toulmin's Theory of Man and the Earth in the Light of the Development of British Geology." *Ann. Science* 35 (1978): 339–52.

James Ussher

BIOGRAPHY

Ussher, James. *The Whole Works of the Most Rev. James Ussher, D. D.*, vol 1, with a life of the author and an account of his writings by Charles Richard Elrington. Dublin: Hodges and Smith, 1847.

Paul Villard

FURTHER READING

Lelong, Benoît. "Paul Villard, J.-J. Thomson et la composition du rayonnement cathodique." *Rev. Hist. Sci.* 50 (1997): 89–130.

Allessandro Volta

BIOGRAPHY

Arago, François. "Alexandre Volta." In *Œuvres completes*, 1:188–240. Paris: Gide and Baudry, 1854.

Dibner, Bern. *Allessandro Volta and the Electric Battery.* New York: F. Watts, 1964.

Alfred Wallace

BIOGRAPHY

George, Wilma. *Biologist Philosopher: A Study of the Life and Writings of Alfred Russel Wallace*. New York: Abelard and Schuman, 1964.

FURTHER READING

Kottler, Malcom Jay. "Alfred Russel Wallace, the Origin of Man, and Spiritualism." *Isis* 65 (1974): 145–92.

Sarton, George. "Discovery of the Theory of Natural Selection." *Isis* 14 (1930): 133–54.

Abraham Gottlob Werner

BIOGRAPHY

Cuvier, Georges. "Éloge historique de Werner." In Cuvier, *Recueil des éloges historiques lus dans les séances publiques de l'Institut de France,* 2:113–34. Paris: Firmin-Didot, 1861.

FURTHER READING

Ospovat, Alex M. "The Distortion of Werner in Lyell's Principles of Geology." *Brit. J. Hist. Sci.* 9 (1976): 190–99.

Thomas Wright (of Durham)

BIOGRAPHY

Gushee, Vera. "Thomas Wright of Durham, Astronomer." *Isis* 33 (1941): 197–218.

Bibliography

Adhémar, Joseph. *Révolutions de la mer, déluges périodiques*. Paris: Lacroix-Comon, 1st ed. 1842, 2nd ed. 1860.

Æpinus, Franz Ulrich Theodosius. *Tentamen theoriae electricitatis et magnetismi*. Translated by R. W. Home and P. J. Connor as *Æpinus's Essay on the Theory of Electricity and Magnetism* (Princeton, NJ: Princeton University Press, 1979).

Ager, Derek V. *The Nature of the Stratigraphical Record*. 3rd ed. New York: Wiley, 1993.

Agricola, Georgius [Georg Bauer]. *De ortu et causis subterraneorum* and *De Natura fossilium*. Basel, 1546. Published in English as *Textbook of Mineralogy*, translated by M. C. Bandy and J. A. Bandy, in *Min. Soc. Amer. Special Paper* 63 (1955).

———. *De re metallica libri XII*. Basel: Froben, 1556. Translated by H. C. Hoover and L. H. Hoover (New York: Dover, 1950).

Aiton, Eric J. *Leibniz: A Biography*. Bristol, Eng.: Adam Hilger, 1985.

Albertus Magnus [Albert of Lavingen]. *Mineralium libri*. Vol. 5 of Albertus, *Opera omnia*, 1–116. Translated with a biography and notes by D. Wyckoff as *Book of Minerals* (Oxford: Clarendon, 1967).

———. *Opera omnia*. Edited by A. Borgnet. 38 vols. Paris: Louis Vivès, 1890–99.

Aldrich, Lyman T., and Alfred O. C. Nier. "Argon 40 in Potassium Minerals." *Phys. Rev.* 74 (1948): 876–77.

Aldrovandi, Ulisse. *Museum metallicum in libros IIII distributum*. Parma: M. A. Bernia, 1648.

Allen, Don C. *The Legend of Noah: Renaissance Rationalism in Art, Science, and Letters*. Urbana: University of Illinois Press, 1963.

Allison, Samuel King. "Arthur Jeffrey Dempster." *Biog. Mem. Nat. Acad. Sci.* 27 (1952): 318–29.

Andrade, Edward Neville Da Costa. "Robert Hooke." *Proc. Roy. Soc.* A201 (1950): 439–73.

Arago, François. "Alexandre Volta." In Œuvres, 1:188–240.

———. "Bailly." In Œuvres, 2:247–426.

———. "Discours sur l'enseignement." In Œuvres, 12:710–11.

———. "Halley." In Œuvres, 3:367–69.

———. "Joseph Fourier." In Œuvres, 1:295–369.

———. "Laplace." In Œuvres, 3:456–515.

———. Œuvres complètes. 17 vols. Paris: Gide and Baudry, 1854–62.

———. "Poisson." In Œuvres, 2:593–689.

Arduino, Giovanni. Due lettere del sig. Giovanni Arduino sopra varie sue osservazioni naturali al Chiariss. Sig. Cavaliere Antonio Vallisnieri. In Nuova Raccolta d'Opuscoli Scientifici e Filologici, 6:xcv–clxxx. 1760.

Aristotle. The Complete Works of Aristotle. 2 vols. Edited by J. Barnes. Princeton, NJ: Princeton University Press, 1984.

———. Meteorology. Translated by E. W. Webster. In Aristotle, Complete Works.

———. On Generation and Corruption. Translated by H. H. Joachim. In Aristotle, Complete Works.

———. On the Heavens. Translated by J. L. Stocks. In Aristotle, Complete Works.

———. Physics. Translated by R. P. Hardie and R. K. Gaye. In Aristotle, Complete Works.

———. Politics. Translated by B. Jowett. In Aristotle, Complete Works.

Armitage, Angus. Edmond Halley. London: Nelson, 1966.

Arrhenius, Svante August. Emmanuel Swedenborg as a Cosmologist. Stockholm: Aftonbladets Tryckeri, 1908.

———. The Life of the Universe, as Conceived by Man from the Earliest Times to the Present Age. 2 vols. London: Harper, 1906.

———. "On the Influence of Carbonic Acid in the Air upon the Temperature of the Ground." Phil. Mag. 41 (1896): 237–76.

Astell, Mary. An Essay in Defense of the Female Sex. London, 1696.

Aston, Francis William. "The Constitution of Atmospheric Neon." Phil. Mag. 39 (1920): 449–55.

———. "The Constitution of Ordinary Lead." Nature 120 (1927): 224.

———. "The Isotopic Constitution and Atomic Weight of Lead from Different Sources." Proc. Roy. Soc. A140 (1933): 535–43.

———. "Isotopic Constitution of Lead from Different Sources." Nature 129 (1932): 649.

———. Mass Spectra and Isotopes. London: Arnold, 1st ed. 1933, 2nd ed. 1942.

———. "The Mass Spectrum of Uranium Lead and the Atomic Weight of Protactinium." Nature 123 (1929): 313.

———. "A New Mass Spectrograph and the Whole Number Rule." Proc. Roy. Soc. A115 (1927): 487–515.

———. "A Positive Ray Spectrograph." Phil. Mag. 38 (1919): 707–14.

Astruc, Jean. Conjectures sur les mémoires originaux dont il paroît que Moyse s'est servi pour composer le livre de la Genèse. Brussels: Fricx, 1753.

———. *Mémoires pour l'histoire naturelle de la province de Languedoc*. Paris: Guillaume Cavelier, 1740.

Atkinson, Robert d'Escourt, and Friedrich Georg Houtermans. "Zur Frage der Aufbaumöglichkeit der Elemente in Sternen." *Z. Phys.* 54 (1929): 656–65.

Aubuisson des Voisins, Jean-François d'. "Sur les travaux de M. Werner en minéralogie." *Ann. Chim.* 69 (1809): 225–48.

———. "Tableau de la classification des minéraux par M. Werner, accompagné de quelques observations sur les travaux de ce minéralogiste." *J. Phys. Chim. Hist. Nat.* 60 (1805): 171–78, 329–39.

———. *Traité de géognosie, ou exposé des connaissances actuelles sur la constitution physique et minérale du globe terrestre*. 3 vols. Paris: F. G. Levrault, 1828.

Aude, Chevalier d'. *Vie privée du comte de Buffon, suivie d'un recueil de poésies, dont quelques pièces sont relatives à ce grand homme*. Lausanne, 1788.

Aufrère, Louis. *De Thalès à Davis: Le relief et la sculpture de la Terre (auteurs, textes, doctrines, ambiances). IV. La fin du XVIIIᵉ siècle, I. Soulavie et son secret, un conflit entre l'actualisme et le créationisme, le temps géomorphologique*. Paris: Hermann, 1952.

Augustine [Aurelius Augustinius]. *The City of God* and *Confessions*. Translated by M. Dods. Chicago: Encyclopedia Britannica, 1952).

———. *On Genesis : Two Books on Genesis against the Manichees; and, On the Literal Interpretation of Genesis, an Unfinished Book*. Translated by R. J. Teske. Washington, D.C.: Catholic University of America Press, 1990.

———. *De natura boni contra Manicheos Liber I*. In *S. Aur. Augustini Hipponensis Episcopi opera omnia*, 8:774–98 Paris: Gaume, 1837.

———. *Letter to Jerome*. Translated by J. G. Cunningham. In *Nicene and Post-Nicene Fathers*, ser. 1, vol. 1. Grand Rapids, MI: Eerdmans, 1994.

Auvray, Paul. *Richard Simon (1638–1712)*. Paris: PUF, 1974.

Avicenna [Ibn Sînâ]. *Avicenna's Treatise on Logic*. Translated by F. Zabeeh. The Hague: Nijoff, 1971.

———. *De congelatione et conglutinatione lapidum, being sections of the Kitâb Al-Shifâ, the Latin and Arabic texts*. Translated by and with notes by E. J. Holmyard and D. C. Mandeville. Paris: Librairie orientaliste Paul Geuthner, 1927.

Bacon, Roger. *Opus majus*. Translated by R. B. Burke. 2 vols. Philadelphia: University of Pennsylvania Press, 1928.

———. *Opus tertium*. In *Fr. Rogeri Bacon opera quædam hactenus inedita*, vol. 1. London: Longman, 1859.

Badash, Lawrence. *Rutherford and Boltwood: Letters on Radioactivity*. New Haven, CT: Yale University Press, 1969.

Bailly, Jean-Sylvain. *Histoire de l'astronomie ancienne depuis son origine jusqu'à l'établissement de l'école d'Alexandrie*. Paris: Debure, 1775.

———. *Histoire de l'astronomie indienne et orientale*. Paris: Debure, 1787.

Barbo, Loïc. *Curie: Le rêve scientifique*. Paris: Belin, 1999.

Barrell, Joseph. "Rhythms and the Measurements of Geologic Time." *Bull. Geol. Soc. Amer.* 28 (1917): 745–904.

———. "The Strength of the Earth's Crust." *J. Geol.* 22 (1914): 655–83.

Basil the Great. *Homilies on the Hexaemeron.* Translated by B. Jackson. In *Nicene and Post-Nicene Fathers,* ser. 2, vol. 8. Grand Rapids, MI: Eerdmans, 1996.

Bayle, Pierre. *Lettre à M.L.A.D.C. Docteur de Sorbonne, Où il est prouvé par plusieurs raisons tirées de la Philosophie, et de la Theologie, que les Cometes ne sont point le presage d'aucun malheur.* Cologne: P. Marteau, 1682.

———. *Pensées diverses sur la comète de 1680.* Edited by A. Prat. 2 vols. Paris: E. Cornely, 1911–12). Published in English as *Various Thoughts on the Occasion of a Comet,* translated by R. C. Bartlett (Albany: New York State University Press, 2000).

Beck, Frederick [pseud.], and W. Godin [pseud.]. *Russian Purge and the Extraction of Confessions.* Translated from German by Eric Mosbacher and David Porter. New York: Viking, 1951. [better to put this ref. in Houterman's Further reading]

Becker, George Ferdinand. "Age of a Cooling Globe in Which the Initial Temperature Increases Directly as the Distance from the Surface." *Science* 27 (1908): 227–33.

———. "Relations of Radioactivity to Cosmogony and Geology." *Bull. Geol. Soc. Amer.* 19 (1908): 113–46.

Becquerel, Antoine, and Edmond Becquerel. *Eléments de physique terrestre et de météorologie.* Paris: Firmin Didot, 1847.

Becquerel, Henri. "Émissions de radiations nouvelles par l'uranium métallique." *C. R. Acad. Sci.* 122 (1896): 1086–88.

———. "Note sur le rayonnement de l'uranium." *C. R. Acad. Sci.* 130 (1900): 1583–85.

———. "Sur la radioactivité de l'uranium." *C. R. Acad. Sci.* 133 (1901): 977–80.

———. "Sur les radiations émises par phosphorescence." *C. R. Acad. Sci.* 122 (1896): 420–21.

———. "Sur les radiations invisibles émises par les corps phosphorescents." *C. R. Acad. Sci.* 122 (1896): 501–3.

———. "Sur quelques propriétés nouvelles des radiations invisibles émises par divers corps phosphorescents." *C. R. Acad. Sci.* 122 (1896): 559–64.

Beer, Gavin de. *Charles Darwin: Evolution by Natural Selection.* New York: Doubleday, 1964.

———. "The Volcanoes of Auvergne." *Ann. Sci.* 18 (1963): 49–61.

Bensaude-Vincent, Bernadette. *Lavoisier: Mémoires d'une révolution.* Paris: Flammarion, 1993.

Bethe, Hans. "Energy Production in Stars." *Phys. Rev.* 55 (1939): 434–56.

———. "Recent Evidence on the Nuclear Reactions in the Carbon Cycle." *Ap. J.* 98 (1940): 118–21.

Bischof, Gustav. *Lehrbuch der chemischen und physikalischen Geologie.* 3 vols. Bonn: Adolph Marcus, 1st ed., 1850, 2nd ed. 1863.

Blondlot, René. *Rayons N.* Paris, 1904. Translated by J. Garcin as *"N" Rays* (London: Longmans, 1905).

Blumenbach, Johann Friedrich. *Handbuch der Natur-Geschichte.* Göttingen, Germany: Dieterich, 1779–1830 (12 eds.).

Boltwood, Bertram. "On the Ultimate Disintegration Products of the Radio-Active Elements." *Am. J. Sci.* 20 (1905): 253–67.

———. "On the Ultimate Disintegration Products of the Radio-Active Elements. Part II. The Disintegration Products of Uranium." *Am. J. Sci.* 23 (1907): 77–88.

Bottéro, Jean. *Lorsque les dieux faisaient l'homme.* Paris: Gallimard, 1989.

———. *Naissance de Dieu: La Bible et l'historien.* 2nd ed. Paris: Gallimard, 1992. Published in English as *The Birth of God: The Bible and the Historian,* trans. K. W. Bolle (University Park: Pennsylvania State University Press, 2000).

Boubée, Nérée. *Cours abrégé de géologie ou développement du tableau de l'état du globe à ses différents ages.* 4th ed. Paris: Bulletin d'Histoire Naturelle de Paris, 1834.

Bowler, Peter J. "Hugo De Vries and Thomas Hunt Morgan: The Mutation Theory and the Spirit of Darwinism." *Ann. Sci.* 35 (1978): 55–73.

Brahe, Tycho. *De Mundi Ætheri recentioribus Phænomenis* and *De Cometa anni 1577.* In Brahe, *Opera omnia,* 4:1–396.

———. *De nova Stella 1573.* In Brahe, *Opera omnia,* 1:1–142.

———. *Tychonis Brahe Dani opera omnia.* Edited by J. L. E. Dreyer. 15 vols. Copenhagen: Gyldendalske Bohandel, 1913. Reprint, Amsterdam: Swets and Zeitlinger, 1972.

Brongniart, Alexandre. "Sur les caractères zoologiques des formations, avec l'application de ces caractères à la détermination de quelques terrains de craie." *Ann. Mines* 6 (1821): 537–72.

Brothers of Purity. Encyclopedia. Arabic ed. in *Die Abhandlungen der Ichwân Es-Safâ,* edited by F. Dieterici. Vol. 9 of *Philosophie der Araber im IX. und X. Jahrundert N. Chr. aus der Theologie des Aristoteles, desn Abhandlungen Alfarabis und den Schriften der Lautern Brüder.* Leipzig: J. C. Hinrichs'sche Buchhandlung, 1886.

Brown, Harrison. "An Experimental Method for the Estimation of the Age of the Elements." *Phys. Rev.* 72 (1947): 348.

Buckland, William. The Discovery of Coprolites, or Fossil Fæces, in the Lias at Lyme Regis, and in Other Formations." *Trans. Geol. Soc. London* 3 (1829): 223–36.

———. *Geology and Mineralogy Considered with Reference to Natural Theology.* 2 vols. The sixth *Bridgewater Treatise on the Power, Wisdom, and Goodness of God as Manifested in Creation.* London: W. Pickering, 1837.

———. "Notice on the Megalosaurus or Great Fossil Lizard of Stonesfield." *Trans. Geol. Soc. London* 1 (1822): 390–96.

Buffon, Georges-Louis Leclerc. *Correspondance inédite de Buffon.* Edited by H. Nadault de Buffon. Paris: Hachette, 1860.

———. *Époques de la nature.* Paris, 1779. Critical ed., with an introduction and notes by J. Roger, *Mém. Mus. Nat. Hist. Nat.* 10 (1962): v–cli, 1–316; reprint, Paris, 1988. Page references are to the 1962 edition.

————. *Expériences sur la pesanteur du feu, et sur la durée de l'incandescence.* Supplement to *Histoire naturelle.* Paris: Imprimerie royale, 1776.

————. *Expériences sur le progrès de la chaleur dans les corps.* Supplement to *Histoire naturelle.* Paris: Imprimerie royale, 1774.

————. *Histoire naturelle, générale et particulière.* Paris: Imprimerie royale, 1749–67. Reprinted in *Œuvres complètes de Buffon,* edited by J. L. Lanessan, 14 vols. (Paris: A. Le Vasseur, 1884–85).

————. *Introduction à l'histoire des minéraux. Partie expérimentale.* Paris: Imprimerie Royale, 1778.

————. *Œuvres philosophiques.* Edited by J. Piveteau. Paris: PUF, 1954.

————. *Premier discours: De la manière d'étudier et de traiter l'Histoire naturelle.—Second discours: Histoire et théorie de la Terre. Preuves de la théorie de la Terre.* Paris: Imprimerie royale, 1749.

————. *Recherches sur le refroidissement de la Terre et des Planètes et Fondemens des recherches précédentes sur la température des Planètes.* Supplement to *Histoire naturelle.* Paris: Imprimerie Royale, 1776.

————. *Suite des expériences sur le progrès de la chaleur dans les différentes substances minérales.* Supplement to *Histoire naturelle.* Paris: Imprimerie royale, 1774.

Burchfield, Joe D. "Darwin and the Dilemma of Geological Time." *Isis* 65 (1974): 301–21.

————. *Lord Kelvin and the Age of the Earth.* 2nd ed., with a new postface. Chicago: University of Chicago Press, 1990.

Buridan, Jean. "The Compatibility of the Earth's Diurnal Rotation with Astronomical Phenomena." Translated by M. Clagett. In *A Source Book in Medieval Science,* ed. E. S. Grant, 500–503. Cambridge, MA: Harvard University Press, 1974.

————. "The Impetus Theory of Projectile Motion." Translated by M. Clagett. In *A Source Book in Medieval Science,* ed. E. S. Grant, 275–84. Cambridge, MA: Harvard University Press, 1974.

————. "On the motion of the earth's center of gravity and the formation of mountains." Translated by E. S. Grant. In *A Source Book in Medieval Science,* ed. E. S. Grant, 621–24. Cambridge, MA: Harvard University Press, 1974.

————. *Quaestiones super libris quattuor De Cælo et Mundo.* Cambridge: Mediaeval Academy of America, 1942.

Burkhardt, Richard W., Jr. *The Spirit of System: Lamarck and Evolutionary Biology.* Cambridge, MA: Harvard University Press, 1977).

Burnet, Thomas. *The Sacred Theory of the Earth, Containing an Account of the Original of the Earth, and of all the general changes which it hath already undergone, or is to undergo till the consummation of all things. The first two books concerning the Deluge and concerning the Paradise.* 2nd ed. London, 1691. Reprint, Carbondale: Southern Illinois University Press, 1965.

Campbell, Thomas. *Life and Letters of Thomas Campbell.* Edited by W. Beattie. 3 vols. London: Hall, Virtue, 1850.

Caneva, Kenneth L. *Robert Mayer and the Conservation of Energy.* Princeton, NJ: Princeton University Press, 1993.

Cardano, Girolamo. *De subtilitate.* Edited by E. Nenci. Milan: F. Anfeli, 2004. Published in French as *Les livres de Hieronime Cardanus medecin milannois, intitules de la Subtilité, et subtiles inuentions, ensemble les causes occultes, et raisons d'icelles,* translated by Richard le Blanc (Paris: Guillaume le Noir, 1556).

———. *De vita propria liber.* Published in English as *The Book of My Life,* translated by J. Stoner. London: J. M. Dent, 1931.

Cardwell, Donald S. L. *James Joule: A Biography.* Manchester, Eng.: Manchester University Press, 1989.

Carnot, Sadi. *Réflexions sur la puissance motrice du feu et sur les machines propres à développer cette puissance.* Paris: Bachelier, 1824. Published in English as *Reflections on the Motive Power of Fire,* translated by R. Fox (New York: Manchester University Press, 1986).

Carton, Raoul. *L'expérience physique chez Roger Bacon: Contribution à l'étude de la méthode et de la science expérimentales au XIIIᵉ siècle.* Paris: Vrin, 1924.

Caspar, Max. *Johannes Kepler.* Munich: V. Kohlhammer, 1948. Published in English as *Johannes Kepler,* translated by C. Doris Hellman (New York: Abelard-Schuman, 1959).

Cassini, Dominique. "Observations astronomiques faites en divers endroits du royaume de France." *Mém. Acad. Roy. Sci.* 2 (1684): 349–75.

Cassini, Jacques. *De la grandeur et de la figure de la Terre.* Paris: Imprimerie Royale, 1720.

Cavendish, Henry. "Experiments with Air." *Phil. Trans. Roy. Soc. London* 75 (1785): 372–84.

Celsus. *On the True Doctrine: A Discourse against the Christians.* Translated by R. J. Hoffmann. Oxford: Oxford University Press, 1987.

Chamberlin, Thomas Chrowder. "The Age of the Earth from the Geological Viewpoint." *Proc. Amer. Phil. Soc.* 61 (1922): 247–71.

———. "Diastrophism and the Formative Processes. XIII. The Bearings of the Size and Rate of Infall of Planetisimals on the Molten or Solid State of the Earth." *J. Geol.* 28 (1920): 665–701.

———. "Fundamental Problems of Geology." *Carnegie Inst. Washingon Yearbook* 3 (1904): 195–258.

———. "A Group of Hypotheses Bearing on Climatic Changes." *J. Geol.* 5 (1897): 653–83.

———. "Lord Kelvin's Address on the Age of the Earth as an Abode Fitted for Life, I and II." *Science* 9 (1899): 889–901; 10 (1899): 11–18.

———. *The Origin of the Earth.* Chicago: University of Chicago Press, 1916.

Chamberlin, Thomas Chrowder, and R. D. Salisbury. *Geology.* 3 vols. New York: H. Holt, 1904–6.

Champollion-Figeac, Aimé-Louis. *J. J. Fourier et Napoléon: L'égypte et les cent jours.* Paris: Firmin Didot, 1844.

Chandrasekhar, Subrahmanyan. *Eddington: The Most Distinguished Astrophysicist of His Time*. Cambridge: Cambridge University Press, 1983.

Charpentier-Morize, Micheline. *Jean Perrin: Savant et homme politique, 1870–1942*. Paris: Belin, 1997.

Chateaubriand, François-Auguste-René de. *Le génie du Christianisme*. 4 vols. Paris: Migneret, 1802. Published in English as *The Genius of Christianity*, translated by C. I. White (New York: Howard Fertig, 1976).

Châtelet, Gabrielle-Émilie du. *Dissertation sur la nature et la propagation du feu*. Paris: Prault, 1744.

Church, Thomas M. "Clair C. Patterson (1922–1995)." *EOS* 77 (1996): 306.

Clarke, Frank Wigglesworth. "William Francis Hillebrand 1853–1925." *Biog. Mem. Nat. Acad. Sci.* 12 (1929): 42–70.

Clarke, John Mason. "The Age of the Earth from the Paleontological Viewpoint." *Proc. Am. Phil. Soc.* 61 (1922): 272–82.

Clausius, Rudolph. "On the Second Fundamental Theorem of the Mechanical Theory of Heat." *Phil. Mag.* 35 (1868): 405–19.

———. "Ueber verschiedene für die Anwendung bequeme Formen der Hauptgleichungen der mechanischen Wärmtheorie." *Ann. Phys. Chem.* 125 (1865): 400.

Cohen, I. Bernard. "Roemer and the First Determination of the Velocity of Light" (1676). *Isis* (1939): 327–79.

Cohen, I. Bernard, and Richard Westfall, eds. *Newton*. New York: Norton, 1995.

Colding, Ludvig August. "On the History of the Principle of the Conservation of Energy." *Phil. Mag.* 27 (1864): 56–64.

———. "On the Universal Powers of Nature and Their Mutual Dependence." *Phil. Mag.* 42 (1871): 1–21.

Coleman, William. *Georges Cuvier, Zoologist: A Study in the History of Evolution Theory*. Cambridge, MA: Harvard University Press, 1964.

Comte, Auguste. *Cours de philosophie positive*. Vol. 2, lecture 19. Paris: Bachelier, 1835. Condensed version published in English as *The Positive Philosophy of Auguste Comte*, translated by H. Martineau (London: George Bell and Sons, 1896).

Condorcet. "Éloge de M. Guettard." *Hist. Acad. Roy. Sci.* (1786): 47–62.

Conti, Antonio Schinella. *Lettre de M. au sujet d'un petit Ecrit intitulé: Réponse aux Observations sur la Chronologie de Mr. Newton*. Paris: Noel Pissot, 1726.

Copernicus, Nicolaus. *Commentariolus*. Translated by E. Rosen. *Isis* 3 (1937): 123–41.

———. *De revolutionibus orbium clestium, libri vi*. Nuremberg: Apud Ioh. Petreium, 1543. Published in English as *On the Revolutions of Heavenly Stars*, translated by C. G. Wallis (Amherst, NY: Prometheus Books, 1995).

Cordier, Louis. "Essai sur la température de l'intérieur de la Terre." *Mém. Acad. Roy. Sci.* 7 (1827): 473–555. Reprint, *Mém. Mus. Nat. Hist. Nat.* 15 (1827): 161–244.

———. "Mémoire sur les substances minérales dites *en masse,* qui entrent dans la composition des roches volcaniques de tous les âges." *J. Phys.* 83 (1816): 135–63, 285–307, 352–87.

Cousin, Victor. "Discours de réception à l'Académie française et notes addition-nelles." In Cousin, Œuvres, 4:1–101. Paris, 1849.

Craig, Gordon Y., Donald B. McIntyre, and Charles D. Waterston, *James Hutton's Theory of the Earth: The Lost Drawings*. Edinburgh: Scottish Academic Press, 1978.

Craig, Harmon. "Introduction of Alfred O.C. Nier for the V.M. Goldschmidt Award 1984." *Geochim. Cosmochim. Acta* 49 (1985): 1161–74.

Crawford, Elisabeth. *Arrhenius: From Ionic Theory to the Greenhouse Effect*. Canton, MA: Watson, 1996.

Croll, James. *Climate and Time in Their Geological Relations*. New York: Appleton, 1893.

———. *Discussions on Climate and Cosmology*. Edinburgh: Adam and Charles Black, 1885.

———. "On the Change in the Obliquity of the Ecliptic, Its Influence on the Climate of the Polar Regions and on the Level of the Sea." *Phil. Mag.* 33 (1867): 426–45.

———. "On the Excentricity of the Earth's Orbit." *Phil. Mag.* 31 (1866): 26–28.

———. "On the Influence of the Tidal Wave on the Motion of the Moon." *Phil. Mag.* 33 (1867): 118–31.

———. "On the Physical Cause of the Change of Climate during Geological Epochs."*Phil. Mag.* 28 (1864): 121–37.

———. "On the Physical Causes of the Submergence and Emergence of the Land during the Glacial Epoch." *Phil. Mag.* 31 (1866): 301–6.

Crombie, Alistair Cameron. *The History of Science from Augustine to Galileo*. New York: Dover, 1995.

Crookes, William. "Antoine Henri Becquerel, 1852–1908." *Proc. Roy. Soc. London* A83 (1909): xx–xxiii.

———. "Chemical Science: Opening Address." *Nature* 34 (1886): 423–32.

———. *Crookes and the Spirit World*. Compiled by R. G. Medhurst. Edited by M. R. Barrington. London: Souvenir, 1972.

———. "Radio-Activity of Uranium." *Proc. Roy. Soc.* 66 (1900): 409–23.

———. "Sur la matière radiante." *Ann. Chim. Phys.* 19 (1880): 195–231.

Curie, Eve. *Madame Curie*. Paris: Gallimard, 1938. Published in English as *Madame Curie: A Biography*, translated by V. Sheean (New York: Doubleday, 1939).

Curie, Marie. See also Sklodowska Curie.

———. *Œuvres complètes de Marie Sklodowska Curie*. Edited by Irène Joliot-Curie. Warsaw: Panstwowe Wydawnictwo Naukowe, 1954.

———. *Pierre Curie*. New York: Macmillan, 1923.

———. *Traité de Radioactivité*. 2 vols. Paris: Gauthier-Villars, 1910.

Curie, Maurice. "Sur les écarts de poids atomique obtenus avec le plomb provenant de divers minéraux." *C. R. Acad. Sci.* 158 (1914): 1676–79.

Curie, Pierre. *Les prix Nobel en 1905*. Stockholm: Imprimerie Royale.

———. *Œuvres de Pierre Curie*. With a foreword by Marie Curie. Paris: Gauthier-Villars, 1908.

———. "Sur la mesure absolue du temps" (1902). In P. Curie, *Œuvres de Pierre Curie*, 439.

———. "Sur la radioactivité induite et sur l'émanation du radium." *C. R. Acad. Sci.* 136 (1903): 223–26.

Curie, Pierre, and Marie Curie. "Sur une substance nouvelle radio-active." *C. R. Acad. Sci.* 127 (1898): 175–78.

Curie, Pierre, Marie Curie, and Gustave Bémont. "Sur une nouvelle substance fortement radio-active contenue dans la pechblende." *C. R. Acad. Sci.* 127 (1898): 1215–17.

Curie, Pierre, and Albert Laborde. "Sur la chaleur dégagée spontanément par les sels de radium." *C. R. Acad. Sci.* 136 (1903): 673–75.

Cuvier, Georges. *Discours préliminaire des recherches sur les ossemens fossiles des quadrupèdes: Où l'on rétablit les caractères de plusieurs espèces d'animaux que les révolutions du globe paroissent avoir détruites.* Paris: Deterville, 1812. Published in English as *A Discourse on the Revolutions of the Surface of the Globe and the Changes Thereby Produced in the Animal Kingdom* (Philadelphia: Carey and Lea, 1831).

———. "Éloge historique de Desmarest." In Cuvier, *Recueil des éloges*, 2:137–62.

———. "Éloge historique de M. de Lamarck." In Cuvier, *Recueil des éloges*, 3:179–210.

———. "Éloge historique de Werner." In Cuvier, *Recueil des éloges*, 2:113–34.

———. "Mémoires sur les espèces d'éléphans tant vivantes que fossiles." *Magasin Encyclopédique* 3 (1796): 440–45.

———. *Recueil des éloges historiques lus dans les séances publiques de l'Institut de France.* 3 vols. Paris: Firmin-Didot, 1861.

Cuvier, Georges, and Alexandre Brongniart. "Essai sur la géographie minéralogique des environs de Paris." *Ann. Muséum. Hist. Nat.* 11 (1808): 293–326.

Daniélou, Jean. *Origène.* Paris: Table Ronde, 1948.

Darboux, Gaston. "Discours prononcé aux funérailles de M. Henri Becquerel." *C. R. Acad. Sci.* 147 (1908): 143–51.

Darwin, Charles Robert. "Extract from an Unpublished Work on Species, Consisting of a Portion of a Chapter Entitled, 'On the Variation of Organic Beings in a State of Nature; on the Natural Means of Selection; on the Comparison of Domestic Races and True Species.'" *J. Linnean Soc.* 3 (1859): 46–50; reprinted in Sarton, "Discovery."

———. *Geological Observations on the Volcanic Islands and Parts of South America Visited during the Voyage of H. M. S. "Beagle."* London: Smith, Elder, 1844.

———. *On the Origin of Species by Means of Natural Selection; or The Preservation of Favoured Races in the Struggle for Life.* London: J. Murray, 1859–72 (6 eds.).

Darwin, Francis. *Life and Letters of Charles Darwin.* London: John Murray, 1887. Reprint, New York: Basic Books, 1959.

Darwin, George Howard. "On the Precession of a Viscous Spheroid, and on the Remote History of the Earth." *Phil. Trans. Roy. Soc. London* 170 (1879): 447–530.

———. "Problems Connected with the Tides of a Viscous Spheroid." *Phil. Trans. Roy. Soc. London* 170 (1879): 540–93.

———. "Radioactivity and the Age of the Sun." *Nature* 68 (1903): 496.

———. *Scientific Papers*. Edited by Francis Darwin. 5 vols. Cambridge: Cambridge University Press, 1907–16.

Daubrée, Auguste. *Études et expériences synthétiques sur le métamorphisme et sur la formation des roches cristallines*. Paris: Imprimerie Impériale, 1859.

Davis, Edward Arthur, and Isabel J. Falconer. *J. J. Thomson and the Discovery of the Electron*. London: Taylor and Francis, 1997.

Davis, J. L., "The Research School of Marie Curie in the Paris Faculty, 1907–14." *Ann. Sci.* 52 (1995): 321–55.

Davison, Charles. "Eminent Living Geologists: Rev. Osmond Fisher, M.A., F.G.S." *Geol. Mag.* 7 (1900): 49–54.

Dean, Dennis R. *James Hutton and the History of Geology*. Ithaca, NY: Cornell University Press, 1992.

Debierne, André. "Sur une nouvelle matière radio-active." *C. R. Acad. Sci.* 129 (1899): 593–95.

De Laeter, John, and Mark D. Kurz. "Alfred Nier and the Sector Field Mass Spectrometer." *J. Mass. Spectrosc.* 41 (2006): 447–54.

De Luc, Jean-André. "Dix-huitième lettre à M. Delamétherie sur les agates, les couches calcaires et une classe de couches d'argile." *Observ. Phys. Hist. Nat. Arts* 39 (1791): 453–64.

———. *Traité élémentaire de géologie* (Paris: Courcier, 1809).

Dempster, Arthur Jeffreys. "Isotopic Constitution of Uranium." *Nature* 136 (1935): 180.

Descartes, René duPerron. *Lettres de Mr Descartes, où son traittées plusieurs belles questions touchant la morale, la physique, la médecine, et les mathématiques*. 3 vols. Paris: C. Angot, 1667.

———. *Œuvres de Descartes*, 11 vols., edited by V. Cousin. Paris: F. G. Levrault, 1824–26.

———. *Principia philosophiæ*. Amsterdam: Elzevier, 1644. Published in French as *Principes de la philosophie*, translated by Abbot Picot (Paris: H. Le Gras, 1647); 4th ed. published by C. Adam and P. Tannery; reprint, Paris: Vrin, 1971; published in English as *Principles of Philosophy*, translated by V. R. Miller and R. P. Miller (Dordrecht, Netherlands: D. Reidel, 1988).

Deshayes, Gérard-Paul. *Description de coquilles caractéristiques des terrains*. Paris: G. Levrault, 1831.

———. *Description des coquilles fossiles des environs de Paris*. 3 vols. Paris: Published by author, 1824–37.

Desmarest, Nicolas. "Mémoire sur l'origine et la nature du basalte à grandes colonnes polygones, déterminées par l'histoire naturelle de cette pierre, observée en Auvergne." *Mém. Acad. Roy. Sci.* (1771): 705–75.

———. "Notice sur plusieurs points importans de l'histoire naturelle de la terre." In *Géographie Physique*, by Desmarest, 1:409–31. Paris: Agasse, 1794–95.

Desmond, Adrian. *Huxley: Evolution's High Priest.* London: M. Joseph, 1997.

———. *Huxley: The Devil's Disciple.* London: M. Joseph, 1994.

Desmond, Adrian, and James Moore. *Darwin.* New York: Warner Books, 1991.

Des Vignoles, Alphonse. *Chronologie de l'histoire sainte et des histoires étrangères qui la concernent depuis la sortie d'Egypte juqu'à la captivité de Babylone.* 2 vols. Berlin: A. Haude, 1738.

DeVorkin, David H. *Henry Norris Russell: Dean of American Astronomers.* Princeton, NJ: Princeton University Press, 2000.

De Vries, Hugo. *Die Mutationstheorie, Versuche und Beobachtungen über die Entstehung von Arten im Pflanzereich.* Leipzig: Veit, 1901–3.

———. "The Evidence of Evolution." *Science* 20 (1904): 395–401.

———. *Species and Varieties: Their Origin by Mutation: Lectures Delivered to the University of California.* 3rd ed. Chicago: Open Court, 1912.

Dézallier d'Argenville, Antoine-Joseph. *Histoire naturelle éclaircie dans une de ses parties pricipales, la conchyologie, qui traite des coquillages de mer, de rivière et de terre; ouvrage dans lequel on trouve une nouvelle méthode Latine et Françoise de les diviser: Augmentée de la zoomorphose, ou représentation des animaux à coquille, avec leurs explications.* Paris: De Bure l'Aîné, 1757.

Dhombres, Jean, and Jean-Bernard Robert. *Fourier: Créateur de la physique mathématique.* Paris: Belin, 1998.

Dibner, Bern. *Allessandro Volta and the Electric Battery.* New York: F. Watts, 1964.

———. *Galvani-Volta.* Norwalk, CT: Burndy Library, 1952.

———. *Œrsted and the Discovery of Electromagnetism.* New York: Blaisdell, 1963.

Diderot, Denis, and Jean Le Rond d'Alembert, eds. *Encyclopédie, ou dictionnaire raisonné des sciences, des arts et des métiers.* 28 vols. Paris: Briasson, 1751–72.

Diodorus the Sicilian. *Historical Library.* Translated by G. Booth. London: Davis, 1814.

Donovan, Arthur. *Antoine Lavoisier: Science, Administration, and Revolution.* Oxford: Blackwell, 1993.

Douglas, Allie Vibert. *The Life of Arthur Stanley Eddington.* London: Thomas Nelson, 1956.

Drake, Ellen T. *Restless Genius: Robert Hooke and His Earthly Thoughts.* Oxford: Oxford University Press, 1996.

Drake, Stillman. *Galileo at Work: His Scientific Biography.* Chicago: University of Chicago Press, 1978.

Du Fay, Charles-François de Cisternai. "Quatrième mémoire sur l'électricité." *Mém. Acad. Roy. Sci.* (1733): 457–76.

Duhem, Pierre. "Albert le Grand." In *Système du monde,* 5:412–65.

———. *Études sur Léonard de Vinci, ceux qu'il a lus et ceux qui l'ont lu.* 3 vols. Paris: A. Hermann, 1906.

———. "La cosmologie de Platon." In *Système du monde,* 1:28–101.

———. "La cosmologie des pères de l'Église." In *Système du monde,* 2:394–494.

———. "La géologie de Buridan et l'université de Paris. I. Un adversaire: Nicole Oresme." In *Système du monde,* 9:306–9.

———. "La géologie de Jean Buridan." In *Système du monde*, 9:293–306.

———. "La géologie des Arabes." In *Système du monde*, 9:257–66.

———. "La physique d'Aristote." In *Système du monde*, 1:130–241.

———. "La première tentative concordiste entre le récit de la Genèse et la physique: Jean Philopon." In *Système du monde*, 2:494–501.

———. "Le nombre infini actuel et l'immortalité de l'âme." In *Système du monde*, 7:4.

———. *Le système du monde: Histoire des doctrines cosmologiques de Platon à Copernic*. 10 vols. Paris: Hermann; vols. 1–7, 1913–1919, reprinted in 1958; vols. 8–10, 1956–58.

———. "Nicole Oresme expose l'hypothèse de la rotation terrestre." In *Système du monde*, 9:329–44.

Dunham, Kingsley Charles. "Arthur Holmes 1890–1965." *Biogr. Mem. Fellows Roy. Soc.* 12 (1966): 291–310.

Dupuy, Charles. "Discours de M. Charles Dupuy, ministre de l'instruction publique." In *Jubilé de M. Pasteur*, 7–11. Paris: Gauthier-Villars, 1893.

Easton, Stewart C. *Roger Bacon and His Search for a Universal Science*. New York: Columbia University Press, 1952.

Eddington, Arthur Stanley. "Further Notes on the Radiative Equilibrium of the Stars." *Mon. Not. Roy. Astron. Soc.* 77 (1917): 596–612.

———. *The Internal Constitution of Stars*. Cambridge: University Press, 1926, 1930.

———. "The Internal Constitution of the Stars." *Observatory* 43 (1920): 341–58.

———. "On the Relation between the Masses and the Luminosities of the Stars." *Mon. Not. Roy. Astron. Soc.* 84 (1924): 308–32.

Eliade, Mircea. *Le mythe de l'éternel retour*. Paris: Gallimard, 1949. Published in English as *The Myth of the Eternal Return*, translated by W. R. Trask (Princeton, NJ: Princeton University Press, 1954).

Elie de Beaumont, Léonce. "Note sur le rapport qui existe entre le refroidissement progressif de la masse du globe terrestre et celui de sa surface." *C. R. Acad. Sci.* 19 (1844): 1327–31.

———. *Notice sur les systèmes de montagnes*. 3 vols. Paris: P. Bertrand, 1852.

———. "Researches on Some of the Revolutions Which Have Taken Place on the Surface of the Globe; Presenting Various Examples of the Coincidence between the Elevation of Beds in Certain Systems of Mountains, and the Sudden Changes Which Have Produced the Lines of Demarcation Observable in Certain Stages of the Sedimentary Deposits." *Phil. Mag.* 10 (1831): 241–64.

Ellenberger, François. *Histoire de la géologie*. 2 vols. Paris: Lavoisier, 1988–94. Published in English as *History of Geology*, translated by R. K. Kaula and M. Carozzi (Rotterdam, Netherlands: A. A. Balkema, 1996, 1999).

———. "Le miracle arabe." In Ellenberger, *Histoire*, 1:77–81.

———. "Précisions nouvelles sur la découverte des volcans de France: Guettard, ses prédécesseurs, ses émules clermontois." *Histoire et Nature* 12–13 (1978): 3–42.

Ellenberger, François, and Gabriel Gohau. "A l'aurore de la stratigraphie paléontologique: Jean-André De Luc, son influence sur Cuvier." *Rev. Hist. Sci.* 39 (1981): 217–57.

Elrington, Charles Richard. In Ussher, *Whole Works*, vol. 1.

Elster, Johann, and Hans Geitel. "Recherches sur la conductibilité et la radioactivité de l'atmosphère." In *Les quantités élémentaires d'électricité, ions, électrons, corpuscules*, edited by H. Abraham and P. Langevin., 185–221. Paris: Gauthier-Villars, 1905.

Emmons, Samuel Franklin. "Clarence King. 1842–1901." *Biog. Mem. Nat. Acad. Sci.* 6 (1906): 25–55.

Epstein, Samuel. "Introduction of Clair C. Patterson for the V.M. Goldschmidt Medal 1980." *Geochim. Cosmochim. Acta* 45 (1981): 1383–97.

Eve, Arthur Stewart. *Rutherford: Being the Life and Letters of the Rt Hon. Lord Rutherford, O. M.* New York: Macmillan, 1939.

Fajans, Kasimir. "Die Stellung der Radioelemente im Periodischen System." *Physik. Zeits.* 14 (1913): 136–42.

Faller, Adolf. "Niels Stensen und der Cartesianismus." In Scherz, *Nicolaus Steno*, 140–66.

Faraday, Michael. *Histoire d'une chandelle*. With a biographical notice by H. Sainte-Claire Deville. Translated by W. Hughes. Paris: Hetzel, 1865. Originally published in English as *The Chemical History of a Candle*.

Fermi, Enrico. "Possible Production of Elements of Atomic Number Higher than 92." *Nature* 133 (1934): 898–99.

Fermi, Laura. *Atoms in the Family*. Chicago: University of Chicago Press, 1954.

Figuier, Louis. *La terre avant le Déluge*. Paris: Hachette, 1862.

Fisher, Osmond. "On the Depression of Ice-Loaded Lands." *Geol. Mag.* 9 (1882): 526–27.

———. *Physics of the Earth's Crust*. 2nd ed. London: Macmillan, 1889.

———. "Review of an Estimate of the Geological Age of the Earth by J. Joly." *Geol. Mag.* 7 (1900): 124–33.

Flammarion, Camille. *Astronomie populaire*. Paris: Marpon and Flammarion, 1880. Published in English as *Popular Astronomy*, translated by J. E. Gore (New York: D. Appleton, 1907).

———. *La Planète Mars et ses conditions d'habitabilité*. Paris: Gauthier-Villars, 1892.

———. *Les forces naturelles inconnues*. Paris: Flammarion, 1907.

Fleck, Alexander. "Frederick Soddy." *Biogr. Mem. Fellows Roy. Soc.* 3 (1957): 203–16.

Flourens, Pierre. "Éloge historique de G. Cuvier." In Cuvier, *Recueil des éloges*, 1:i–liii.

Fölsing, Albrecht. *Heinrich Hertz: Eine biographie*. Hamburg: Hoffmann and Campe, 1997.

———. *Wilhelm Conrad Röntgen: Aufbruch ins Innere der Materie*. Munich: C. Hanser, 1995.

Fontenelle, Bernard le Bovier de. "Éloge de M. Cassini." In Fontenelle, *Éloges*, 2:296–334.

———. "Éloge de M. Lémery." In Fontenelle, *Éloges*, 1:356–71.

———. "Éloge de Mr. G.G. Leibniz." In Fontenelle, *Éloges*, 1:447–506.

———. *Éloges des Académiciens de l'Académie royale des sciences, morts depuis l'année 1699*. 2 vols. New ed. Paris: Libraires associés, 1766.

———. *Entretiens sur la pluralité des mondes*. Paris: C. Blageart, 1686. Published in English as *Conversations on the Plurality of Worlds*, translated by H. A. Hargreaves (Berkeley: University of California Press, 1990).

———. "Sur les feux souterrains, les tremblemens de terre, le tonnerre, etc. expliqués chimiquement." *Hist. Acad. Roy. Sci.* (1700): 51–52.

Fourier, Joseph. "Éloge historique de M. le marquis de La Place." In Laplace, *Système du monde*, 6th ed.

———. "Extrait d'un mémoire sur le refroidissement séculaire du globe terrestre."*Ann. Chim. Phys.* 13 (1820): 418–37.

———. *Œuvres*. 2 vols. Paris: Gauthier-Villars, 1888–90.

———. Preface to *L'exposition du Système du monde*, by Laplace. 6th ed. (1836).

———. "Remarques générales sur les températures du globe terrestre et des espaces planétaires." *Ann. Chim. Phys.* 27 (1824): 136–67. Reprinted in *Mém. Acad. Roy. Sci.* 7 (1824): 568–604.

———. *Théorie analytique de la chaleur*. Paris: Firmin Didot, 1822. Published in English as *The Analytical Theory of Heat*, translated by A. Freeman (reprint, New York: Dover, 1955).

"Fourier's Analytical Theory of Heat." *Nature* 18 (1878): 192.

Fournier d'Albe, Edmund Edward. *The Life of Sir William Crookes*. London: T. F. Unwin, 1923.

Franklin, Benjamin. *Benjamin Franklin's Experiments: A New Edition of Franklin's Experiments and Observations on Electricity*. Edited by I. Bernard Cohen. Cambridge, MA: Harvard University Press, 1941.

———. *Œuvres de M. Franklin, docteur ès loix*. Translated into French by M. Barbeu Dubourg. Paris: Quillau l'aîné, 1773.

Fréret, Nicolas. *Défense de la chronologie fondée sur les monumens de l'histoire ancienne, contre le systême chronologique de M. Newton*. Paris: Durand, 1758. Published after Fréret's death by J. P. de Bougainville.

———. *Observations sur la chronologie de M. Newton*. Paris: Guillaume Cavelier, 1725.

Freund, Philip. *Myths of Creation*. New York: Washington Square Press, 1965.

Füchsel, Georg Christian. *Entwurf zu der ältesten Erd und Menschengeschichte, nebst einem Versuch, den Ursprung der Sprache zu finden*. Frankfurt, 1773.

———. "Historia terae et maris, ex historia Thuringae, per montium descriptionem." In *Actorum academae electoralis moguntiae scientiarum utilium quœ erfordae est*, 2:44–254. Erfurt, Germany: Ioannem Fridericum Wreerum, 1761.

Galilei, Galileo. *Le opere di Galileo Galilei, prima edizione completa, condotta sugli autentici manoscritti palatini*. 20 vols. Florence, Italy: G. Barbèra, 1968.

————. *Sidereus nuncius, magna, longeque admirabilia spectacula pandens, sus-piciendaque proponens unicuique, præsertim vero.* Venice: T. Baglionum, 1610. Published in English as *The Sidereal Messenger*, translated by A. van Helden (Chicago: University of Chicago Press, 1989).

Galvani, Luigi. *De viribus electricitatis in motu musculari. Commentarius.* Bologna: Instituti scientiarum, 1791.

Gaukroger, Stephen. *Descartes: An Intellectual Biography.* Oxford: Clarendon, 1995.

Gayon, Jean, ed. *Buffon 88: Actes du colloque international pour le bicentenaire.* Paris: Vrin, 1992.

Geikie, Archibald. *The Founders of Geology.* London: Macmillan, 1905. Reprint, New York: Dover, 1960.

————. "Geological Change, and Time." *Ann. Rept. Smithsonian Inst.* (1892): 111–31.

————. "On Modern Denudation." *Trans. Geol. Soc. Glasgow* 3 (1868–69): 153–90.

————. *Textbook of Geology.* 2 vols. London, 1903.

Gensanne, Antoine de [Genssane]. *Histoire naturelle de la province du Languedoc, partie minéralogique et géoponique.* 5 vols. Montpellier: Rigaud, Pons, and Cie, 1776–79.

George, Wilma. *Biologist Philosopher: A Study of the Life and Writings of Alfred Russel Wallace.* New York: Abelard and Schuman, 1964.

Gerling, Erich Carlovich. "Age of the Earth according to Radioactivity Data." *Doklady Akad. Nauk.* 34 (1942): 259–61. Published in English by C. T. Harper in *Benchmark Papers in Geology* (Stroudsburg, PA: Dowden, Hutchinson, and Ross, 1973), 5:121–23.

————. *Doklad. Akad. Nauk. URSS* 61 (1948): 297–300.

Gesner, Konrad. *De rerum fossilium, lapidum et Gemmarum maximè, figuris & simil-itudinibus Liber: non solùm Medicis, sed omnibus rerum Naturæ ac Philologiæ studiosis, utilis & iucundus futurus.* Zürich: Tiguri, 1565.

Gillispie, Charles Coulson, Robert Fox, and Ian Grattan-Guinness. *Pierre-Simon Laplace (1749–1827): A Life in Exact Science.* Princeton, NJ: Princeton University Press, 1997.

Glasser, Otto. *Wilhelm Conrad Röntgen und die Geschichte der Röntgenstrahlen.* Berlin, 1933. Published in English as *William Conrad Röntgen and the Early History of the Roentgen Rays* (Baltimore: Charles C. Thomas, 1934; reprint, San Francisco: Norman, 1993).

Gohau, Gabriel. *Les sciences de la terre aux XVIIe et XVIIIe siècles: Naissance de la géologie.* Paris: Albin Michel, 1990.

Goldich, Samuel Stevin. "Alfred O. Nier, 1911–1994." *EOS* 75 (1994): 66.

Goldman, Martin. *The Demon in the Aether: The Story of James Clerk Maxwell.* Edinburgh: Paul Harris, 1983.

Goodchild, John. "Some Geological Evidence regarding the Age of the Earth." *Proc. Roy. Phys. Soc. Edinburgh* 14 (1897): 259–308.

Goodchild, Peter. *Oppenheimer: Shatterer of Worlds.* Boston: Houghton Mifflin, 1981.

Gould, Stephen Jay. *Time's Arrow, Time's Cycle: Myth and Metaphor in the Discovery of Geological Time.* Cambridge, MA: Harvard University Press, 1987.

Grafton, Anthony T. *Joseph Scaliger: A Study in the History of Classical Scholarship.* Oxford: Oxford University Press, 1983.

———. "Joseph Scaliger and Historical Chronology: The Rise and Fall of a Discipline." *History and Theory* 14 (1975): 156–85.

Gray, Andrew. "Famous Scientific Workshops. I. Lord Kelvin's Laboratory in the University of Glasgow;" *Nature* 55 (1897): 486–92.

———. *Lord Kelvin: An Account of His Scientific Life and Work.* London: J. M. Dent, 1908.

Greenberg, John L. "Isaac Newton and the Problem of the Earth's Shape." *Arch. Hist. Exact Sci.* (1996): 371–91.

Gregory, William K. "William Diller Matthew, Paleontologist (1871–1930)." *Science* 72 (1930): 642–45.

Greig, James. *Silvanus P. Thompson Teacher.* London: H. M. Stationery Office, 1979.

Guédès, Michel. "Les revenus de Lamarck." *Histoire et Nature* 21 (1982): 49–60.

Guettard, Jean-Etienne. "Mémoire et carte minéralogique sur la nature et la situation des terreins qui traversent la France et l'Angleterre." *Mém. Acad. Roy. Sci.* (1746): 363–92.

———. "Mémoire sur quelques montagnes de la France qui ont été des volcans." *Mém. Acad. Roy. Sci.* (1752): 27–59.

Guinguéné, P.-L. "L'abbé Antoine Schniella Conti." In *Biographie universelle*, 9(1854): 122–24.

Gushee, Vera. "Thomas Wright of Durham, Astronomer." *Isis* 33 (1941): 197–218.

Hahn, Otto. "Ein neues Zwischenprodukt im Thorium." *Chem. Ber.* 40 (1907): 1462–69.

———. *Mein Leben.* Munich: Bruckmann, 1968. Published in English as *My Life: The Autobiography of a Scientist*, translated by E. Kaiser and E. Wilkins (New York: Herder and Herder, 1970).

———. "Über ein neues, die Emanation des Thoriums gebendes radioaktive Element." *Jahrb. Radioaktiv. Elektron.* 2 (1905): 233–66.

———. *Vom Radiothor zur Uranspaltung.* Brunswick, Germany: F. Vieweg, 1962. Published in English as *Otto Hahn: A Scientific Autobiography*, translated by W. Ley (New York: Scribner's, 1966).

———. *Was lehrt uns die Radioaktivität über die Geschichte der Erde?* Berlin: J. Springer, 1926.

Hahn, Otto, and Fritz Strassmann. "Nachweis der Entstehung activer Bariumisotope aus Uran und Thorium durch Neutronenbestrahlung: Nachweis weiketer aktiver Bruchtucke bei der Uranspaltung." *Naturwiss.* 27 (1939): 89–95.

Hahn, Otto, and Ernest Walling. "Über die Möglichkeit geologischer Altersbestimmungen rubidiumhaltiger Mineralien und Gesteine." *Z. Anorg. Allg. Chem.* 236 (1938): 78–82.

Hall, James. "Account of a Series of Experiments, Shewing the Effects of Compression in Modifying the Action of Heat." *Trans. Roy. Soc. Edinb.* 6 (1812): 71–184.

———. "Curious Circumstances upon Which the Vitreous or the Stony Character of Whinstone and Lava Respectively Depend; with Other Facts." *J. Natural Philos. Chem.* 2 (1798–99): 285–88.

———. "Experiments on Whinstone and Lava." *Trans. Roy. Soc. Edinb.* 5 (1805): 42–75.

Hall, Trevor. H. *The Spiritualists: The Story of Florence Cook and William Crookes.* New York: Helix Press, 1963. Reprinted as *The Medium and the Scientist* (Buffalo, NY: Prometheus Books, 1984).

Halley, Edmond. "An Estimate of the Quantity of Vapour Raised out of the Sea by the Warmth of the Sun; Derived from an Experiment Shown before the Royal Society, at One of Their Meetings." *Phil. Trans.* 16 (1687): 366–70.

———. "An Historical Account of the Trade Winds and Mansoons, Observable in the Seas between and Near the Tropics; with an Attempt to Assign Their Physical Causes." *Phil. Trans.* 16 (1686): 153–68.

———. "Of the Infinity of the Sphere of Fixed Stars," *Phil. Trans.* 31 (1720): 22.

———. "On the Cause of the Change in the Variation of the Magnetic Needle; with an Hypothesis of the Structure of the Internal Parts of the Earth." *Phil. Trans.* 17 (1692): 563–78.

———. "On the Cause of the Saltness of the Ocean, and of the Several Lakes That Emit no Rivers; with a Proposal, by Means Thereof, to Discover the Age of the World." *Phil. Trans.* 29 (1715): 296–300. Long excerpts reprinted in G. F. Becker, "Halley on the Age of the Ocean," *Science* 31 (1910): 459–61.

———. "On the Height of the Mercury in the Barometer at Different Elevations above the Surface of the Earth; and on the Rising and Falling of the Mercury on the Change of Weather." *Phil. Trans.* 16 (1686): 104–16.

———. "Remarks upon Some Dissertations Lately Publish'd at Paris, by the Rev P. Souciet, against Sir Isaac Newton's Chronology." *Phil. Trans.* 34 (1726–27): 205–10.

———. "Some Account of the Ancient State of the City of Palmira, with Remarks on the Inscription Found There." *Phil. Trans.* 19 (1695): 160–75.

———. "Some Considerations about the Cause of the Universal Deluge," *Phil. Trans.* 33 (1724): 118.

———. "Some Farther Remarks on P. Souciet's Differations against Sir Isaac Newton's Chronology, in a Letter to Dr. Jurin." *Phil. Trans.* 35 (1727–28): 296–300.

———. "Some Further Thoughts upon the Same Suject, Delivered on the 19th of the Same Month?" *Phil. Trans.* 33 (1724–25): 118–25.

———. "That the Obliquity of the Ecliptic and the Elevation of the Pole Continue Unaltered." *Phil. Trans.* 16 (1687): 403.

Hampton, John. *Nicolas-Antoine Boulanger et la science de son temps.* Geneva: Droz, 1955.

Harkins, William Draper, and R. E. Hall. "The Evolution of the Elements and the Stability of Complex Atoms. I. A New Periodic System Which Shows a Relation between the Abundance of the Elements and the Structure of the Nuclei of Atoms." *J. Amer. Chem. Soc.* 39 (1917): 856–79.

———. "The Periodic System and the Properties of the Elements." *J. Amer. Chem. Soc.* 38 (1916): 169–221.

Harkins, William Draper, and Ernst D. Wilson. "The Changes of Mass and Weight Involved in the Formation of Complex Atoms." *J. Amer. Chem. Soc.* 37 (1915): 1367–83.

———. "Recent Work on the Structure of the Atom." *J. Amer. Chem. Soc.* 37 (1915): 1396–1421.

———. "The Structure of Complex Atoms. The Hydrogen-Helium System." *J. Amer. Chem. Soc.* 37 (1915): 1383–96.

Harwood, John T. "Rhetoric and Graphics in *Micrographia*." In *Robert Hooke, New Studies*, edited by M. Hunter and S. Schaffer. Woodbridge, Eng.: Boydell Press, 1989.

Haughton, Samuel. *Manual of Geology*. London: Longman, 1871.

———. "Physical Geology. A Geological Proof That the Changes of Climate in Past Times Were Not Due to Changes in the Position of the Pole; with an Attempt to Assign a Minor Limit to the Duration of Geological Time." *Nature* 18 (1878): 266–68.

Havard, Jean-Alexandre. *Voltaire et Mme du Châtelet: Révélations d'un serviteur attaché à leurs personnes*. Paris: E. Dentu, 1863.

Hazard, Paul. *La pensée européenne au XVIIIe siècle, de Montesquieu à Lessing*. Paris: Boivin, 1946. Published in English as *European Thought in the Eighteenth Century*, translated by J. Lewis May (Harmondsworth, Eng.: Penguin, 1965).

Heilbron, John L. *Elements of Early Modern Physics*. Berkeley: University of California Press, 1982.

———. *H. G. J. Moseley: The Life and Letters of an English Physicist*. Berkeley: University of California Press, 1974.

Heim, Roger, ed. *Buffon*. Paris: Muséum National d'Histoire Naturelle, 1952.

Helmholtz, Hermann von. "On the Interaction of Natural Forces." *Phil. Mag.* 11 (1856): 489–518. Reprinted in *Popular Lectures*, translated by E. Atkinson (London: Longmans, 1903), along with a short autobiography; and in *Science and Culture: Popular and Philosophical Essays*, edited by D. Cahan. (Chicago: University of Chicago Press, 1995).

———. "Ueber die Erhaltung der Kraft. Eine physikalische Abhandlung." In *Wissenschaftliche Abhandlungen von Hermann Helmholtz*, 3 vols., 1:12–75. Leipzig: J. A. Barth, 1882.

Hérault de Séchelles, Marie Jean. *Voyage à Montbar, contenant des détails très-intéressans sur le caractère, la personne et les écrits de Buffon*. Paris: Solvet, 1800. Reprint by Y. Gaillard, *Buffon: Biographie imaginaire et réelle*. Paris: Hermann, 1979.

Herivel, John. *Joseph Fourier: The Man and the Physicist*. Oxford: Clarendon Press, 1975.

Herodotus. *History*. Translated by G. Rawlinson. New York: Tudor, 1936.

Héron de Villefosse, Antoine-Marie. "Considérations sur les fossiles, et particulièrment ceux que présente le pays de Hanovre; ou extrait raisonné d'un ouvrage de M. Blumembach, ayant pour titre: *Specimen Archeologia telluris, terrarum que imprimis Hannoveranarum*." *J. Mines* 16 (1804): 5–36.

Herrmann, Dieter B. *Ejnar Hertzsprung: Pionier der Sternforschung*. Berlin: Springer, 1994.

Herschel, William. "Astronomical Observations Relating to the Sidereal Part of the Heavens, and Its Connection with the Nebulous Part: Arranged for the Purpose of a Critical Examination." *Phil. Trans.* 104 (1814): 248–84.

———. "On the Construction of the Heavens." *Phil. Trans.* 75 (1785): 213–51.

———. "On the Remarkable Appearances at the Polar Regions of the Planet Mars, the Inclination of Its Axis, the Position of Its Poles, and Its Spheroidal Figure; with a Few Hints Relating to Its Real Diameter and Atmosphere." *Phil. Trans.* 74 (1784): 233–73.

———. *The Scientific Papers of Sir William Herschel*. 2 vols. London: Royal Society, 1912.

Hertz, Heinrich. *Untersuchungen über die Ausbreitung der elektrischen Kraft*. Leipzig: Johann Ambrosius Barth, 1892. Published in English as *Electric Waves*, translated by D. E. Jones (New York: Macmillan, 1962).

Hertzsprung, Ejnar. "Zur Strahlung der Sterne. I." *Z. Wissenschaftliche Photog.* 3 (1905): 429–42.

———. "Zur Strahlung der Sterne. II." *Z. Wissenschaftliche Photog.* 5 (1907): 86–107.

Hillebrand, William Francis. "On the Occurrence of Nitrogen in Uraninite and on the Composition of Uraninite in General." *Bull. U.S. Geol. Survey* 78 (1890): 43–78.

Hippolytus Antipope. *Refutatio omnium haeresium*. Berlin: W. De Gruyter, 1986.

Hoffmann, Klaus. *Schuld und Verantwortung: Otto Hahn, Konflikte eines Wissenschaftlers*. Berlin: Springer, 1993. Published in English as *Otto Hahn: Achievement and Responsibility*, translated by J. M. Cole (New York: Springer, 2001).

Holbach, Paul Henri Thiry [M. de Mirabaud, pseud.]. *Système de la nature, ou des loix du monde physique & du monde moral*. 2 vols. London, 1770. Reprint, Paris: Fayard, 1990. Published in English as *The System of Nature; or, Laws of the Moral and Physical World*, translated by H. D. Robinson (Boston: J. P. Mendum, 1889). Page references are to the 1990 edition.

Holmes, Arthur. *The Age of the Earth*. London: Harper and Brothers, 1913.

———. *The Age of the Earth*. London: Ernest Benn, 1927.

———. "The Association of Lead with Uranium in Rock-Minerals, and Its Application to the Measurement of Geological Time." *Proc. Roy. Soc.* 85 (1911): 248–56.

———. "An Estimate of the Age of the Earth." *Nature* 157 (1946): 680–84.

———. Introduction to *The Precambrian*, edited by K. Rankama., 1:xi–xxiii. New York: Interscience, 1963.

———. *Principles of Physical Geology.* 3rd ed. London: Nelson, 1965.

———. "Radioactivity and Geological Time." In *The Age of the Earth*, by A. Knopf, E. W. Brown, A. Holmes, A. F. Kovarik, A. C. Lane, and C. Schuchert, 124–459. Washington, DC: National Research Council, 1931.

———. "A Revised Estimate of the Age of the Earth." *Nature* 159 (1947): 127–28.

———. "A Revised Geological Timescale." *Trans. Edinb. Geol. Soc.* 17 (1959): 183–216.

Homer. *Iliad.* Translated by S. Lombardo. Indianapolis: Hackett, 1997.

Höningschmid, Otto, and Mlle St. Horovitz. "Sur le poids atomique du plomb de la pechblende." *C. R. Acad. Sci.* 158 (1914): 1796–98.

Hooke, Robert. *A Description of Helioscopes and Some Other Instruments.* London: John Martyn, 1676.

———. "Lectures and Discourse of Earthquakes and Subterranean Eruptions." In *The Posthumous Works of Robert Hooke, Containing His Cutlerian Lectures and Other Discourses Read at the Meetings of the Illustrious Royal Society*, edited by R. Waller and R. S. Secret. London: Smith and Walford, 1705. Reprinted with an introduction by R. S. Westfall. (New York: Johnson Reprint, 1969); and by Drake in his *Restless Genius*.

———. *Micrographia; or, Some Physiological Descriptions of Minute Bodies Made by Magnifying Glasses with Observations and Inquiries Thereupon.* London: James Allestry, 1665. Reprint, New York: Dover, 1961.

Hopkins, William. "Researches in Physical Geology—On the Phenomena of Precession and Nutation, Assuming the Fluidity of the Interior of the Earth." *Phil. Trans. Roy. Soc.* 129 (1839): 381–423.

———. "Researches in Physical Geology—Second Series. On Precession and Nutation, Assuming the Interior of the Earth to Be Fluid and Heterogeneous." *Phil. Trans. Roy. Soc.* 130 (1840): 193–208.

———. "Researches in Physical Geology—Third Series. On the Thickness and Constitution of the Earth's Crust." *Phil. Trans. Roy. Soc.* 132 (1842): 43–55.

Hoskin, Michael A. "The Cosmology of Thomas Wright of Durham." *J. Hist. Astron.* 1 (1970): 44–52.

———. *William Herschel and the Construction of the Heavens.* With notes by D. W. Dewhirst. London: Oldbourne, 1963.

Houtermans, Friedrich Georg. "Das Alter des Urans." *Z. Naturforsch.* 29 (1947): 322–28.

———. "Determination of the Age of the Earth from the Isotopic Composition of Meteoritic Lead." *Nuevo Cimento* 10 (1953): 1623–33.

———. "Die Isotopenhäufigkeiten im natürlichen Blei und das Alter des Urans." *Naturwiss.* 33 (1946): 185–86.

———. "History of the K-Ar Method of Geochemistry." In *Potassium-Argon Dating*, compiled by O. A. Schaeffer and J. Zähringer, 1–6. New York: Springer, 1966.

Hoyt, William Graves. *Lowell and Mars*. Tucson: University of Arizona Press, 1976.
———. "Vesto M. Slipher, 1875–1969." *Biog. Mem. Nat. Acad. Sci.* 52 (1980): 410–49.
Hubble, Edwin Powell. "The Law of Red-Shifts." *Mon. Nat. R. A. S.* 113 (1953): 658–66.
———. "A Relation between Distance and Radial Velocity among Extra-Galactic Nebulae." *Proc. Nat. Acad. Sci.* 15 (1929): 168–73.
Hubble, Edwin Powell, and Milton L. Humason. "The Velocity-Distance Relation among Extra-Galactic Nebulae." *Ap. J.* 74 (1931): 43–80.
Huggins, William. "On the Spectra of Some of the Chemical Elements." *Phil Trans.* 154 (1864): 139–60.
Huggins, William, and W. A. Miller. "On the Lines of the Spectra of Some of the Fixed Stars." *Proc. Roy. Soc.* 12 (1863): 444–45.
———. "On the Spectra of Some of the Fixed Stars." *Phil. Trans.* 154 (1864): 413–35, 436–44.
Humboldt, Alexander von. *Essai géognostique sur le gisement des roches dans les deux hémisphères*. Paris: G. Levrault, 1823.Translated as *A Geognostical Essay on the Superposition of Rocks in both Hemispheres* (London: Longman, 1823).
———. *Voyage aux régions équinoxiales du Nouveau Continent, fait en 1799, 1800, 1801, 1802, 1803 et 1804 par Al. de Humboldt et A. Bonpland*. Vol. 1, *Relation historique*. Paris: F. Schoell, 1814.
Huot, Jean-Jacques Nicolas. *Nouveau cours élémentaire de géologie*. 2 vols. Paris: Roret, 1837–39.
———. *Nouveau manuel complet de géologie*. Paris: Roret, 1840.
Hurwic, Anna. *Pierre Curie*. Paris: Flammarion, 1995.
Hutton, James. *Abstract of a Dissertation Read in the Royal Society of Edinburgh*. 1785.
———. "Theory of the Earth; or an Investigation of the Laws Observable on the Composition, Dissolution, and Restoration of Land upon the Globe." *Trans. Roy. Soc. Edinb.* 1 (1788): 209–304. Reprinted with the *Observation of Granite* and Playfair's "Biographical account of Hutton" (Darien, CT: Hafner, 1970).
———. *Theory of the Earth, with Proofs and Illustrations*. 2 vols. Edinburgh: W. Creech, 1795. The 3rd vol. was published by A. Geikie (London: Geological Society, 1899) and reprinted and edited by D. R. Dean as *James Hutton in the Field and in the Study, Being an Augmented Reprinting of Vol. 3 of Hutton's Theory of the Earth as First Published by Sir Archibald Geikie. (1899)* (New York: Delmar, 1997).
Huxley, Thomas. "The Anniversary Address of the President." *Quart. J. Geol. Soc.* 25 (1869): xxviii–lii.
Imbrie, John, and Katharine P. Imbrie. *Ice Ages: Solving the Mystery*. London: Macmillan, 1979.
Ioannis Buridani expositio et quæstiones in Aristotelis De cælo. Edited by B. Patar. Louvain: Peeters, 1996.
J. L. "Lord Kelvin." *Proc. Roy. Soc.* A81 (1908): iii–lxxvi.

Jahn, Melvin E., and Daniel J. Woolf. *The Lying stones of Dr. Johann Bartholomew Adam Beringer being his Lithographiæ Wirceburgensis.* Berkeley: University of California Press, 1963.

Jameson, Robert. *The Wernerian Theory of the Neptunian Origin of Rocks: A Facsimile Reprint of Elements of Geognosy, 1808.* New York: Hafner Press, 1976.

Janssen, Jules. "Indication de quelques uns des résultats obtenus à Cocanada, pendant l'éclipse du mois d'août dernier, et à la suite de cette éclipse." *C. R. Acad. Sci.* 67 (1868): 838–39.

Jauncey, George E. M. "The Early Years of Radioactivity." *Amer. J. Phys.* 14 (1946): 226–41.

Jeans, James. "The Ages and Masses of the Stars." *Nature* 114 (1924): 828–29; 118 suppl. (1926): 29–40; and 122 suppl. (1928): 689–700.

———. *Astronomy and Cosmogony.* Cambridge: University Press, 1929.

———. "The Physics of the Universe." *Nature* 122 (1928) suppl.: 689–700.

———. "Recent Developments of Cosmical Physics." *Nature* 118 (1926) suppl.: 29–40.

———. "A Suggested Explanation of Radio-Activity." *Nature* 70 (1904): 101.

———. *The Universe around Us.* Cambridge: Cambridge: University Press, 1929.

Jerome. *The Letters of Saint Jerome.* Translated by C. C. Mierow. New York: Newman Press, 1963.

"John George Goodchild, F.G.S." *Geol. Mag.* 33 (1906): 189–90.

"John Joly 1857–1903." *Obit. Not. Roy. Soc. London* 1 (1934): 259–86.

John Philoponus. *Against Aristotle: On the Eternity of the World.* Translated by C. Wildberg. Ithaca, NY: Cornell University Press, 1987.

Jolly, W. P. *Marconi.* New York: Stein and Day, 1972.

Joly, John. "The Age of the Earth." *Nature* 109 (1922): 480–85.

———. "The Age of the Earth." *Phil. Mag.* 22 (1911): 357–80.

———. "An Estimate of the Geological Age of the Earth." *Trans. Roy. Dublin Soc.* 7 (1899): 23–66.

———. "A Geological Description of Glen Tilt." *Trans. Geol. Soc. London* 3 (1816): 259–337.

———. "Pleiochroic Halos." *Phil. Mag.* 6 (1907): 381–83.

———. *Radioactivity and Geology: An Account of the Influence of Radioactive Energy on Terrestrial History.* London: A. Constable, 1909.

———. "Radium and the Geological Age of the Earth." *Nature* 68 (1903): 526.

———. "Radium and the Sun's Heat." *Nature* 68 (1903): 572.

———. *The Surface-History of the Earth.* Oxford: Clarendon Press, 1925, 1930.

Joly, John, and E. Rutherford. "The Age of Pleochroic Haloes." *Phil. Mag.* 25 (1913): 644–57.

Jones, Jean, Hugh S. Torrens, and Eric Robinson. "The Correspondence between James Hutton (1726–1797) and James Watt (1736–1819) with Two Letters from Hutton to George Clerk Maxwell (1715–1784): Part II." *Ann. Sci.* 52 (1995): 357–82.

Jonsson, Inge. *Emanuel Swedenborg.* Translated by C. Djurklou. New York: Twayne, 1971.

Josephus, Flavius. *The Works of Flavius Josephus, the Learned and Authentic Jewish Historian and Celebrated Warrior, Containing Twenty Books of the Jewish Antiquities, Seven Books of the Jewish Wars, and the Life of Josephus, Written by Himself.* Translated by W. Whiston. Philadelphia: J. Grigg, 1832.

Joule, James Prescott. *The Scientific Papers of James Prescott Joule.* 2 vols. London, 1884, 1887.

———. "Sur l'équivalent mécanique du calorique." *C. R. Acad. Sci.* 28 (1849): 132–35.

Julius Africanus [Sextius Julius Africanus]. "The Extant Fragments of the Five Books of the Chronography of Julius Africanus." Edited by A. Cleveland Coxe. In *The Ante-Nicene Fathers,* vol. 6. New York: Christian Literature, 1890.

Kant, Immanuel. *Allgemeine Naturgeschichte und Theorie des Himmels, oder Versuch von der Verfassung und dem mechanischen Ursprunge des ganzen Weltgebäudes nach Newtonischen grundsätzen abgehandelt.* Königsberg, Germany: J. F. Petersen, 1755. Published in English as *Universal Natural History and Theory of the Heavens,* translated by S. L. Jaki (Edinburgh: Scottish Academic Press, 1981).

Kardec, Allan. *Le livre des esprits.* 2nd ed. Paris: Didier, 1860. Published in English as *Book on Mediums; or, A Guide for Mediums and Invocators: Containing the Special Instructions of the Spirits on the Theory of All Kinds of Manifestations, the Development of Mediumship; the Difficulties and the Dangers That Are to Be Encountered in the Practice of Spiritism,* translated by E. A. Wood (Kila, MT: Kessinger, 1996).

———. *Spiritisme expérimental: Le livre des médiums ou guide des médiums et des évocateurs contenant l'enseignement spécial des esprits sur la théorie de tous les genres de manifestation, les moyens de communiquer avec le monde invisible, le développement de la médiumnité, les difficultés et les écueils que l'on peut rencontrer dans la pratique du spiritisme, pour faire suite au livre des esprits.* Paris: Didier, 1867.

Keesom, Willem H. *Helium.* Amsterdam: Elsevier, 1942.

Keferstein, Christian. "Notice sur Füchsel et ses ouvrages." *J. Géologie* 2 (1830): 191–97.

Kelvin. See William Thomson.

Kepler, Johannes. *Astronomia nova* (1609). In vol. 3 of *Gesammelte Werke,* by Kepler. Published in English as *New Astronomy,* translated by W. H. Donahue (Cambridge: Cambridge University Press, 1992).

———. *De vero Anno, quo æternus Dei Filius humanam naturam in Utero benedictæ Virginis Mariæ assumpsit* (1614). In vol. 5 of Kepler, *Gesammelte Werke.*

———. *Gesammelte Werke,* edited by M. Caspar and F. Hammer. 21 vols. Munich: Beck, 1938–93). Vol. 1 published in English as *The Secret of the Universe,* translated by A. M. Duncan (New York: Abaris Books, 1981).

———. *Harmonice Mundi Libri V* (1619). In vol. 6 of Kepler, *Gesammelte Werke.* Published in English as *The Harmony of the World*, translated by E. J. Aiton, A. M. Duncan, and J. V. Field (Philadelphia: American Philosophical Society, 1997).

———. *Mysterium cosmographicum* (1596). Vol. 1 of Kepler, *Gesammelte Werke.*

Khlopin, Vitali G., and Eric Carlovich Gerling. "New Method of Age Determination in Minerals." *Doklad. Akad. Nauk. URSS* 58 (1947): 1415–17.

Khriplovich, Iosif B. "The Eventful Life of Fritz Houtermans." *Physics Today*, July 1992, 29–37.

King, Clarence. "The Age of the Earth." *Amer. J. Sci.* 45 (1893): 1–20.

King, Elizabeth. *Lord Kelvin's Early Home.* London: Macmillan, 1909.

Kircher, Athanasius. *Mundus Subterraneus in xii Libros digestus; quo Divinum Subterrestris Mundi Opificium, mira Ergasteriorum Naturæ in eo distributio, verbo παντα μορφον Protei Regnum, Universæ denique Naturæ Majestas et divitiæ summa rerum varietate exponentur. Abditorum effectuum causæ acri indagine inquisitæ demonstrantur; cognitæ per Artis et Naturæ conjugium ad humanæ vitæ necessarium usum vario experimentorum apparatu, necnon novo modo, & ratione applicatur.* 2 vols. Amsterdam: Joannem Janssonium and Elizeum Weyerstraten, 1665.

Kirsch, Gerhard. *Geologie und Radioaktivität: Die radioaktiven Vorgänge als geologische Uhren und geophysikalische Energiequellen.* Berlin: Springer, 1928.

Klaproth, Martin Heinrich. "Chemische Untersuchung des Uranits, einer neuentdeckten metallischen Substanz." *Ann. Chem. (Crell)* 2 (1789): 387–403.

Klickstein, Herbert S. *Wilhelm Conrad Röntgen on a New Kind of Rays.* Vol. 1. Philadelphia: Mallinckrodt Classics of Radiology, 1966.

Knott, Cargill Gilston. *The Life and Scientific Work of Peter Guthrie Tait.* Cambridge: University Press, 1911.

Knox, R. Buick. *James Ussher: Archbishop of Armagh.* Cardiff: University of Wales Press, 1967.

Köningsberger, Leo. *Hermann von Helmholtz.* Brunswick, Germany: Vieweg, 1911.

Kottler, Malcom Jay. "Alfred Russel Wallace, the Origin of Man, and Spiritualism." *Isis* 65 (1974): 145–92.

Koyré, Alexandre. *La révolution astronomique.* Paris: Hermann, 1961. Published in English as *The Astronomical Revolution*, translated by R. E. W. Maddison (New York: Dover, 1992).

———. *Newtonian Studies.* Cambridge, MA: Harvard University Press, 1965.

———. "Paracelse." *Rev. Hist. Philos. Relig.* 13 (1933): 46–75, 145–63.

———. "An Unpublished Letter of Robert Hooke to Isaac Newton." *Isis* 43 (1952): 312–37. Reprinted in Koyré, *Newtonian Studies.*

Kuhn, Thomas S. *The Copernican Revolution.* Cambridge, MA: Harvard University Press, 1957.

Lamarck, Jean-Baptiste de Monet. *Hydrogéologie.* Paris, 1802. Published in English as *Hydrogeology*, translated by A. V. Carozzi (Urbana: University of Illinois Press, 1964).

———. *Philosophie zoologique*. Paris: Dentu, 1809. Published in English as *Zoological Philosophy: An Exposition with Regard to the Natural History of Animals*, translated by H. Elliot (Chicago: University of Chicago Press, 1984).

La Mettrie, Julien Offray de. *L'homme machine*. Leiden: E. Luzac, 1748. Published in English as *Machine Man and Other Writings*, translated by A. Thomson (Cambridge: Cambridge University Press, 1996).

La Peyrère, Isaac de. *Prae-adamitæ, sive Exercitatio super versibus duodecimo, decimotertio & decimoquarto, capitis quinti Epistolæ D. Pauli ad Romanos, quibus inducuntur Primi Homines ante Adamum conditi*. Amsterdam, 1655.

———. *Systema theologicum ex prae-adamitarum hypothesis, pars prima*. Amsterdam, 1655. Published in English as *A Theological Systeme upon that Presupposition that Men were before Adam*. London, 1655.

Laplace, Pierre-Simon. *Essai philosophique sur les probabilités*. 6th ed. Paris: Bachelier, 1840. Published in English as *Philosophical Essay on Probabilities*, translated by A. I. Dale (New York: Springer, 1995).

———. *L'Exposition du Système du monde*. 6th ed. Paris: Bachelier, 1836. Earlier edition published in English in two volumes as *The System of the World*, translated by H. H. Harte (London: Longmans, 1830).

———. *Œuvres complètes*. 14 vols. Paris: Gauthier-Villars, 1878–1912.

Lapparent, Albert-Auguste Cochon de. "De la mesure du temps par les phénomènes de sédimentation." *Bull. Soc. Géol. France* 18 (1889–90): 351–55.

———. *Traité de Géologie*. 5th ed. 3 vols. Paris: Masson, 1906.

Launay, Louis de. *La science géologique, ses méthodes, ses résultats—ses problèmes, son histoire*. Paris: Armand Colin, 1905.

———. *L'histoire de la Terre*. Paris: Flammarion, 1920.

———. *Une grande famille de savants, les Brongniart*. Paris: G. Rapilly, 1940.

Lavisse, Ernest. *Histoire de France*. Vol. 9. Paris: Hachette, 1910.

Lavoisier, Antoine-Laurent de. "Observations générales sur les couches modernes horizontales qui ont été déposées par la mer et sur les conséquences qu'on peut tirer de leurs dispositions relativement à l'ancienneté du globe." *Mém. Acad. Roy. Sci.* (1789): 351–71. Reprinted in Lavoisier, *Œuvres*, vol. 5 (1892).

———. *Œuvres complètes*. 6 vols. Paris: Imprimerie Nationale, 1864–93.

Lavoisier, Antoine-Laurent de, and Pierre-Simon Laplace. "Mémoire sur la chaleur" (1780). In Lavoisier, *Œuvres*, 2:283.

Lawren, William. *The General and the Bomb*. New York: Dodd and Mead, 1988.

Le Bon, Gustave. "La lumière noire." *C. R. Acad. Sci.* 122 (1896): 188–90.

———. "La photographie à la lumière noire." *C. R. Acad. Sci.* 122 (1896): 233–35.

———. *Psychologie des foules*. Paris: Alcan, 1895. Published in English as *The Crowd: A Study of Popular Mind* (Atlanta: Cherokee, 1982).

Lehmann, Johann Gottlob. *Versuch einer Geschichte von Flötz-Gebürgen betreffend deren Entstehung, Lage, darinne befindliche Metallen, Mineralien und Foßilien, gröstentheils aus eigenen Wahrnehmungen, chymischen und physicalischen*

Bersuchen, und aus denen Grundsäken der Natur-Lehre bergeleitet, und mit nöthigen Kupfern. Berlin, 1756.

Leibniz, Gottfried Wilhelm. *Protogæa, sive de prima facie telluris et antiquissimæ historiæ vestigiis in ipsis naturæ monumentis dissertatio, ex schedis manuscriptis viri illustris in lucem edita a Chritiano Ludovico Scheidio.* In *Opera Omnia*, by Leibniz, 6 vols., 2:181–240. Geneva: Fratres de Tournes, 1768. Published in French, with complements and notes by J. M. Barrande, as *Protogée: De l'aspect primitif de la terre et des traces d'une histoire très ancienne que renferment les monuments même de la nature*, translated by B. de Saint-Germain (Toulouse: Presses Universitaires du Mirail, 1993).

Lelong, Benoît. "Paul Villard, J.-J. Thomson et la composition du rayonnement cathodique." *Rev. Hist. Sci.* 50 (1997): 89–130.

Lémery, Nicolas. "Explication physique et chymique des feux souterrains, des tremblemens de terre, des ouragans, des éclairs et du tonnerre." *Mém. Acad. Roy. Sci.* (1700): 101–10.

Lenard, Philip. *Deutsche Physik.* Munich: J. F. Lehmanns Verlag, 1938.

———. *Über Kathodenstrahlen.* Leipzig: J. A. Barth, 1906 and 1920.

Leonardo da Vinci. *The Notebooks of Leonardo da Vinci, Compiled and Edited from the Original Manuscripts.* Edited by Jean Paul Richter. 2 vols. New York: Dover, 1970.

Lewis, Cherry. *The Dating Game: One Man's Search for the Age of the Earth.* Cambridge: Cambridge University Press, 2000.

Lhwyd, Edward. "Of the Origin of Marine Fossils; and of Mineral Leaves, Branches, Etc." In *Life and Letters of Edward Lhwyd*, vol. 14 of *Early Science in Oxford*, edited by R. T. Gunther. Oxford: University Press, 1945.

Lindemann, Frederick Alexander. "The Age of the Earth." *Nature* 95 (1915): 203, 372.

Lindsay, Robert Bruce. *Lord Rayleigh—The Man and His Work.* Oxford: Pegamon, 1966.

Locke, John. *The Works.* Vol. 3. London: A. Bettesworth, 1727.

Lockyer, Norman. *Inorganic Evolution.* London: Macmillan, 1900.

———. "Lettre à M. Warren de la Rue sur une méthode pour observer en temps ordinaire le spectre des protubérances signalées dans les éclipses totales de soleil" (23 October 1868). *C. R. Acad. Sci.* 67 (1868): 836–38.

———. *The Meteoritic Hypothesis.* London, 1890.

———. "The Story of Helium." *Nature* 53 (1896): 319–22, 342–46.

Lockyer, T. M., and W. L. Lockyer. *Life and Work of Sir Norman Lockyer.* London: Macmillan, 1928.

Lodge, Oliver. "Becquerel Memorial Lecture." *J. Chem. Soc.* 101 (1912): 2005–42.

———. "The University Aspect of Psychical Research." In *The Case For and Against Psychical Belief, by Sir Oliver Lodge*, edited by Carl Murchison, 3–23. Worcester, MA: Clark University, 1927.

Lods, Adolphe. "Histoire de la critique du Pentateuque." In *Histoire de la littérature hébraïque et juive*, 83–118. Paris: Payot, 1950.

————. *Jean Astruc et la critique biblique au XVIIIe siècle*. Strasbourg: Librairie Istra, 1922.

Lowell, Percival. *The Evolution of Worlds*. New York: Macmillan, 1909.

————. *Mars*. London: Longmans, Green, 1896.

————. *Mars and Its Canals*. New York: Macmillan, 1911.

————. *Mars as the Abode of Life*. New York: Macmillan, 1908.

Lucretius [Titus Lucretius Carus]. *De Natura rerum*. Published in English as *On the Nature of Things*, translated by J. S. Watson (Amherst, NY: Prometheus Books, 1997).

Lumière, Auguste, and Louis Lumière. "A propos de la photographie à travers les corps opaques." *C. R. Acad. Sci.* 122 (1896): 463–465.

————. "Recherches photographiques sur les rayons de Röntgen." *C. R. Acad. Sci.* 122 (1896): 382–83.

Luther, Martin. *The Table Talk of Martin Luther*. Translated by W. Hazlitt. London: G. Bell, 1902.

Lyell, Charles. "Award of the Wollaston Medal and Donation Fund." *Quart. J. Geol. Soc.* 6 (1850): xxiii–xxvi.

————. *Life, Letters, and Journals of Sir Charles Lyell, Bart*. Edited by Mrs Lyell [Sir Charles's sister-in-law]. 2 vols. London: J. Murray, 1881.

————. *Principles of Geology, Being an Attempt to Explain the Former Changes of the Earth's Surface, by Reference to Causes Now in Operation*. London: J. Murray, 1830–75 (12 eds.). Reprint of 1st ed., Chicago: University of Chicago Press, 1990.

MacCulloch, John. "A Geological Description of Glen Tilt." *Trans. Geol. Soc. London* 3 (1816): 259–337.

Maillet, Benoît de. *Telliamed, ou entretiens d'un philosophe indien avec un missionnaire françois sur la diminution de la mer*. Amsterdam: L'Honoré and Son, 1748. 2nd ed. published in English, with a short biography, as *Telliamed or Conversations between an Indian Philosopher and a French Missionary on the Diminution of the Sea, the Origin of Man, etc.*, 1755; translated by A. V. Carozzi (Urbana: University of Illinois Press, 1968).

Mairan, Jean-Jacques Dortous de. *Dissertation sur la glace, ou explication physique de la formation de la glace, & de ses divers phénomènes*. 4th ed. Paris: Imprimerie Royale, 1749.

————. "Éloge de M. Halley." *Mém. Acad. Roy. Sci.* (1742): 172–88. Reprinted in *Correspondence and Papers of Edmond Halley*, edited by E. F. MacPike. (Oxford: Clarendon, 1932). Page references are to the reprint.

————. "Mémoire sur la cause générale du froid en hiver, et de la chaleur en été." *Mém. Acad. Roy. Sci.* (1719): 104–35.

————. "Nouvelle recherche sur la cause générale du chaud en été et du froid en hiver, en tant qu'elle se lie à la chaleur interne et permanente de la terre." *Mém. Acad. Roy. Sci.* (1765): 143–266.

————. "Recherches géométriques sur le diminution des degrés terrestres en allant de l'équateur vers les poles. Où l'on examine les consequences qui en résultent, tant à l'égard de la figure de la Terre, que de la pesanteur des corps, et de l'accourcissement du pendule." *Mém. Acad. Roy. Sci.* (1720): 231–77.

Mandonnet, P. *Siger de Brabant and l'Averroïsme latin au XIIIeme siècle.* Vol. 1, *Etude critique*, 2nd ed. (Louvain: Institut supérieur de philosophie de l'université, 1911). Vol. 2, *Textes inédits* (Louvain: Institut supérieur de philosophie de l'université, 1908).

Manuel, Frank E. *Isaac Newton, Historian.* Cambridge, MA: Belknap Press of Harvard University Press, 1963.

———. *A Portrait of Isaac Newton.* Cambridge, MA: Belknap Press of Harvard University Press, 1968.

Marconi, Guglielmo. "Wireless Telegraphy." *Proc. Instit. Elec. Eng.* 28 (1899): 273–91.

Marrou, Irénée. *Saint Augustin et l'augustinisme.* Paris: Editions du Seuil, 1956. Published in English as *St. Augustine and His Influence through the Ages*, translated by P. Hepburne-Scott (New York: Harper Touchbooks, 1957).

Matthew, William Diller. "Time Ratios in the Evolution of Mammalian Phyla: A Contribution to the Problem of the Age of the Earth." *Science* 40 (1915): 232–35.

Maupertuis, Pierre-Louis Moreau de. *La Figure de la terre.* Paris: Imprimerie Royale, 1738. Published in English as *The Figure of the Earth, Determined from Observations Made by Order of the French King, at the Polar Circle, by Messrs. de Maupertuis, Clairaut, Camus, Le Monnier, of the Royal Academy of Science, and the abbé Outhier, and Mr. Celsius.* London: T. Cox, 1738.

———. "Lettre sur la comète qui paroissoit en 1742." In *Œuvres de Mr. de Maupertuis*, 4 vols., 3:209–56. Lyon: J. M. Bruyset, 1756.

Maxwell, James Clerk. "Introductory Lecture on Experimental Physics." In Maxwell, *Scientific Papers*, 2:241–55.

———. *The Scientific Papers of James Clerk Maxwell.* 2 vols. Cambridge: University Press, 1890. Reprint, New York: Dover, 1965.

Mayer, Jean. "Portrait d'un chimiste: Guillaume-François Rouelle." *Rev. Hist. Sci. Appl.* 23 (1970): 305–32.

Mayer, Julius Robert. "On Celestial Dynamics. I. Introduction." *Phil. Mag.* 25 (1863): 241–48, 387–409.

———. "On Celestial Dynamics. IX. The Heat of the Interior of the Earth." *Phil. Mag.* 25 (1863): 417–28.

———. "Remarks on the Mechanical Equivalent of Heat." *Phil. Mag.* 25 (1863) suppl.: 493–522.

———. "Sur la transformation de la force vive en chaleur." *C. R. Acad. Sci.* 27 (1848): 385–87.

McKee, D. Rice. "Isaac de la Peyrère, a Precursor of Eighteenth-Century Critical Deists." *PMLA* 59 (1944): 456–85.

Meitner, Lise, and Otto Frisch. "Disintegration of Uranium by Neutrons: A New Type of Nuclear Reactions." *Nature* 143 (1939): 239–40.

Merricks, Linda. *The World Made New: Frederick Soddy, Science, Politics, and Environment*. Oxford: Oxford University Press, 1997.

Millin, Aubin-Louis. *Discours sur l'origine et les progrès de l'histoire naturelle en France*. Paris, 1792.

Milne, Edward Arthur. *Sir James Jeans*. Cambridge: Cambridge University Press, 1952.

Mirabaud, Jean-Baptiste de. *Le monde, son origine et son antiquité*. London, Paris: Briasson, 1751.

Moissan, Henri. *Traité de chimie minérale*. 5 vols. Paris: Masson, 1904–6.

Molière [Jean-Baptiste Poquelin]. *Les femmes savantes*. Act 4, scene 3 published in English as *The Clever Women*, translated by M. Slater (Oxford: Oxford University Press, 2001).

Montaigne, Michel Eyquem de. *Essais*. Bordeaux: S. Millanges, 1580. Published in English as *The Complete Essays of Montaigne*, translated by D. M. Frame (Stanford, CA: Stanford University Press, 1958).

More, Louis Trenchard. *Isaac Newton: A Biography, 1642–1727*. London: Scribner's, 1934.

Moréri, Louis. *Grand dictionnaire historique de Moréri*. New ed. Vol. 8. Paris, 1749.

Moro, Anton Lazzaro. *De Crostacei e degli altri marini corpi che si truovano su' monti, libri due*. Venice: Stefano Monti, 1740.

Morse, Edward S. *Mars and Its Mystery*. Boston: Little, Brown, 1906.

Moseley, Henry Gwinn Jeffreys. "The High-Frequency Specra of the Elements." *Phil. Mag.* 26 (1914): 1025–34; 27:703–13.

Muraoka, Hanichi. "Das Johanniskäferlicht." *Ann. Phys.* 59 (1896): 773–71.

Muraoka, Hanichi, and M. Kasuya. "Das Johanniskäferlicht und die Wirkung der Dämpfe von festen und flüssigen Körpen auf photographische Platten." *Ann. Phys.* 64 (1898): 186–92.

Needham, John Turberville. *Nouvelles recherches physiques et métaphysiques sur la Nature et la Religion, lettre à M. de Buffon*. Paris: Lacombe, 1769.

Newton, Isaac. *An Abstract of Sir Isaac Newton's Chronology of Ancient Kingdoms*. 2nd ed. Vol. 1. London: W. Innys, 1732.

———. *The Chronology of Ancient Kingdoms Amended, to which is Prefix'd a Short Chronicle from the First Memory of Things in Europe, to the Conquest of Persia by Alexander the Great*. London: J. Tonson, 1728. Translated into French by F. Granet as *La Chronologie des anciens royaumes, corrigée. A laquelle on a joint une chronique abrégée, qui contient ce qui s'est passé anciennement en Europe, jusqu'à la conquête de la Perse par Alexandre le Grand*. (Paris: G. Martin, 1728).

———. *The Correspondence*. Edited by H. W. Turnbull. Cambridge: Royal Society, 1960.

———. *Philosophiae naturalis principia mathematica*. 3rd ed. London, 1726. Published in English as *Sir Isaac Newton's Mathematical Principles of Natural Philosophy and His System of the World* (London, 1727); translated by A. Motte, with notes by F. Cajori, 2 vols. (Berkeley: University of California Press, 1962).

————. *Principes mathématiques de la philosophie naturelle.* Translated into French by Gabrielle-Émilie du Châtelet. 2nd ed. 2 vols. Paris: Desaint et Saillant, 1759. Reprint, Paris: A. Blanchard, 1966.

————. "Remarks on the Observations Made on a Chronological Index of Sir Isaac Newton, Translated into French by the Observator, and Published at Paris." *Phil. Trans.* 33 (1725): 315–21.

Niderst, Alain. *Fontenelle à la recherche de lui-même (1657–1702).* Paris: Nizet, 1972.

Niepce de Saint-Victor, Abel. "Cinquième mémoire sur une action de la lumière inconnue jusqu'ici." *C. R. Acad. Sci.* 53 (1861): 33–35.

————. "Suite du deuxième mémoire sur une nouvelle action de la lumière." *C. R. Acad. Sci.* 46 (1858): 489–91.

Nier, Alfred O. C. "The Isotopic Constitution of Uranium and the Half-Lives of the Uranium Isotopes." *Phys. Rev.* 55 (1939): 150–53.

————. "Some Reminiscences of Isotopes, Geochronology, and Mass Spectrometry." *Ann. Rev. Earth Planet. Sci.* 9 (1981): 1–17.

————. "Some Reminiscences of Mass Spectrometry and the Manhattan Project." *J. Chem. Educ.* 66 (1989): 385–88.

————. "Variations in the Relative Abundances of the Isotopes of Common Lead from Various Sources." *J. Amer. Chem. Soc.* 60 (1938): 1571–76.

Nier, Alfred O. C., Eugene T. Booth, John R. Dunning, and Aristid Von Grosse. "Nuclear Fission of Separated Uranium Isotopes." *Phys. Rev.* 57 (1940): 546.

Nier, Alfred O. C., and Michael B. McElroy. "Composition and Structure of Mars' Upper Atmosphere: Results from the Neutral Mass Spectrometers on Viking 1 and 2." *J. Geophys. Res.* 82 (1977): 4341–49.

Nier, Alfred O. C., Robert W. Thomson, and Byron F. Murphey. "The Isotopic Constitution of Lead and the Measurement of Geological Time. III." *Phys. Rev.* 60 (1941): 112–16.

Noddack, Irma Z. "Über das Element 93." *Ang. Chem.* 47 (1934): 653.

Nye, Mary-Jo. "Gustave LeBon's Black Light: A Study in Physics and Philosophy in France at the Turn of the Century." *Hist. Stud. Phys. Sci.* 4 (1974): 163–95.

Oersted, Hans Christian. "Experiments on the Effect of a Current of Electricity on the Magnetic Needle." Reprinted in *Œrsted and the Discovery of Electromagnetism,* by Bern Dibner. New York: Blaisdell, 1963.

O'Hara, James G., and Willibad Pricha. *Hertz and the Maxwellians: A Study and Documentation of the Discovery of Electromagnetic Wave Radiation, 1873–1894.* London: Peter Peregrinus, 1987.

Oldenburg, Henry. "Of the *Mundus Subterraneus* of Athanasius Kircher." *Phil. Trans.* 6 (1665): 109–17.

Oliphant, Mark. *Rutherford: Recollections of the Cambridge Days.* Amsterdam: Elsevier, 1972.

Olmsted, John W. "The Scientific Expedition of Jean Richer to Cayenne." *Isis* 34 (1942–43): 117–28.

Oppenheimer, Robert. *Letters and Recollections.* Edited by A. C. Smith and C. Weiner. Cambridge, MA: Harvard University Press, 1980.

Oresme, Nicole. *Le livre du ciel et du monde*. Edited by A. D. Menut and A. J. Denomy. Translated by and with an introduction by A. D. Menut. Madison: University of Wisconsin Press, 1968.

Origen. *Contra Celsus*. In *The Ante-Nicene Fathers*, vol. 4. New York: Scribner's, 1903.

———. *On First Pinciples*. Translated by F. Crombie. In *The Ante-Nicene Fathers*, vol. 4. New York: Scribner's, 1926.

Ospovat, Alex M. "The Distortion of Werner in Lyell's Principles of Geology." *Brit. J. Hist. Sci.* 9 (1976): 190–99.

Outram, Dorinda. *Georges Cuvier: Vocation, Science, and Authority in Post-Revolutionary France*. Manchester, Eng.: Manchester University Press, 1984.

Palassou, Pierre-Bernard. *Essai sur la minéralogie des Monts-Pyrénées*. 2nd ed. Paris: Didot, 1784.

Palissy, Bernard. *Discours Admirables*. Edited by P. A. Cap. Paris : Dubochet et Cie, 1844. Published in English as *The Admirable Discourses of Bernard Palissy*, translated by A. La Rocque (Urbana: University of Illinois Press, 1957).

———. *Œuvres de Bernard Palissy*. Edited by B. Faujas de Saint-Fond and N. Gobet. Paris: Ruault, 1777.

Pallas, Simon. *Reise durch verschiedene Provinzen des russische Reichs*. Graz: Akademische Drück, 1967. Partially translated as *Travels through the Southern Provinces of the Russian Empire in the Years 1793 and 1794* (London: T. N. Longman and O. Rees, 1802).

Paneth, Friedrich Adolf. "Meteorites and the Age of the Solar System." *Nature* 149 (1942): 235–38.

———. *The Origin of Meteorites*. Oxford: Clarendon, 1940.

Paneth, Friedrich Adolf, D. Urry, and W. Koeck. "Zur Frage des Ursprungs der Meteorite." *Z. Elektrochem.* 36 (1930): 727–32.

Paracelsus [Theophrastus Bombastus von Hohenheim]. *Samtliche Werke von Theophrast von Hohenheim*. Edited by Karl Sudhoff and Wilhelm Matthiessen. 12 vols. Hildesheim, Germany: Olms, 1971.

———. *Selected Writings*. Translated by Norbert Guterman. Bollinger series, no. 28. Princeton, NJ: Princeton Universtity Press, 1942.

Parkinson, James. *Organic Remains of a Former World: An Examination of the Mineral Remains of the Vegetables and Animals of the Antediluvian World Generally Termed Extraneous Fossils*. 3 vols. London: Sherwood, Neely, and Jones, 1804–11.

Patterson, Clair C. "Acceptance Speech for the V.M. Goldschmidt Medal." *Geochim. Cosmochim. Acta* 45 (1981): 1385–88.

———. "Age of Meteorites and the Earth." *Geochim. Cosmochim. Acta* 10 (1956): 230–37.

———. "Contaminated and Natural Lead Environments of Man." *Arch. Environm. Health* 11 (1965): 344–60.

———. "Delineation of Separate Brain Regions Used for Scientific versus Engineering Modes of Thinking." *Geochim. Cosmochim. Acta* 58 (1994): 3321–27.

———. "Historical Changes in Integrity and Worth of Scientific Knowledge." *Geochim. Cosmochim. Acta* 58 (1994): 3141–43.

Patterson, Clair C., M. Murozumi, and Tsaihwa J. Chow. "Chemical Concentrations of Pollutant Lead Aerosols, Terrestrial Dusts, and Sea Salts in Greenland and Antartic Snow Strata." *Geochim. Cosmochim. Acta* 33 (1969): 1247–94.

Patterson, Clair C., George Tilton, and Mark Inghram. "Age of the Earth." *Science* 121 (1955): 69–75.

Pécaut, Catherine. "L'Oeuvre géologique de Leibniz." *Rev. Gén. Sci.* 58 (1951): 282–96.

Péligot, Eugène. "Recherches sur l'uranium; deuxième mémoire." *C. R. Acad. Sci.* 18 (1844): 682–87.

———. "Sur le poids atomique de l'urane." *C. R. Acad. Sci.* 12 (1841): 735–37.

———. "Sur le poids atomique de l'uranium." *C. R. Acad. Sci.* 22 (1846): 487–94.

Pera, Marcello. *La rana ambigua.* Turin: G. Einaudi, 1986.

Perrin, Jean. "Matière et lumière: Essai de synthèse de la mécanique chimique." *Ann. Physique* 11 (1919): 5–108.

———. "Nouvelles propriétés des rayons cathodiques." *C. R. Acad. Sci.* 121 (1895): 1130–34.

———. "Quelques propriétés des rayons de Röntgen." *C. R. Acad. Sci.* 122 (1896): 186–88.

Perry, John. "On the Age of the Earth." *Nature* 51 (1895): 224–27, 582–85.

Peteau, Denys. *Opus de doctrina temporum.* 3rd ed. Antwerp: J. Harduin, 1627.

Petit, Pierre. *Dissertation sur la nature des comètes au Roy. Avec un discours sur les Prognostiques des Eclipses & autres Matieres curieuses.* Paris: L. Billaine, 1665.

Pezron, Paul. *L'antiquité des temps rétablie et defenduë contre les Juifs & les Nouveaux Chronologistes.* Amsterdam: H. Desbordes, 1687.

Phillips, John. *Life on the Earth: Its Origin and Succession.* Cambridge: Macmillan, 1860.

———. *Memoirs of William Smith, LL.D.* London, 1844. Reprint, New York: Arno Press, 1978.

Philo. *Allegorical Interpretation of Genesis.* Translated by F. H. Colson and G. H. Whittaker. Cambridge, MA: Harvard University Press, 1962.

Philoponus. See John Philoponus.

Piché, David. *La condamnation parisienne de 1277.* Paris: Vrin, 1999.

Piéron, Henri. "Grandeur et décadence des rayons N: Histoire d'une croyance." *Année Psychologique* 13 (1907): 143–69.

Piggot, Charles Snowden. "Isotopes and the Problem of Geologic Time." *J. Amer. Chem. Soc.* 52 (1930): 3161–64.

———. "Lead Isotopes and the Problem of Geologic Time." *J. Wash. Acad. Sci.* 18 (1928): 269–73.

Piggot, Charles Snowden, and Clarence Norman Fenner. "The Mass-Spectrum of Lead from Bröggerite." *Nature* 123 (1929): 793–94.

Plato. *Timaeus.* Translated by D. J. Zeyl. In *Plato: Complete Works.* Indianapolis: Hackett, 1997.

Playfair, John. "Biographical Account of the Late Dr James Hutton, F. R. S. Edin." *Trans. Roy. Soc. Edinb.* 5 (1805): 39–99.

———. *Illustrations of the Huttonian Theory of the Earth.* Edinburgh: William Creech, 1802. Reprint, with an introduction by G. W. White, Urbana: University of Illinois Press, 1956.

Pliny the Elder [Caius Plinius Secundus]. *Natural History.* Translated by H. Rackham. Cambridge. MA: Harvard University Press, 1938.

Plotinus. *The Enneads.* Translated by S. MacKenna. New York: Burdett Larson, 1992.

Poincaré, Henri. "Les rayons cathodiques et les rayons Röntgen." *Rev. Gen. Sci.* 7 (1896): 52–59.

Poirier, Jean-Pierre. *Antoine-Laurent de Lavoisier, 1743–1794.* Paris: Pygmalion, 1995. Published in English as *Lavoisier: Chemist, Biologist, Economist,* translated by R. Balinski (Philadelphia: University of Pennsylvania Press, 1996).

Poisson, Siméon Denis. "Mémoire sur les températures de la partie solide du globe, de l'atmosphère, et du lieu de l'espace où la Terre se trouve actuellement." *C.R. Acad. Sci.* 4 (1837): 137–66.

———. *Traité mathématique de la chaleur.* Paris: Bachelier, 1835.

Polybius. *The Histories of Polybius.* Book 1. Translated by F. Hultsch and E. S. Shuckburgh. Bloomington: Indiana University Press, 1962.

Porter, Roy. "George Hoggart Toulmin's Theory of Man and the Earth in the Light of the Development of British Geology." *Ann. Science* 35 (1978): 339–52.

Pouillet, Claude. *Eléments de physique expérimentale et de météorologie.* Paris: Hachette, 1827–56 (7 eds.).

———. "Mémoire sur la chaleur solaire, sur les pouvoirs rayonnants et absorbants de l'air atmosphérique, et sur la température de l'espace." *C. R. Acad. Sci.* 7 (1838): 24–65.

Poulsen, Jacob E., and Egill Snorrason, eds. *Nicolaus Steno, 1638–1686: A Reconsideration by Danish Scientists.* Gentofte, Denmark: Nordisk Insulinlaboratorium, 1986.

Prout, William. "Correction of a Mistake in the Essay on the Relation between the Specific Gravities of Bodies in Their Gaseous State and the Weight of their Atoms." *Ann. Philos.* 7 (1816): 111–113.

———. "On the Relation between the Specific Gravities of Bodies in Their Gaseous State and the Weight of their Atoms." *Ann. Philos.* 6 (1815): 321–30.

Ptolemy, Claude. *Almagest.* Translated by G. J. Toomer. New York: Springer, 1984.

———. *Tetrabiblos.* Translated by F. E. Robbins. Cambridge, MA: Harvard University Press, 1998.

Puech, H. C. *En quête de la gnose.* Paris: Gallimard, 1978.

Ramsay, William. "Annual General Meeting." *J. Chem. Soc.* 67 (1895): 1107–9.

———. "Argon and Its Companions." *Phil. Trans. Roy. Soc. London* A197 (1901): 47–89.

———. "L'hélium." *Ann. Chim. Phys.* 13 (1898): 433–81.

———. "Nouvelles recherches de M. Ramsay sur l'argon et sur l'hélium." *C. R. Acad. Sci.* 120 (1895): 660–62;

———. "On a Gas Showing the Spectrum of Helium, the Reputed Cause of D_3, One of the Lines in the Coronal Spectrum: Preliminary Note." *Proc. Roy. Soc.* 58 (1895): 65–67.

Ramsay, William, and Frederick Soddy. "Experiments in Radio-Activity and the Production of Helium from Radium." *Nature* 68 (1903): 354–55.

———. "Gases Occluded by Radium Bromide." *Nature* 68 (1903): 246.

Rappaport, Rhoda. "Fontenelle Interprets the Earth's History." *Rev. Hist. Sci.* 44 (1991): 281–300.

———. "G. F. Rouelle: An Eighteenth-Century Chemist and Teacher." *Chymia* 6 (1960): 68–102.

———. "Lavoisier's Geologic Activities, 1763–1792." *Isis* 58 (1968): 375–84.

Raven, Charles E. *John Ray Naturalist: His Life and Works.* 2nd ed. Cambridge: Cambridge University Press, 1950. Reprint, 1986.

Ray, John. *Observations Topographical, Moral, & Physiological; made in a Journey Through part of the Low-Countries, Germany, Italy, and France: with A Catalog of Plants not Native of England, Found Spontaneously Growing in those Parts, and their Virtues.* London: J. Martyn, 1673.

———. *The Wisdom of God Manifested in the Works of Creation; in two Parts.* 4th ed. London: Samuel Smith, 1704.

Rayleigh, Lord (John William Strutt), and William Ramsay. "Argon, a New Constituent of the Atmosphere." *Proc. Roy. Soc.* 57 (1895): 265–87.

Redondi, Pietro. *Galileo eretico.* Turin: G. Einaudi, 1983. Published in English as *Galileo Heretic*, translated by R. Rosenthal (Princeton, NJ: Princeton University Press, 1987).

Reid, Robert. *Marie Curie.* New York: E. P. Dutton, 1974.

Reilly, P. Connor. *Athanasius Kircher: Master of a Hundred Arts, 1602–1680.* Wiesbaden, Germany: Edizioni del Mondo, 1974.

Revelle, Roger. "Harrison Brown, September 26, 1917–December 8, 1986." *Biog. Memoir Nat. Acad. Sci.* 65 (1994): 41–55.

"Rev. Osmond Fisher, M.A., F.G.S." *Geol. Mag.* 1 (1914): 383–84.

Reyment, Richard. "John Phillips (1800–1874)." *Terra Nova* 8 (1996): 5–7.

———. "William Smith (1769–1839)—The Father of English Geology." *Terra Nova* 8 (1996): 662–64.

Richards, Theodore Williams. "The New Outlook in Chemistry." *Science* 26 (1907): 297–305.

Richards, Theodore Williams, and L. P. Hall. "The Atomic Weight of Uranium Lead and the Age of an Especially Ancient Uraninite." *J. Amer. Chem. Soc.* 48 (1926): 704–8.

Richards, Theodore Williams, and M. E. Lembert. "The Atomic Weight of Lead of Radioactive Origin." *J. Amer. Chem. Soc.* 36 (1914): 1329–44.

Richer, Jean. "Observations astronomiques et physiques faites en l'isle de Caïenne." *Mém. Acad. Roy. Sci.* 7 (1666–99): 3–94.

Richet, Charles. *Traité de métapsychique.* Paris: Alcan, 1922. Published in English as *Thirty Years of Psychical Research: Being a Treatise of Metapsychics*, translated by S. De Brath (New York: Macmillan, 1923).

Roger, Jacques. *Buffon: Un philosophe au Jardin du Roi.* Paris: Fayard, 1989. Translated by S. L. Bonnefoi as *Buffon: A Life in Natural History* (Ithaca, NY: Cornell University Press, 1996).

———. "The Cartesian Model and Its Role in Eighteenth-Century 'Theory of the Earth.'" In *Problems of Cartesianism*, ed. T. M. Lennon, J. M. Nicholas, and J. W. Davis, 95–112. Montreal: McGill–Queen's University Press, 1982.

———. "La théorie de la terre au XVIIe siècle." *Rev. Hist. Sci.* 26 (1973): 23–48.

Romé de L'Isle, Jean-Baptiste-Louis. *Cristallographie, ou description des formes propres à tous les corps du regne minéral, dans l'état des combinaison saline, pierreuse ou métallique.* 4 vols. Paris: Imprimerie de Monsieur, 1783.

———. *Essai de cristallographie, ou description des figures géométriques, propres à différens corps du regne minéral, connus vulgairement sous le nom de cristaux.* Paris: Didot, 1772.

———. *L'action du feu central démontrée nulle à la surface du Globe, contre les assertions de MM. le Comte de Buffon, Bailly, de Mairan, &c.* 2nd ed. Stockholm, 1781.

Römer, Ole. "Démonstration touchant le mouvement de la lumière trouvé par M. Römer de l'Académie Royale des Sciences." *J. Sçavants* (1676): 233–36.

Ronan, Colin A. *Edmond Halley: Genius in Eclipse.* New York: Doubleday, 1969.

Ronsard, Pierre de. *Les Amours.* In *Œuvres complètes*, edited by J. Céard, D. Ménager, and M. Simonin. Paris: Gallimard, 1993.

Röntgen, Wilhelm Conrad. "On a New Kind of Rays." *Nature* 53 (1896): 274–76.

———. "Ueber eine neue Art von Strahlen." *Ann. Phys. Chem.* NF 64 (1898): 1–11.

Rossi, Paolo. *I segni del tempo: Storia delle terra e delle nazioni da Hooke a Vico.* Milan: Feltrinelli, 1979.

Rouelle, Guillaume François. "Theory of the Earth." In N. Desmarest, *Géographie Physique*, vol. 1. Paris: Agasse, 1794.

Royou, Thomas Marie. *Le monde de verre réduit en poudre, ou analyse et réfutation des Epoques de la nature de M. le comte de Buffon* (Paris: Mérigot le jeune, 1780).

Rudwick, Martin J. S. *Georges Cuvier, Fossil Bones, and Geological Catastrophes: New Translations and Interpretations of Primary Texts.* Chicago: University of Chicago Press, 1997.

Rupke, Nicholas A. *The Great Chain of History: William Buckland and the English School of Geology (1814–1849).* Oxford: Clarendon, 1983.

Russell, Henry Norris. "Relation between the Spectra and Other Characteristics of the Stars." *Nature* 93 (1914): 227–30, 252–58, 281–86.

———. "A Superior Limit to the Age of the Earth's Crust." *Proc. Roy. Soc.* A99 (1921): 84–86.

Rutherford, Ernest. *The Collected Papers of Lord Rutherford of Nelson.* Edited by J. Chadwick. 3 vols. London: Allen and Unwin, 1962–65.

———. "Les problèmes actuels de la radioactivité." *Arch. Sci. Phys. Nat.* 19 (1905): 31–59, 125–50.

———. "The Mass and Velocity of the α-Particles Expelled from Radium and Actinium." *Phil. Mag.* 12 (1906): 348–71.

———. "Origin of Actinium and the Age of the Earth." *Nature* 123 (1929): 313–14.

———. "The Radiation and Emanation of Radium." *Technics*, July 1904, 11–16.

———. *Radioactivity.* Cambridge: University Press, 1st ed. 1904, 2nd ed. 1905.

———. "Radium—the Cause of the Earth's Heat." *Harper's Magazine*, February 1905, 390–96.

———. "Uranium Radiation and the Electrical Conduction Produced by It." *Phil. Mag.*, 1899. In Rutherford, *Collected Papers*, 1:175.

Rutherford, Ernest, and Howard Turner Barnes. "Heating Effects of the Radium Emanation." *Nature* 68 (1903): 622.

Rutherford, Ernest, and Bertram Boltwood. "The Relative Proportion of Radium and Uranium in Radio-Active Minerals." *Amer. J. Sci.* 20 (1905): 55–57.

———. "The Relative Proportion of Radium and Uranium in Radio-Active Minerals." *Amer. J. Sci.* 22 (1906): 1–3.

Rutherford, Ernest, and Frederick Soddy. "The Cause and Nature of Radioactivity." *Phil. Mag.* 4 (1902): 370–96.

———. "A Comparative Study of the Radioactivity of Radium and Thorium." *Phil. Mag.* 5 (1903): 445–57.

———. "Radioactive Change." *Phil. Mag.* 5 (1903): 576–91.

———. "The Radioactivity of Uranium." *Phil. Mag.* 5 (1903): 441–57.

Saint-Pierre, Jacques-Henri Bernardin de. *Etudes de la Nature.* 5 vols. Paris, 1784. Published in English as *Studies of Nature*, translated by H. Hunter (London: J. Mawman, 1805).

Sarton, George. "Discovery of the Theory of Natural Selection." *Isis* 14 (1930): 133–54.

———. "Moseley. The Numbering of the Elements." *Isis* 9 (1927): 96–111.

Scaliger, Joseph-Juste. *Autobiography of Joseph Scaliger, with Autobiographical Selections from His Letters, His Testament, and the Funeral Orations by Daniel Hensius and Dominicus Baudius.* Translated by G. W. Robinson. Cambridge: Harvard University Press, 1927.

———. *De emendatione temporum, octo libris distinctum in quo praeter dierum ciuilium, mensium, annorum & epocharum cognitionem exactam, doctrinam accuratam, priscorum temporum methodiis, ac nouorum annorum forma, aut ipsorum veterum emendatio examinanda & dignoscenda acute proponitur.* Frankfurt: Ioannem Wechelum, 1593.

———. *Thesaurus temporum. Eusebii Pamphili Caesarae Palestinae Episcopi. Chronicum canonum omnimodæ historiæ libri duo, interprete Hieronymo, ex fide vetustissimorum Codicum castigati. Item auctores omnes derelicta ab Eusebio, & Hieronymo continuantes. Eivsdem Evsebii Vtriusque partis Chronicorum Canonum reliquiæ Græcæ, quæ colligi potuerunt, antehac non editæ.* Lyons: T. Basson, 1606.

Schaffer, Simon. "Halley's Atheism and the End of the World." *Notes Records Roy. Soc.* 32 (1978): 17–40.

Scherz, Gustav, ed. *Nicolaus Steno and His Indice.* Copenhagen: Munksgaard, 1958.

Schmidt, Gerhard C. "Ueber die von den Thorverbindungen und einigen anderen Subtanzen ausgehende Strahlung." *Ann. Physik* 65 (1898): 141–51.

Schmolz, Helmut, and Hubert Weckbach. *Robert Mayer: Sein Leben und Werk in Dokumenten.* Heilbronn, Germany: Anton H. Konrad, 1964.

Schuchert, Charles. "Biographical Memoir of John Mason Clarke." *Biog. Mem. Nat. Acad. Sci.* 12 (1929): 182–244.

———. "Biographical Memoir of Joseph Barrell, 1869–1919." *Biog. Mem. Nat. Acad. Sci.* 12 (1929): 1–40.

———. "Geochronology or the Age of the Earth on the Basis of Sediments and Life." In *The Age of the Earth*, edited by A. E. Knopf, W. Brown, A. Holmes, A. F. Kovarik, A. C. Lane, and C. Schuchert., 10–64. Washington, DC: National Research Council, 1931.

Schuster, Arthur. *Biographical Fragments.* London: Macmillan, 1932.

———. *The Progress of Physics during 33 years (1875–1908): Four Lectures Delivered to the University of Calcutta during March 1908.* Cambridge: University Press, 1911.

Severinus, Petrus [Peder Sørenson]. *Idea Medicinæ Philosophicæ, fundamenta continens totius doctrinæ Paracelsicæ, Hippocraticæ & Galenicæ.* Basel, 1571.

Sévigné, Marie de Rabutin-Chantal. *Lettres de Mme de Sévigné à sa fille et à ses amis, 2 janvier 1681 au comte de Bussy.* Paris: Bossange and Masson, 1818.

Sharlin, Harold Issadore. *Lord Kelvin: The Dynamic Victorian.* University Park: Pennsylvania State University Press, 1979.

Sharov Alexander S., and Igor D. Novikov. *Edwin Hubble: The Discoverer of the Big Bang Universe.* Translated from the Russian by V. Kisin. Cambridge: Cambridge University Press, 1993.

Simon, Gérard. *Kepler astronome astrologue.* Paris: Gallimard, 1979.

Simon, Richard. *Histoire critique du Vieux Testament.* Rotterdam, Netherlands, 1685. Reprint, Frankfurt: Minerva, 1967.

Sklodowska Curie, Marie. "Rayons émis par les composés de l'uranium et du thorium." *C. R. Acad. Sci.* 126 (1898): 1101–3; and 127 (1898): 175–78, 1215–17.

Slipher, Vesto Melvin. "The Radial Velocity of the Andromeda Nebula." *Lowell Obs. Bull.* 58 (1913): 56–57.

———. "A Spectrographic Investigation of Spiral Nebulae." *Proc. Amer. Philos. Soc.* 56 (1917): 403–10.

Smith, Crosbie, and M. Norton Wise. *Energy and Empire: A Biographical Study of Lord Kelvin.* Cambridge: Cambridge University Press, 1989.

Smith, Kirk R., Fereidun Fesharaki, and John P. Holdren, eds. *Earth and the Human Future: Essays in Honor of Harrison Brown.* Boulder: Westview Press, 1986.

Smith, William. *Strata Identified by Organized Fossils.* London: W. Arding, 1816.

Soddy, Frederick. "The Chemistry of Mesothorium." *Trans. Chem. Soc.* 99 (1911): 72–83.

———. "Discussion on Isotopes." *Proc. Roy. Soc.* A99 (1921): 87–104.

———. *Radioactivity and Atomic Theory . . . Annual Progress Reports on Radioactivity 1904–1920 to the Chemical Society by Frederick Soddy F.R.S.* Compiled, with commentary, by T. J. Trenn. London, 1975.

———. *Radioactivity: An Elementary Treatise from the Standpoint of the Disintegration Theory.* London, 1904.

Soddy, Frederick, and H. Hyman. "The Atomic Weight of Lead from Ceylon Thorite." *Trans. Chem. Soc.* 105 (1914): 1402–8.

Sollas, William Johnson. "The Age of the Earth." *Nature* 51 (1895): 533–34.

———. "The Age of the Earth." *Nature* 108 (1921): 281–83.

———. *The Age of the Earth and Other Geological Stories.* New York: E. P. Dutton, 1905.

Sorabji, Richard. "Fear of Death and Endless Recurrence." In *Time, Creation, and the Continuum,* by Sorabji, 174–90. London: Duckworth, 1983.

———. *Philoponus and the Rejection of Aristotelian Science.* London: Duckworth, 1987.

Souciet, Etienne. *Cinq dissertations contre la chronologie de M. Newton, suivies d'une Dissertation sur une médaille singulière d'Auguste.* Paris: Rollin, 1726.

Soulavie, Jean-Louis Giraud. *Chronologie physique des éruptions des volcans éteints de la France méridionale, depuis celles qui avoisinent la formation de la terre, jusques à celles qui sont décrites dans l'Histoire.* Vol. 4 of Soulavie, *Histoire naturelle.*

———. *Histoire naturelle de la France méridionale.* 7 vols. Paris: Quillot et al., 1780–84.

Spinoza, Baruch. *Tractatus Theologico-Politicus, Continens Dissertationes aliquot, Quibus ostenditur Libertatem Philosophandi non tantum salva Pietate, & Reipublicæ Pace posse concedi: sed eandem nisi cum Pace Reipublicæ, ipsaque Pietate tolli non posse.* Hamburg: H. Künrath, 1670 (actually published in Amsterdam by C. Conrad). Published in English as *Theologico-Political Treatise,* translated by S. Shirley (Boston: Brill Academic, 1997).

Steno, Nicolaus [Niels Stensen]. *Canis Carchariæ Dissectum Caput.* Translated by A. Garboe as *The Earliest Geological Treatise* (London: Macmillan, 1958).

———. *De solido intra solidum naturaliter contento dissertationis prodromus.* Florence, 1669. Published in English as *The Prodromus of Nicolaus Steno's Dissertation concerning a Solid Body Enclosed by Process of Nature within a Solid,* translated by J. G. Winter (London: Macmillan, 1916; reprint, New York: Hafner, 1968).

———. *Discours de Monsieur Sténon, sur l'anatomie du cerveau.* Paris: Robert de Ninville, 1669.

———. *Elementorum myologiæ specimen, seu musculi descriptio geometria. Cui accedunt Canis carchariæ dissectum caput, et Dissectus piscis ex canum genere.* Florence, 1667.

———. *In nomine Iesu CHAOS: A Danish Student in His Chaos-Mansucript 1659.* Edited by H. D. Schpelern. Copenhagen: University Library, 1987.

Stimson, Dorothy. *Scientists and Amateurs: A History of the Royal Society.* New York: H. Schuman, 1948.

Stoney, George Johnstone. "On Dr Johnstone Stoney's Logarithmic Law of Atomic Weights." *Proc. Roy. Soc.* A85 (1911): 471–73.

———. "On the 'Electron,' or Atom of Electricity." *Phil. Mag.* 38 (1894): 418–20.

Strabo. *Geography.* Translated by H. L. Jones. Cambridge, MA: Harvard University Press, 1960.

Strutt, Guy. "Robert John Strutt, Fourth Baron Rayleigh." *Appl. Opt.* 3 (1964): 1105–12.

Strutt, John William. See Lord Rayleigh.

Strutt, Robert John (fourth Baron Rayleigh). *John William Strutt, Third Baron Rayleigh.* London: E. Arnold, 1924. Reprinted as *Life of John William Strutt, Third Baron Rayleigh, O.M., F.R.S.* (Madison: University of Wisconsin Press, 1968).

———. *The Life of Sir J. J. Thomson.* Cambridge: University Press, 1943.

———. "Measurements of the Rate at Which Helium Is Produced in Thorianite and Pitchblende, with a Minimum Estimate of Their Antiquity." *Proc. Roy. Soc.* 84 (1910): 379–88.

———. "On the Distribution of Radium in the Earth's Crust, and on the Earth's Internal Heat." *Proc. Roy. Soc.* A77 (1906): 472–85.

———. "On the Radio-Active Minerals." *Proc. Roy. Soc.* A76 (1905): 88–101.

———. "Some Reminiscences of Scientific Workers of the Past Generation, and Their Surroundings." *Proc. Phys. Instit.* 48 (1936): 217–46.

Susskind, Charles. *Heinrich Hertz: A Short Life.* San Francisco: San Francisco Press, 1995.

Swedenborg, Emmanuel. *Principia.* Published in English as *The Principia; or, The First Principles of Natural Things, Being New Attempts toward a Philosophical Explanation of the Elementary World,* translated by A. Clissold, 2 vols. (London: W. Newbery, 1846).

Tait, Peter Guthrie. *Lectures on Some Recent Advances in Physical Science, with a Special Lecture on Force.* 3rd ed. London: Macmillan, 1885.

Taton, René, ed. *Roemer et la vitesse de la lumière.* Paris: Vrin, 1978.

Taylor, Kenneth L. "Nicolas Desmarest and Geology in the Eighteenth Century." In *Toward a History of Geology,* edited by C. J. Schneer. Cambridge, MA: MIT Press, 1969.

Tempier, Etienne. "Condemnation of 219 Propositions." Translated by E. L. Fortin and P. D. O'Neill. In *Medieval Political Philosophy: A Sourcebook,* edited by R. Lerner and M. Mahdi, 335–54. Ithaca, NY: Cornell University Press, 1984.

Termier, Pierre. *À la gloire de la terre.* Paris: Desclée de Brouwer, 1922.

Tertullian. *The Apology.* In *The Ante-Nicene Fathers,* vol. 3. New York: Scribner's, 1903.

Theophilus of Antioch. *Theophilus to Autolycus.* Translated by M. Dods. In *The Ante-Nicene Fathers,* vol. 2. New York: Scribner's, 1903.

Thompson, Jane Smeal, and Helen G. Thompson. *Silvanus Phillips Thompson: His Life and Letters*. London: T. F. Unwin, 1920.

Thompson, Silvanus P. *The Life of William Thomson, Baron Kelvin of Larg*. 2 vols. London: Macmillan, 1910.

———. "On Hyperphosphorescence." *Phil. Mag.* 42 (1896): 103–7.

Thomson, George Paget. *J. J. Thomson and the Cavendish Laboratory in His Day*. New York: Doubleday, 1965.

Thomson, Joseph John. "Cathode Rays." *Phil. Mag.* 44 (1897): 293–316.

———. "Positive Rays of Electricity." *Nature* 91 (1913): 362.

———. "Rays of Positive Electricity." *Proc. Roy. Soc.* A89 (1913): 1–20.

———. *Recollections and Reflections*. New York: Macmillan, 1937.

Thomson, Joseph John, and E. Rutherford. "On the Passage of Electricity through Gases Exposed to Röntgen Rays." *Phil. Mag.* 42 (1896): 392–407.

Thomson, William (Baron Kelvin of Largs). "The Age of the Earth." *Nature* 51 (1895): 438–40.

———. "The Age of the Earth as an Abode Fitted for Life." *Science* 9 (1899): 665–74.

———. "Contribution by Lord Kelvin to the Discussion on the Nature of the Emanations from Radium Which Was Opened by Professor E. Rutherford at the Meeting of the British Association Last September." *Phil. Mag.* 7 (1904): 220–22.

———. "Inaugural Address of Sir William Thomson, LL.D., F.R.S., President." *Nature* 4 (1871): 262–70.

———. "Lord Kelvin and His First Teacher in Natural Philosophy." *Nature* 68 (1903): 624–25.

———. *Mathematical and Physical Papers* (MPP). 6 vols. Cambridge: University Press, 1911.

———. "On a Universal Tendency in Nature to the Dissipation of Mechanical Energy." *Phil. Mag.* 3 (1852): 304–6.

———. "On Geological Dynamics" (1869). In Thomson, *PLA*, 2:73–131.

———. "On Geological Time" (1868). In Thomson, *PLA*, 2:10–72.

———. "On the Mechanical Energies of the Solar System." *Phil. Mag.* 8 (1854): 409–30.

———. "On the Rigidity of the Earth." *Phil. Trans.* 153 (1863): 573–82.

———. "On the Secular Cooling of the Earth" (1862). In Thomson, *MPP*, 3:295–311.

———. "On the Use of Observations of Terrestrial Temperature for the Investigation of Absolute Dates in Geology" (1855). In Thomson, *MPP*, 2:175–77.

———. "Plan of an Atom to Be Capable of Storing an Electrion with Enormous Energy for Radio-Activity." *Phil. Mag.* 10 (1905): 695–98.

———. *Popular Lectures and Addresses* (PLA). 3 vols. London: Macmillan, 1894.

———. "Presidential Address to the British Association, 1871." In Thomson, *PLA*, 2:132–205.

———. *Reprints of Papers on Electrostatics and Magnetism*. London: Macmillan, 1872.

Thoren, Victor E., and John Robert Christianson, *The Lord of Uraniborg*. Cambridge: Cambridge University Press, 1990.

Thorndike, Lynn. *A History of Magic and Experimental Science*. Vol. 3. New York: Columbia University Press, 1933.

Tilden, William A. *Sir William Ramsay: Memorials of His Life and Work*. London, 1918.

Tilton, George R. "Charles Snowden Piggot, 1892–1973." *Biog. Memoirs Nat. Acad. Sci.* 66 (1995): 2–20.

Toulmin, George Hoggart. *The Antiquity and Duration of the World*. London: T. Cadell, 1780.

Toulouse, Dr. "Les rayons N existent-ils?" *Revue scientifique* 2 (1904): 545–52, 590–91, 620–23, 656–60, 682–86.

Tournefort, Joseph Pitton de. "Description du labirinthe de Candie, avec quelques observations sur l'accroissement et sur la generation des pierres." *Mém. Acad. Roy. Sci.* (1702): 224–41 (2nd ed., 1720).

Turnor, Edmund. *Collections for the History of the Town and Soke of Grantham. Containing Authentic Memoirs of Sir Isaac Newton*. London: W. Miller, 1806.

Ussher, James. *Annales Veteris Testamenti, a prima mundi origine deducti: Una cum rerum asiaticarum and ægyptiacarum chronico, a temporis historici principio usque ad Maccabaicorum initia producto*. 2 vols. London, 1650–54. Reprinted in Ussher, *Whole Works*, vol. 8.

———. *The Whole Works of the Most Rev. James Ussher, D. D.*, 16 vols., with a life of the author and an account of his writings by R. Elrington. Dublin: Hodges and Smith, 1847.

Vaccari, Ezio. *Giovanni Arduino (1714–1795) Il contributo di uno scienzato veneto al dibattito settecentesco sulle scienze della Terra*. Florence: L. S. Olschki, 1993.

Vaillot, René. *Voltaire et son temps: Avec Madame du Châtelet*. Oxford: Voltaire Foundation, 1988.

Vallisnieri, Antonio, *Lezione accademica intorno all'origine delle fontane*. Venezzia: Gabbrielle Ertz, 1715.

Van Doren, C. *Benjamin Franklin*. New York: Viking Press, 1938.

Vico, Giambattista. *Principi di una scienza nuova d'intorno alla comune natura delle nazioni*. Naples: F. Mosca, 1730. Published in English as *The New Science of Giambattista Vico*, translated by T. G. Bergin and M. H. Fisch (Ithaca, NY: Cornell University Press, 1984).

Villard, Paul. "Sur le rayonnement du radium." *C. R. Acad. Sci.* 130 (1900): 1178–79.

———. "Sur la réflexion et la réfraction des rayons cathodiques et des rayons déviables du radium." *C. R. Acad. Sci.* 130 (1900): 1010–12.

Vinci. See Leonardo da Vinci.

Volta, Allessandro. *L'Opera di Alessandro Volta*. Milan: V. Hoepli, 1927.

———. "On the Electricity Excited by the Mere Contact of Conducting Substances of Different Kinds." *Phil. Trans.* 90 (1800): 403–31.

Voltaire [Jean-Marie Arouet]. *Correspondance*. Ed. T. Besterman. Paris: Gallimard, 1977.

———. *Des singularités de la nature*. Basel, 1768.

———. *Dissertation sur les changements arrivés dans notre Globe, et sur les pétrifications qu'on prétend en être encore le témoignage*. Dresden, 1748.

———. *Éléments de la philosophie de Newton*. Amsterdam: E. Ledet, 1738. Reprinted in Voltaire, *Œuvres complètes*, vol. 15 (Oxford: Voltaire Foundation, 1992). Published in English as *The Elements of Sir Isaac Newton's Philosophy* (London: S. Austeen, 1738).

———. *Essai sur la nature du feu et de sa propagation*. 1738. Reprinted in Voltaire, *Œuvres complètes*, vol. 22 (Garnier, Paris, 1879).

———. *Lettres philosophiques*. Translated by Voltaire as *Letters concerning the English Nation*, ed. Nicholas Cronk (1734; reprint, Oxford: Oxford University Press, 1994).

———. *Mélanges de philosophie*. Geneva, 1757.

———. *Quatrième discours en vers sur l'homme*. (1737). Reprint, Paris: Garnier, 1877.

Wade, Ira O. *The Clandestine Organization and Diffusion of Philosophic Ideas in France from 1700 to 1750*. Reprint, New York: Octagon Books, 1967.

Wallace, Alfred Russel. *Contributions to the Theory of Natural Selection: A Series of Essays*. London: Macmillan, 1870.

———. *Darwinism: An Exposition of the Theory of Natural Selection with Some of Its Applications*. London: Macmillan, 1889.

———. *Island Life; or, The Phenomena and Causes of Insular Faunas and Floras Including a Revision and Attempted Solution of the Problem of Geological Climates*. 2nd ed. London: Macmillan, 1892.

———. *Is Mars Habitable?* London: Macmillan, 1907.

———. "The Measurement of Geological Time I." *Nature* 1 (1870): 399–401.

———. "The Measurement of Geological Time II." *Nature* 1 (1870): 452–55.

———. "On the Tendency of Varieties to Depart Indefinitely from the Original Type." *J. Linnean Soc.* 3 (1859): 53–62; reprinted in Sarton, "Discovery."

Weart, Spencer R. "The Discovery of the Risk of Global Warming." *Physics Today*, January 1997, 34–40.

Weisheipl, James A., ed. *Albertus Magnus and the Sciences: Commemorative Essays*. Toronto: Pontifical Institute of Mediæval Studies, 1980.

Weissberg, Alexander. *The Accused*. New York: Simon and Schuster, 1951.

Weizsäcker, Carl Friedrich von. "Über die Möglichkeit dualen β-Zerfalls von Kalium." *Phys. Z.* 38 (1937): 623–24.

———. "Über Elementumwandlungen im Innern des Sterne. I." *Phys. Z.* 38 (1937): 176–91.

———. "Über Elementumwandlungen im Innern des Sterne. II." *Phys. Z.* 39 (1938): 633–46.

Werner, Abraham Gottlob. *Kurze Klassifikation und Beschreibung der verschiedenen Gebirgsarten*. Dresden, Germany, 1787. Published in English as *A Short*

Classification and Description of the Various Rocks, translated by A. M. Ospovat (New York: Hafner, 1971).

———. "Observations sur les roches volcaniques, et sur le basalte." *Obs. Phys. Hist. Nat. Arts* 38 (1791): 409–20.

———. *Von den aüsserlichen Kennzeichen der Fossilien*. Leipzig, 1774. Reprint, Amsterdam: Asher, 1965. Published in English as *On the External Classification of Minerals*, translated by A. V. Carozzi: (Urbana: University of Illinois Press, 1962).

Westfall, Richard. *Never at Rest*. Cambridge: Cambridge University Press, 1980.

Whiston, William. *Astronomical Principles of Religion, Natural and Reveal'd*. 2nd ed. London: J. Senex and W. Taylor, 1725.

———. *Memoirs of the Life and Writings of Mr. William Whiston, Containing Memoirs of several of his Friends also. Written by himself*. 3 vols. London: 1749–50.

———. *A New Theory of the Earth, from its Original to the Consummation of all Things: Wherein The Creation of the World in Six Days, The Universal Deluge, and General Conflagration, As laid down in the Holy Scriptures, Are shewn to be perfectly Agreeable to Reason and Philosophy. With a Large Introductory Discourse Concerning the Genuine Nature, Style, and Extent of the Mosaick History of the Creation*. London: 1696–1752 (6 eds.).

White, Michael. *Leonardo: The First Scientist*. London: Little, Brown, 2000.

Whitrow, Gerald James. *Time in History: Views of Time from Prehistory to the Present Day*. Oxford: Oxford University Press, 1988.

Wilson, David. *Rutherford: Simple Genius*. Cambridge, MA: MIT Press, 1983.

Wilson, Leonard G. *Charles Lyell: The Years to 1841: A Revolution in Geology*. New Haven, CT: Yale University Press, 1972.

Wilson, W. E. "Radium and Solar Energy." *Nature* 68 (1903): 222.

With, Etienne. *L'écorce terrestre*. Paris: Plon, 1874.

Wood, Robert W. "The *N*-Rays." *Nature* 71 (1904): 530–31.

Woodward, A. Smith, and William Whitehead Watts. "William Johnson Sollas 1849–1936." *Obit. Not. Roy. Soc. London* 2 (1938): 265–81.

Woodward, John. *An Essay toward a Natural History of the Earth and Terrestrial Bodies, Especially Minerals: as also of the Seas, Rivers, and Springs. With an Account of the Universal Deluge: and of the Effects that it had upon Earth*. London: R. Wilkin, 1695. Reprint, New York: Arno Press, 1978.

Wright (of Durham), Thomas. *An Original Theory or New Hypothesis of the Universe, Founded upon the laws of Nature and solving by mathematical principles the general phænomena of the visible creation; and particularly the Via Lactea Compris'd in Nine Familiar Letters from the Author to his friend. And Illustrated with upwards of Thirty Graven and Mezzotinto Plates, by the Best Masters*. London: H. Chapelle, 1750. Reprint, with notes by M. A. Hoskin, London: Macdonald, 1971.

———. *Second or Singular Thoughts upon the Theory of the Universe*. Edited by M. A. Hoskin. London: Dawsons of Pall Mall, 1968.

Wurtzelbaur, Johann Philipp von. "An account of some observations lately made at Nuremberg, shewing that the Latitude of that place has continued without sensible alteration for 200 years last past; as likely the Obliquity of the Ecliptick; by comparing them with what was observed by Bernard Walther in the Year 1487." *Phil. Trans.* 16 (1687): 403–6.

Wyckoff, Dorothy. "Albertus Magnus on Ore Deposits." *Isis* 49 (1958): 109–22.

Young, Thomas. *A Course of Lectures in Natural Philosophy and the Mechanical Arts.* 2 vols. London: J. Johnson, 1807.

General Reference Works

Biographical Memoirs of the Fellows of the Royal Society. London: Royal Society, in progress since 1955.

Biographical Memoirs of the National Academy of Sciences. Published annually by the National Academy of Sciences, Washington, DC.

Biographie universelle ancienne and moderne, ou histoire, par ordre alphabétique, de la vie publique and privée de tous les hommes qui se sont faits remarquer par leurs écrits, leurs actions, leurs talents, leurs vertus ou leurs crimes. Edited by M. Michaud. New ed. 45 vols. Paris: Mme. C. Desplaces, 1854–65.

Dictionary of Scientific Biography. Edited by C. C. Gillespie. 16 vols. New York: Scribner's, 1970–80. Continued by F. L. Holmes, 2 supplements (1990).

Index biographiques des membres and correspondants de l'Académie des sciences du 22 décembre 1666 au 15 décembre 1967. Paris: Gauthier-Villars, 1968.

Les prix Nobel. Stockholm: Imprimerie Royale.

Obituary Notices. London: Royal Society, 1932–54.

Poggendorf Biographisch-literarisches Handwörterbuch zur Geschichte der exakten Wissenschaften. Leipzig: J. A. Barth, 1863–1904; Verlag Chemie, 1905–6; Berlin: Verlag Chemie, 1936–39; Akademie Verlag, in progress since 1956.

Index